Applied and Systemic-Structural Activity Theory

Human Activity: Efficiency, Safety, Complexity, and Reliability of Performance
Series Editor: Gregory Z. Bedny
Research Associate at Evolute, Inc., Louisville, Kentucky

Work Activity Studies Within the Framework of Ergonomic, Psychology and Economics
Gregory Z. Bedny and Inna S. Bedny

For more information about this series, please visit: https://www.crcpress.com/Human-Activity/book-series/CRCHUMACEFSACP

Applied and Systemic-Structural Activity Theory

Advances in Studies of Human Performance

Edited by

Gregory Z. Bedny and Inna S. Bedny

CRC Press
Taylor & Francis Group
Boca Raton London New York

CRC Press is an imprint of the
Taylor & Francis Group, an **informa** business

CRC Press
Taylor & Francis Group
6000 Broken Sound Parkway NW, Suite 300
Boca Raton, FL 33487-2742

First issued in paperback 2023

ISBN 13: 978-1-03-257046-4 (pbk)
ISBN 13: 978-1-138-60672-2 (hbk)

DOI: 10.1201/9780429466311

This book contains information obtained from authentic and highly regarded sources. Reasonable efforts have been made to publish reliable data and information, but the author and publisher cannot assume responsibility for the validity of all materials or the consequences of their use. The authors and publishers have attempted to trace the copyright holders of all material reproduced in this publication and apologize to copyright holders if permission to publish in this form has not been obtained. If any copyright material has not been acknowledged, please write and let us know so we may rectify this oversight in any future reprint.

Library of Congress Cataloging-in-Publication Data

Names: Bedny, Gregory Z., editor. | Bedny, Inna, editor.
Title: Applied and systemic-structural activity theory : advances in studies of human
    performance / edited by Gregory Z. Bedny and Inna S. Bedny.
Description: Boca Raton, FL : CRC Press, [2019] | Series: Human activity:
    Efficiency, safety, complexity, and reliability of performance | Includes index.
Identifiers: LCCN 2019004434| ISBN 9781138606722 (hardback : alk. paper) |
    ISBN 9780429466311 (e-book)
Subjects: LCSH: Performance standards. | Performance technology. |
Management—Psychological aspects. | Quantitative research.
Classification: LCC HF5549.5.P35 A77 2019 | DDC 658.3/125—dc23
LC record available at https://lccn.loc.gov/2019004434

Publisher's Note
The publisher has gone to great lengths to ensure the quality of this reprint but points out that some imperfections in the original copies may be apparent.

Visit the Taylor & Francis Web site at
http://www.taylorandfrancis.com

and the CRC Press Web site at
http://www.crcpress.com

Dedication

_____

To the memory of my beloved
husband Gregory Bedny
-Inna Bedny

Contents

SECTION I The Concept of Self-Regulation in Studies of Human Activity

SECTION II Efficiency of Performance and Quantitative Methods of Its Assessment

SECTION III   Management and Education

Foreword

Gregory Bedny was my lifelong friend. We became friends in our early teens. I would like to share my memories of him.

Gregory did not have a formal education in Applied Psychology. Psychology fields like human factors/ergonomics, engineering psychology, social psychology, and industrial/organizational psychology were relatively new in the former Soviet Union during the early 1970s. None of the universities in the country had a major in these fields. Some universities were in the process of establishing applied psychology departments, but there were no qualified professors to teach the subjects, and the textbooks were just being worked on. For Soviet citizens, going abroad to study was impossible because it was a closed country. We tried to obtain some knowledge in these fields piecewise from books and articles translated into Russian, and from any other possible source we could get our hands on. Gregory was an excellent example of that approach. His only formal education was the study of technical subjects in college, and subsequent study in pedagogy, anatomy, and physiology at the University of South Ukraine (Odessa, Ukraine).

We studied together at the university. We also shared our love for artistic gymnastics. We trained hard, participated in many competitions, and both achieved the level of "Master of Sports" (an elite level). This athletic title was established in 1935. The sport was government subsidized and very popular. In spite of that, when we were awarded the title in 1963, there were only some thirty thousand masters of sports in the entire country combined because it took a lot of dedication and talent to achieve it. It was pretty cool that we got it! Each holder of the title received a hardcopy picture ID and a small square chest badge. It was a rare distinction. People always looked with admiration at those with a master of sports badge on their jacket lapel.

After graduation, Gregory taught some technical subjects in a junior college, and began working on his Candidates of Science dissertation (PhD) on his own. He successfully defended it in 1969. Gregory was the first one in Odessa to obtain a PhD in industrial organizational psychology. He created and published courses in engineering psychology, transportation ergonomics, and the ergonomic basis of work safety, and taught related courses to students at the Civil Engineering University and at the Merchant Maritime University.

A few years later, again on his own, he prepared his Doctor of Science dissertation, the highest advanced degree possible in the USSR. A Doctor of Science was awarded to a select group of scholars; fewer than 10% of people who had PhDs, also earned a Doctor of Science for their achievement in science. Such a degree automatically qualified one for full professor positions at any university of higher education. Gregory successfully defended his Doctor of Science in 1987. His knowledge of technical subjects gave him an edge in the field and a unique character to his work. In the following two decades he was able to produce several monographs in the fields of ergonomics and engineering psychology.

In the US, Gregory's road to science was a very difficult journey. He immigrated in 1990 with no English and no connections in the scientific community. But he didn't abandon his passion for science, and he succeeded spectacularly. His first book in English *The Russian Theory of Activity* was published in 1997, just seven (!) years after his arrival in the country. Despite the obstacles, he went on to publish a tremendous body of work. He created systemic-structural activity theory (SSAT), produced several fundamental scholarly monographs on human factors/ergonomics, engineering psychology, and on some theoretical issues, as well as numerous articles published in respected journals in many countries. Every time his newest weighty book came out, Gregory would joke, "Another brick into the foundation of science."

To the scientific community he was the respected Dr. Gregory Bedny. Several years ago, he was invited to give lectures in two universities in the UK. After he finished his presentation of the first one, he was driven to the second location. There he noticed some faces of those who listened to his presentation the day before. He referred to them with a smile, "Why are you here? I am not a movie star." They responded, "We want to listen to you again." Gregory often received requests from students of domestic and foreign universities to lead their PhD dissertations based on his scientific work. For his theoretical research, the Russian Academy of Aviation and Aeronautics awarded Gregory the title of Honorary Academician and the Academy Medal.

Gregory participated in many International Conferences on Applied Human Factors and Ergonomics (AHFE). He chaired the SSAT section and presented his own work. At the conference in Krakow, Poland, where Gregory's wife Inna and I also participated, we witnessed a funny episode. A young scientist from Switzerland approached Gregory and pronounced with a happy face, "Doctor Bedny, Doctor Bedny, I can't believe I am talking to you. I read many of your works…" And he did not let Gregory go for about half an hour. But Gregory did not mind. As the Russian proverb says, "Don't feed him, just let him do what he loves."

The 9th AHFE Conference was held in Florida in July of 2018. He was preparing to go there, but unfortunately his health condition did not allow it. Gregory passed away on July 21, 2018, just a day before his section at the conference was supposed to start. His legacy as a friend is in his personality. He was a man of giving, never expecting anything in return, and was genuinely happy for his friends' successes. Gregory was an excellent professor and his students loved him. His legacy as a scientist is in his works. Over the course of his life, he wrote nineteen books (in Russian and English). In recent years, when he began working on another book, he used to say, "This is my last one." Well, now we can say with sorrow that this *is* his last book. In this collection of articles Gregory gathered authors in different fields, whose research is within the frame of the systemic-structural activity theory. In the book readers can see that SSAT found its reflection in a variety of scientific and applied areas of research and practice.

Apart from his legacy as a scholar, he left behind a loving wife, daughter, and granddaughter as well as many loving friends and relatives whose lives he changed for the better. His life was not easy but it was brilliantly successful.

Fred Voskoboynikov

Preface

This collection of articles presents data obtained in applied activity theory (AAT) and in systemic-structural activity theory (SSAT) that can be used in the study of human performance.

The principles developed in AAT and in SSAT can be utilized in the study of various types of occupations. Specifically, recently obtained data can be utilized for the study of extremely complex human-machine and human-computer systems, and for evaluation of efficiency, complexity, and reliability of such systems.

Advanced principles of standardized and unified description of human performance are suggested in this book. We believe that readers will find these principles useful in their own studies.

The examples given in the articles emphasize the possibility for practitioners to apply the described methods in their work.

We also believe that the material in this book will be useful not only for psychologists, ergonomists, and industrial engineers, but also for students of graduate and undergraduate programs specializing in these fields.

The strength of the book lies in the careful selection of works of different authors that complement each other and presenting a holistic view of the field of human performance.

The material offered in the book is written in clear and easy-to-read language. By striking a compelling balance between the presented materials in various fields of human performance study on the one hand and their interdependence on the other hand, this will be a useful resource for readers interested in the field.

The book consists of three sections.

Section I is called "The Concept of Self-Regulation in Studies of Human Activity." This section consists of six chapters with an emphasis on the self-regulation of activity in the study of human performance.

The concept of self-regulation of activity was developed in the framework of SSAT, and it is well adapted to the task analysis.

Activity self-regulative principles are utilized for uncovering the most effective strategies of task performance. The term strategy is one of the basic concepts in the analysis of activity self-regulation. Strategy is dynamic and adaptive, enabling changes in goal attainment during task performance. Selection and design of the most efficient strategies of task performance in different domains is the main purpose of the first chapter of this section written by Gregory Bedny, Inna Bedny, and Waldemar Karwowski.

In this chapter there is a brief comparative analysis of general, applied, and systemic-structural activity theories. It gives readers a good overview for the further understanding of the topics described in the following chapters. In contrast to applied activity theory, which presents useful data that is not integrated in one theoretically based approach, the SSAT is a unified theory. In this chapter, it is demonstrated that SSAT has rigorous standardized terminology and procedures. Analysis of some basic terminology and methods of study are given.

The authors of the second chapter are Gregory Bedny and Inna Bedny. The topic of this chapter is the analysis of variable tasks with probabilistic structure and reliability of their performance. This paper suggests a new method of analysis of variable tasks with complex logical organization requiring multiple decisions during their performance. Professional ergonomists encounter difficulties when analyzing the performance of such tasks. The scale of the subjective probability evaluation of events is suggested. The probabilistic characteristics of activity structure can be described based on the obtained data. Consequently, this data can be utilized in the analytical methods of task complexity evaluation and reliability of performance.

Chapter three is collaboration between Julian Vince and Gregory Bedny. This chapter is dedicated to the ontological nature of autonomous technologies from an activity theory perspective in terms of their role in socio-technical systems. This work highlights human-autonomy interaction. The SSAT is utilized to describe the systemic nature of behavior and cognition in human-autonomy interaction through the perspective of goal-directed, tool mediated activity.

The fourth chapter, written by Alexander Yemelyanov, provides an analysis of the existing viewpoint on decision-making under risk and uncertainty with the focus on the motivational approach from the SSAT perspective. This chapter covers assessment of the level of motivation; rules of motivation and self-regulation; and the self-regulation model of decision-making. The proposed Performance Evaluation Process (PEP), in which the goal, selection criteria, and mental model are formed during decision-making is based on this model. Express Decision (ED, a mobile application for quick everyday decision-making, is developed.

The fifth chapter, presented by Gregory Bedny and Inna Bedny, offers models of cognitive processes such as perception and memory. These processes are considered as complex goal-directed self-regulative systems. The models provide the framework for an analysis of cognitive processes when studying man-machine or human-computer interaction systems. They can serve as a useful tool for conducting task analysis when perception or memories are critical factors.

The last chapter of this section is written by the same authors and is called "The Study of Work Motivation and the Concept of Activity Self-Regulation." The authors discuss the relationship between the concepts of goal and motives, the stages of motivation and their role in the activity of self-regulation, and regulatory aspects of motivation and stages of work motivation. Motivational aspects of self-regulation, their specific features, and their contradictions are also analyzed in this chapter. New viewpoints on the stages of work motivation and their role in the activity of self-regulation, and regulatory aspects of motivation are offered as well.

The second section of the book is called "Efficiency of Performance and Quantitative Methods of Its Assessment."

Written by Alexander Yemelyanov and Alina Yemelyanov, the seventh chapter presents psychological experiments for analyzing emotional motivational mechanisms and their role in regulation of pilots' work activity in stressful situations. The authors of this chapter describe the experiments that were conducted by Mikhail Kotik at Tartu University, Estonia, in the period of 1966 to 1994. The method that

Kotik used allowed assessment of not only conscious emotional manifestations, but their subconscious motives as well. The method also makes it easier to implement and interpret the resulting data. In addition to experiments characterized by a negative valence, further experiments evaluated people's attitude toward attractive or appealing events and their social, material, and physical aspects.

Ephraim Suhir presents the eighth chapter. It deals with the application of mathematics in psychology and ergonomics outside of traditional statistical methods. Limitations of existing quantitative methods are considered. The main theoretical factor in quantitative analysis is a consideration of the relationship between the human capacity factor and the existing mental workload. The taken approach, with the appropriate modifications and generalizations, can be applicable to various practical situations. The author also discusses the "Miracle on the Hudson" accident that took place in 2008.

The ninth chapter, authored by Gregory Bedny, discusses a method of analyzing production processes in construction industry by using technological units of analysis. In SSAT there are two types of analysis units. One type describes elements of work activity during performance of various tasks utilizing technical terminology or common language. The other method uses standardize psychological terminology to describe elements of activity. This combination provides a clear understanding of task description, and helps to create analytical models of task performance. When physical work dominates, the first type of analysis unit should be used. The new method of chronometrical study that is based on this description is suggested, and a quantitative assessment of the efficiency of task performance is presented in this chapter as well.

Iuriy Solonin presents the last chapter of this section. The issues of evaluation and optimization of physical workload and stress during task performance is considered in this chapter. The influence of such factors as an interaction of physical workload and microclimate characterized by extreme heat is shown. Recommendations on optimizing the physical state of operators in hard environmental conditions are suggested.

The last section of the book is called "Management and Education."

Fred Voskoboynikov authors chapter eleven. Activity theory distinguishes two types of activity: "object-oriented" and "subject-oriented." The former refers to a subject using tools on material objects with the goal of completing the task and evaluating the results. The latter refers to social interaction between people, which is the most important element in management. The chapter is dedicated to methods for increasing the effectiveness of managerial work. The influence of the group structure on individual and group performance, particularly on operators' performance efficiency, is discussed in this chapter, as well as the phenomenon of psychological compatibility. Some aspects of decision-making in management and management styles are also considered.

Fred Voskoboynikov also authors chapter twelve. It covers the systemic approach of applying psychological factors to the practice of management. The emphasis is on taking into account people's individual personality features in the process of managing their activity with the goal of improving their performance and preserving their health. The influence of a group environment on individuals' behavior and performance, and phenomenon of compatibility are analyzed as well. The importance of systematic communication between the manager and subordinates, and with other contacts managers have to deal with on a regular basis, are considered here.

In the final chapter of this volume, author Mohammed-Aminu Sanda presents an analysis of the concept of activity from the cultural-historical, general, and systemic-structural activity theory perspectives. Different models of activity are analyzed here. The structure of activity is considered from different theoretical perspectives. Task complexity and emotionally motivational factors in the operator's performance are described using examples from the study of miners' work.

The editors of this book present their own theories and methods along with a variety of other studies of human performance to demonstrate different approaches with the mutual goal of improving its reliability and efficiency.

Editor Biographies

Gregory Z. Bedny taught psychology at Essex County College. Dr. Bedny earned a PhD in industrial organizational psychology and a post-doctorate degree (ScD) in experimental psychology. He was also a board certified ergonomist. He authored and co-authored seven original scholarly books in English published between 1997 and 2018 by Lawrence Erlbaum Associates, Inc., Taylor & Francis Group and CRC Press. Dr. Bedny developed systemic-structural activity theory, which is a high-level generality theory that offers unified and standardized methods for the study of human work. He applied his theoretical studies in the field of human-computer interaction, manufacturing, merchant marines, aviation and transportation, robotic systems, work motivation, training, and reducing fatigue. CRC Press published his latest book *Work Activity Studies within the Framework of Ergonomics, Psychology, and Economics* in 2018.

Inna S. Bedny is a computer professional with a PhD in experimental psychology. She conducts research applying SSAT to human-computer interaction. She has authored and co-authored a number of scientific publications in the field of human-computer interaction. In 2015, she co-authored the book *Applying Systemic-Structural Activity Theory to Design of Human-Computer Interaction Systems* published by CRC Press. She has also taught math and physics at a high school level, and has trained computer professionals. Dr. I. Bedny has worked in the information technology department of UPS for the last two decades.

Section I

The Concept of Self-Regulation in Studies of Human Activity

1 Applied and Systemic-Structural Activity Theories

Gregory Z. Bedny
Essex County College, New Jersey, US

Inna Bedny
United Parcel Service, Atlanta, US

Waldemar Karwowski
University of Central Florida, Florida, US

CONTENTS

1.1 INTRODUCTION

In this chapter we present a brief overview of the applied and systemic-structural activity theories with specific focus on the SSAT. General activity theory and the works of its founders Rubinshtein, Vygotsky, and Leont'ev' will not be analyzed in great detail here because it was discussed in previous publications (Bedny and Meister, 1997; Bedny and Karwowski, 2007).

General activity theory was introduced by Rubinshtein (1959) and later received further development in the works of Leont'ev (1977/1978) and other scientists. General philosophical and psychological principles for the study of personality in psychology in accordance with the concept of activity theory (AT) were first formulated by Rubinshtein (1934). Human development, according to these principles, is the result of interaction of material and social practice with human individuality. Personality is developed through a person's participation in activity, which depends on the relationship between a subject, a situation, and their social interaction. This principle eliminates the contradiction between social and intra individual aspects of human development. The difference between Rubinshtein's and Leont'ev's schools of activity theories has been discussed in Bedny and Karwowski (2007). The Leont'eve's concept of activity theory

3

was introduced to English-speaking readers through the translation of Leont'ev's book (1978) and collections of articles edited by Wertsch (1981). However, general activity theory is not applicable to the study of human performance. An analysis of the interpretation of basic concepts of activity theory in the West demonstrates an unfortunate failure in attempts to capture the original meaning of terminology in this field. It is not surprising that the attempts to adapt general activity theory to the task analysis in general and to human-computer interaction specifically were ineffective. Even Leont'ev, who was considered the leading psychologist and philosopher in the field of general activity theory, made mistakes when he tried to apply it to work studies. For instance, in his book (Leont'ev, 1978; 1977, p. 107) he explained the difference between cognitive or motor actions and psychological operations as their components.

To demonstrate the meaning of the term *psychological operations*, Leont'ev considered the motor action that consists of motor operations as smaller units of analysis in comparison with motor actions. He analyzed the locksmith task as follows: "It is possible to divide metal object physically utilizing different instruments each of which defines how motor actions are performed. In some conditions cutting operation would be adequate, and in other cases sawing operation would be more adequate. It is assumed that a worker should be able to use the corresponding tools, like a knife, or a saw, or others." The errors in this interpretation of psychological operations is obvious because he mixed the production task with psychological operations as components of a motor action. When a worker is cutting a metal or wood object, they perform various motor actions that include several motor operations. For example, a worker moves a hand and grasps an instrument. This first motor action includes two motions or psychological operations "move hand to the instrument" and "grasp it." Similarly, a worker moves their hand with an instrument and installs it into the specific position, and so on. A motor action integrates several motions, which should be considered psychological operations, by the goal of action. Each motor action can be decomposed into motions or psychological operations. Similarly, dividing a material object by means of a saw is a technical operation or a task. As can be seen, Leont'ev's understanding of psychological operation can be confused with terms such as technical operation, cutting operation, etc. In SSAT the term "motion" is used for the psychological operation of motor actions. The concept of motion is also central in time and motion studies (Barnes, 1980). However, the concept of motor action as a goal directed component of human activity does not exist in this field.

Let's now consider cognitive action. Cognition is not simply a system of cognitive processes; it is a system of cognitive actions. A standardized description of cognitive and behavioral actions is necessary in order to describe activity structure. Comparison of activity structure and configuration of equipment or interface is the basic principle of design in SSAT (Bedny, 2015; Bedny, Karwowski, and I. Bedny, 2015; Bedny and I. Bedny, 2018). Cognitive actions usually have short durations and are completed when the goal of these actions is achieved. Cognitive actions are classified based on the dominant cognitive processes similar to motor actions' classification. For example, they are goal directed, they have a beginning and an end, and can be considered self-regulative elements of activity. Cognitive actions transform not material objects but information, and mental images according to the goal of action. There are two methods of describing motor and cognitive actions. One method involves utilizing technological terms or terms that describe actions by common language. For example, taking a reading from a digital display is an example of a perceptual action that utilize common language or technological terminology. Depending on the illumination, the distance of observation, and the constructive feature of display, the content of an action and its duration can vary. Such description does not facilitate the clear understanding of what action a subject performs. In addition, the standardized psychological description of action or psychological units of analysis should be introduced for this purpose. If we describe the above mentioned action as the simultaneous perceptual action with a duration of 0.35 sec in addition to technological units of analysis, it would be possible to clearly understand what action is performed by a subject. Such a method of an actions' description has been developed in SSAT. This method is critically important for design in engineering psychology, ergonomics, and for labor economics.

1.2   BASIC CONCEPTS OF SYSTEMIC-STRUCTURAL ACTIVITY THEORY

Activity is the basic concept in activity theory. In general activity theory *activity* is understood as a purposeful interaction of the subject with the world, a process in which mutual transformations between the poles of "subject-object" are accomplished (Leont'ev, 1978). The SSAT definition of activity has a practical application. We understand activity as a self-regulated system that integrates cognitive, behavioral, and emotional-motivational components, and is directed toward achieving a conscious goal of activity (Bedny, Karowowski, 2011a, Figure 1.1). The relationship between external and internal components of activity in SSAT is also described in Bedny, et al (2011b) Activity in SSAT is described as a system consisting of sub-systems and smaller elements that are in a specific relationship with other elements of activity.

In activity theory there are concepts such as object and subject-object relationship. An object of activity that can be material or mental (symbols, images, etc.) is something that can be modified by a subject according to the goal of the activity. There is also subject ↔ subject interaction when subjects interact with each other by using speech, material, and mental objects. Three basic terms of activity theory are hierarchically organized basic activity elements: *activity* → *cognitive* or *motor actions* → *psychological operations*. Activity is an object of study, and cognitive and

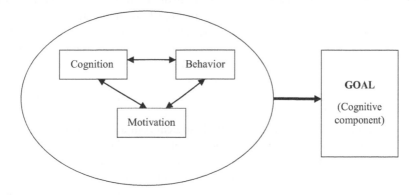

FIGURE 1.1   Simplified schema of activity as a system.

behavioral actions and psychological operations are basic units of activity analysis. These notions will be used in future discussions.

Our lives can be conceptualized as s continuing performance of various tasks. When studying human performance, "task" should be considered the basic component of work activity. The main purpose of job analysis is the description and optimization of tasks that are involved in human activity, discovering their organization, and conducting the task analysis. From the SSAT perspective, task analysis involves describing the structure of activity during task performance. For this purpose, we use formalized qualitative and quantitative methods. Production operation can be considered synonymous to the term task when studying traditional types of work. Job and task analysis have their unique meaning for the equipment design, and methods of task performance, as well as for analysis of safety, training, and education. In a production environment, tasks are performed in a particular order. Each work process consists of a number of tasks. From an SSAT perspective, a task consists of logically organized cognitive and behavioral actions that are directed to achieve the goal of task (Bedny, Meister, 1997; Bedny, Karwowski, 2007). Therefore, general hierarchical scheme of work activity can be presented as *work activity, task, cognitive and behavioral actions,* and *operations.*

Cognitive psychology treats cognition only as a process, making it difficult to study human cognition as a system. SSAT describes cognition and external behavior as a structurally organized system. Cognition is not merely a process or a mental picture of the world, but is also a system of mental actions and operations intimately related to external motor actions (Bedny, et al., 2001). As in physics, where light has both wave and particle characteristics, in SSAT, cognition is understood both as a process and as a system of actions or other functional information units. Cognition incorporates both a process and a structure.

Concepts such as goal and task are critically important in SSAT. These basic concepts are understood differently in SSAT as they are in cognitive psychology and even in general activity theory. The concept of goal is utilized in various fields of science: in psychology, cybernetic, management, engineering, and so on. Here we describe the concept of goal from an SSAT point of view. A goal is one of the most important anticipatory mechanism of activity. Goal is a mental model of a desired future result that is formulated by a subject during their activity. It plays a critical role in task performance. It is not accidental that Leont'ev stated that a task is a situation requiring achievement of the goal in the specific conditions (1977/1978). More specifically, Leont'ev meant the goal of task. However, there are also goals of actions that separate one action from another. A goal of a task organizes all task elements into a holistic system. Goal of activity and of a task can be modified during performance. If a subject changes the goal completely, they are performing a new activity or a new task. In SSAT, the goal is just a cognitive component of activity that includes conscious imaginative and verbally-logical components. Therefore, the goal should be at least partly verbalized. Without awareness of the goal, there is no goal of human activity. Existence of the conscious goal is one of the most important factors that differentiates reactive animal behavior and human activity.

Even highly automated actions should be distinguished from reactive behavior. For example, a very quick response to an emergency signal looks like a reactive response. However, this is not reaction but a meaningful and purposeful action because it has a specific goal or a desired future result. In task analysis we distinguish an overall goal of task, a partial or intermediate goal of actions, and sub-goals of a tasks. A goal cannot be presented to a subject in a ready form and should be distinguished from requirements of the task. Usually requirements are presented as instructions. Only after interpreting these requirements can the subject transform them into the goal. During this process the subject compares requirements with the past experience and the motivational state. This stage can lead to the acceptance of the goal, its partial modification, or even rejection of the goal. Hence, the objectively formulated requirements (can be named as objectively given goal) are not the same as the subjectively excepted goal.

Understanding and analyzing this difference is the critically important stage of task analysis. A subject can also formulate the goal independently. All these aspects of acceptance and formulation of a goal are ignored in cognitive psychology, where goal has a different meaning. Thus, we can conclude that goal does not exist in a ready form and cannot be considered simply as the end state to which human behavior or activity is directed. A lot of errors occur at the stage of goal interpretation, acceptance, and independent goal formulation.

Emotionally-motivational factors are involved in goal interpretation, acceptance, and in its independent formulation. The desired future result emerges as a goal only when it is accompanied by motivation and a subject is involved in an activity for achieving the result that matches the goal. For example, a subject wants to earn money, but in order to reach this goal they have to begin working. Therefore, an imaginative future result is transformed into a goal only when a desired future result that is accompanied by motivation, desires, and wishes becomes motives of the subject's activity. Goals and motives are often interpreted incorrectly. For example, in goal setting theory, a goal integrates two primary attributes: content and intensity (Lee et al., 1989). Intensity is considered a motivational factor. According to this theory, the more intensive the goal is, the more the subject desires to achieve it. However, goal is just a cognitive mechanism that should be distinguished from motivational factors. The goal can be precise or imprecise, correspond to the given requirements or not, but it cannot be intensive. Leont'ev (1978) interpreted a motive as a material or ideal object of need and stated that needs are *objectified*. An object cannot be a motive, but is rather a source of motives or motivation. On the other hand, needs can turn into motives when the goal of an activity is to satisfy these needs. For instance, a boy has a natural ability for gymnastics. He has the need to be a good gymnast but he also wants to spend time with his friends. That is, he is confronted with the choices of to be involved in an intense training process, or spend time with his friends and play videogames. If he chooses the first scenario and dedicates his time to a serious and systematic workout in the gymnastics club, he formulates the goal to be a good gymnast. If he chooses the second option, he formulates the goal to enjoy good time with his friends. Only if he starts to practice hard, can we say that his need to be a good gymnast has transformed into his motive. The needs are transformed into the motives only when a subject connects the motives to the goal. We can say that motives are connected to the goal and metaphorically can be presented as a vector *motive → goal*. Leont'ev (1978) said that in some situations the motive can be shifted into the goal. We cannot agree with this interpretation because in this case the vector *motive* → *goal* would be transformed into a dot and the activity would lose its goal directedness.

Some theorists (Locke, Latham, 1984; Pervin, (Ed), 1989; Lee et al., 1989) postulate that the goal has both cognitive and affective features. They state that the goal can be weak or intense and consider the goal as a motivational factor. According to them the more intense the goal is, the more one strives to reach it. Hence, the goal "pulls" the activity. They, and some other scientists, consider goal a source of inducing behavior. In other words, the goal and motives or motivation are not distinguished in these theories.

From the SSAT point of view the more intense the motives are, the more a person will expend their efforts to reach the goal. We can say that the more intensive the motives are, the more they push a subject to reach the goal. The goal should be considered a conscious mental representation of a future result that is connected with motives.

In our discussions, we demonstrated the difference between the goal and motives by using a psychological approach. We will consider this issue from the neuropsychological perspective. According to Luria (1970), the most general functions of the frontal lobe cortex are temporal organization and temporal integration of goal-directed cognitive and behavioral actions. In other words, it is the temporal organization of human speech, reasoning, and behavior. The human frontal lobe provides *representational processing* that allows a subject to interpret reality, to understand past events, and based on them, to predict future events. This part of the brain is responsible for creating the mental model of reality. The mental model should be understood as a construction of dynamic mental scenarios of various situations that can unfold in time and space. A subject can act based on the mental model created in accordance with the goal and on the specifics of the environment. The frontal lobe plays a leading role in temporal integration and regulation of goal-directed activity. This part of the brain is also involved in monitoring and/or self-regulation of motor actions. The data existing in neuropsychology demonstrates that the frontal lobe region is critical in keeping the goal in mind, and directing activity toward the goal.

Emotionally-motivational processes involve different parts of the brain. Their structural relationship plays a leading role in goal directed conscious activity. One part of the brain is known as reticular activating system (RAS) and is responsible for activation of consciousness. Activating system of the brainstem controls awareness and attention. There are other parts of the brain that are responsible for our emotions. The brain areas that are responsible for our emotions include the limbic system, which contains such regions as the amygdala, hippocampus, hypothalamus, and anterior cingulate cortex (ACC). Another subcortical region of the brain that is involved in emotionally-motivational processes is the retrosplenial cortex (RSC), which is a part of the posterior cingulate cortex. One important mechanism of motivation is the dopamine system, which depends on the functioning of the brain's ventral tegmental area (VTA). We are not going to discuss this topic in great details here and address readers to our prior publications (Bedny, 2015). In general, we can say that different parts of the brain are involved in formation of the goal and emotionally-motivational processes. In summary, the goal cannot be considered as a unitary mechanism that combines informational (cognitive) and energetic (emotional and motivational) components of activity. Therefore, the concept of the vector *motive* → *goal* has not only a psychological basis but a neuropsychological basis as well. A detailed discussion of the relationship between a goal and motives from the neuro-psychological perspective is presented in Bedny (2015).

Understanding the concepts of goal and motives is an important factor in the development of goal directed models of self-regulation of activity. The overview of all existing models of self-regulation developed in control theory and cognitive psychology can be found in the work of Bedny, Karwowski, and I. Bedny (2015). We agree with Lock (1994) that the theories of self-regulation outside of SSAT are too mechanistic and should be abandoned altogether. Nevertheless, the concept of self-regulation is critically important in psychology. In SSAT, self-regulation is defined as the influence on a system that derives from the system itself in order to correct its own behavior or activity.

1.3 CONCEPT OF SELF-REGULATION IN SYSTEMIC-STRUCTURAL ACTIVITY THEORY

The concept of self-regulation is critically important in AAT and especially in SSAT (Kotik, 1974; 1978; 1987; Konopkin, 1980; Bedny, 2015; Bedny, Karwowski, I. Bedny, 2015). SSAT offers the original concept of self-regulation that is described by two models of self-regulation of activity: the model of self-regulation of orienting activity and the general model of activity self-regulation. We also developed the self-regulation models of separate cognitive processes. The concept of activity self-regulation in SSAT is adapted to task analysis. The main purpose of studying activity self-regulation is continuing reconsideration of activity strategies or even changes in the goal of activity when internal and external conditions or situation have changed. Strategy is a plan or a program of performance that is responsive to external contingences, as well as to the internal state of the system. Strategy is dynamic and adaptive, enabling changes in goal attainment as a function of the external and internal conditions of a self-regulative system. It is achieved by a systemic qualitative analysis of the task that is performed by a subject.

The psychological self-regulation is a goal directed process. The system can change its own structure based on its experience. Such a system can form its own goals and sub-goals, and its own criteria for an activity evaluation. Psychological self-regulation integrates cognitive, executive, evaluative, and emotional aspects of activity. It includes conscious and unconscious levels of self-regulation that are interdependent. Goal and verbally logical components of activity play the leading role in the conscious level of self-regulation, whereas imagination, intuition, and non-verbalized meaning play this role in an unconscious level of self-regulation set.

When studying work activity, quite often it is not productive to consider such mental processes as sensation, perception, memory, thinking, etc., separately because they are interrelated, perception involves memory, thinking also depends on memory, and so on. This is why the activity during task performance should be studied not only in terms of cognitive processes, but also in terms of functional mechanisms or functional blocks. Functional mechanisms or functional blocks are main units of activity analysis and they also represent the stages of self-regulation.

The functional model of self-regulation consists of various functional blocks. Hence, any function block includes various combinations of cognitive processes. However, their integration can be carried out in various ways depending on the specificity of the task performed by a subject. The content of each function block can also be described in terms of cognitive and behavioral actions. A function block represents a coordinated system of sub-functions that has a specific purpose in activity regulation. For example, there is a function block that is responsible for the process of goal formation; the other one is responsible for creation of the conceptual model of a situation; there are blocks that involve evaluation of significance of activity, evaluation of its difficulty, or formation of the level of motivation, and so on. In the process of each task analysis the most relevant function blocks are selected for each considered task. The relationship between the considered blocks also should be analyzed. Self-regulation can be performed in the internal mental plane when a subject uses internal or mental feedback while performing various cognitive actions. A model of activity self-regulation can be interpreted as interdependent systems of windows (function blocks) from which one can observe task performance from various angles. For example, a researcher can open a window called "Goal" and concentrate on activity aspects such as goal interpretation, goal formation, and goal acceptance. At the next step they can open another box called "subjectively relevant task conditions." Here a researcher would study such aspects of activity as "operative image" and "situation awareness" that partly overlap. There are other function blocks such as "meaning of input information," "assessment of task sense (significance)," "assessment of task difficulty," etc. The interrelation of data obtained using different function blocks should be taken into account during task analysis. In order to use the SSAT models one should be familiar with their theoretical foundation.

In the process of self-regulation, a subject has the ability to formulate goals, and to change the way of achieving the goal while performing various tasks. The process of self-regulation includes not only conscious but also unconscious components. Analysis of activity based on developed models of self-regulation is called functional analysis because the main units of analysis are function blocks or functional mechanisms.

Below we present the model of self-regulation of orienting activity. The orienting activity precedes execution (see Figure 1.2). In cognitive psychology, an activity that precedes execution is known as situation awareness (Endsley, 2000). In our model of self-regulation, it is just one sub-block or mechanism of activity self-regulation.

We are not going into a detailed discussion of this model because it is described in the work of Bedny, Karwowski, and I. Bedny (2015). We only present the basic terminology that is necessary for understanding of this model (the considered blocks are not described in numerical order).

The function block *goal* (2) is the integrative mechanism of self-regulative process that interacts with motivation and creates the vector *motive* → *goal*. This vector gives direction to the self-regulation process. Studies show that two people may have an entirely different understanding of a goal, even if objectively identical

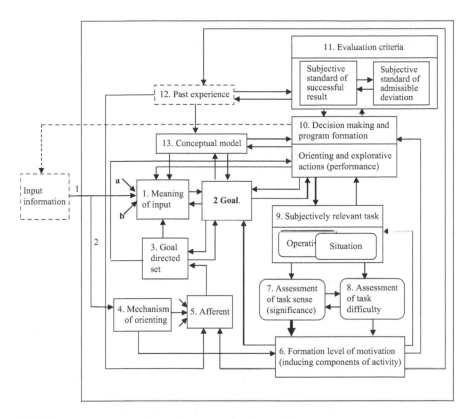

FIGURE 1.2 Model of self-regulation of orienting activity.

requirements or instructions are given. Hence, we distinguish between "subjective" and "objective" interpretation of the goal. As was mentioned above, a goal should not be considered something externally given to the subject in a ready form. A goal is always associated with some phase of activity, and includes stages such as goal recognition, goal interpretation, goal formation, and goal reformulation.

The function block *meaning* (1) is involved in interpretation of input information (can be extracted not only from external data but from memory as well). It reflects the relationship not only between a sign and its referent but also the connection between a sign and activity. The function block "meaning" in its study of the relationship between a sign and its meaning considers not only an individual but also the culture created by human activity. The model considers conscious and unconscious meaning. Conscious meaning has verbal and conscious aspects. In these aspects of self-regulation, the function block "meaning" is associated with a conscious goal. When the block "meaning" works together with the function block "set" it is involved in extracting "non-verbalized situational meaning."

The function block *sense* (7) reflects emotionally evaluative aspects of activity and personal significance of its various components. Personal significance within the goal-directed activity leads a person to interpret the meaning of the presented information and transfer it into the subjective sense.

The function block *assessment of task difficulty* (8) covers the awareness of the objective complexity of the task, as well as some intuitive assessment of complexity. The more complex the task is the greater the probability that the task will be difficult for a subject. A subject can evaluate the same task as more or less difficult depending on their past experience or individual features. Therefore, the cognitive effort and inducing motivational components of activity depend on task difficulty. An individual can under- or overestimate the objective complexity of a task and this would influence strategies of task performance. Incorrect assessment of the task difficulty can result in an inadequate personal sense or motivation to sustain the efforts to complete the task.

The function block *orienting reflex* (4) describes the creation of conditions for a heightened receptivity of the body to sudden changes in the situation that is accomplished by development of complex, short-lived, and transitory physiological processes, i.e., the change of an activation level in the neural system with a general inhibition of conquering ongoing activity.

The function block *afferent synthesis* (5) provides analysis, comparison, and synthesis of all data that a body needs in order to attain adoptive responses in the given circumstances. The main stimulus that causes a reaction never exists in isolation. It interacts with supplementary environmental stimuli that influences what information is extracted from memory that is relevant to the response, current motivational state, and the response itself.

The function block *set* (3) is characterized by the role it plays in the formation of purposeful behavior. A set is responsible for creation of the internal state of a body that determines purposefulness of human behavior but this state is not conscious. A set creates a predisposition to processing incoming information in a particular way, or predisposition to perform particular actions. A set can be transferred into a conscious goal and vice versa. Therefore, a set performs similar functions at the unconscious level of self-regulation as a goal does at a conscious level.

The function block *subjectively relevant task conditions* (9) analyzes creation of a "dynamic model of the situation." It is involved in creating a holistic mental model of reality and includes two sub-blocks: *operative image* that to a large extent provides unconscious dynamic reflection of the situation and *situation awareness"*, which includes a logical and conceptual sub-system of dynamic reflection of a situation when a subject is aware of processing information. These two subsystems of dynamical reflection overlap. The overlapping part of the imaginative subsystem facilitates the conscious processing of information. Conscious and unconscious components of dynamic reflection can to some degree be transformed into each other.

The function block *conceptual model* (13) is responsible for developing a broad and relatively stable mental model, which serves as a general framework for understanding various situations relevant to particular professional duties. Although this model is general, and is stored in long term memory, it is more specific than the general past experience. An imaginative component is one of the distinguishing characteristics of this model, and plays an important role in its functioning.

The function block *past experience* (12) analyzes the general background of a subject that also influences strategies of performance and therefore can be considered as a functional mechanism. It includes general and professional knowledge of a subject, knowledge of culturally accepted norms of behavior, and customs that describe how the community functions. Past experience is acquired through activity that evolve over time within the culture. An interaction of past experience and of the new information results in the assessment of the meaning of the immediately inputted information. Past experience includes not only cognitive but also emotionally motivational components and evaluation of task difficulty.

The function block *decision-making and program formation of task performance* (10) involves analysis of decisions made during the task performance and development of the program for execution of actions directed to achieve the accepted goal. The program is developed prior to the task performance and can be modified during the performance.

The function block *motivation* (6) is responsible for the development of inducing components of activity. While the block "sense" refers to emotionally evaluative components of activity, the motivational block determines activity goal-directness and energetic components involved in attaining a specific goal. The function blocks *sense* and *motivation* are intimately connected, but sometimes emotionally evaluative components of motivation can be in conflict with inducing components.

The function block *criteria of evaluation* (11) is responsible for development of subjective criteria for success, which might deviate from objective requirements. This function block has two sub-blocks: subjective standard of a successful result that is responsible for the development of a subjective criteria of success that is often not the same as objective requirement; subjective standard of admissible deviation demonstrates that subjects define which errors are significant and which are not. If these deviations do not exceed subjective tolerance, subjects do not correct their actions. In the general model of self-regulation, where executive motor activity plays an important role, there are two additional blocks: 1) *positive and negative evaluation of result* and 2) *Information about interim and final result*.

Orienting activity precedes execution stage, and based on this characteristic is similar to situation awareness (SA) in cognitive psychology. However, SA is just one sub-block in this model. This sub-block involves verbalized conscious reflection of the situation during task execution. The overlapping part of the sub-block "operative image" is not involved in the unconscious reflection of the situation. The concept of SA is not sufficient for analysis of activity that precedes task execution. During task analysis, each function block determines the range of issues that are connected with this block, and should be considered when the role of this block in activity regulation is examined. It can be determined how functioning of separate blocks influences each other based on analysis of interconnections between these blocks. The detailed analysis of this model can be found in Bedny, Karwowski, and I. Bedny (2015).

Let us consider in an abbreviated manner an experiment that demonstrate the application of the concept of self-regulation to the study of positioning actions. Such characteristics as time, precision, and amplitude are the most important ones when studying such actions. The Fitts's Law is the most famous outcome in this area of study (Fitts, 1954). One specific aspect of Fitts's experiment from which he derived his low was that subjects had to move a metal stick between two targets with the maximum speed or move a metal stick from the starting position to the particular target with the maximum speed. The Fitts's study demonstrated that the duration of

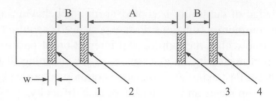

FIGURE 1.3 Layout of four targets: 1–4 targets; A and B—distance between targets and w—width of the targets.

the movement is linearly related to the logarithm of the index of difficulty. This index integrates two characteristics: amplitude and precision. When one tries to use this data in a working environment, the assumption is that each action is performed at the maximum speed, and each action does not depend on either previous or subsequent action. These assumptions contradict all data that has been obtained when studying the coherent activity structure. Actions are logically organized and influence each other. The Fitts's Low reflects a behavioristic approach, when human behavior is considered as a sum of independent reactions that are considered to be independent from each other.

Below we demonstrate in a concise form new data gathered in the study of positioning actions from the perspective of activity self-regulation. In our experiment subjects hit two targets and four targets (see Figure 1.3). The distance between targets 1 and 2, and between targets 3 and 4 was 60 mm. The distance between targets 2 and 3 was 120 mm. The width of each target in the first series was 50 mm and in the second series 7 mm. Subjects had to hit targets with maximum speed and precision the same as it is in Fitts's experiment. The main purpose of the study was to determine the strategies and performance time of positioning actions in various conditions.

The following function blocks are important in the analysis of positioning actions.

1. Goal (block 2).
2. Subjectively relevant task conditions (block 9) that is responsible for the development of stable or dynamic mental model of task or situation.
3. Assessment of task difficulty (block 8).
4. Assessment the sense of task or task significance (block 7).
5. Formation of the level of motivation (block 6).
6. Evaluation criteria block includes two sub-blocks: *subjective standard of successful result* and *subjective standard of admissible deviation* (block 11).

Analysis of these mechanisms is carried out by taking into account their interaction based on feed-forward and feedback connections presented in the model of self-regulation. Let us briefly consider performance of positioning actions from the perspective of activity self-regulation.

Block 2 (*goal*) and block 9 (*subjectively relevant task conditions*) are involved in creating a mental representation of the task or its mental model. Our findings supported the assumption that the instruction "hit the target at the maximum speed and precision" gave the subjects an opportunity to widely vary their subjective mental representation of the task. Some subjects considered precision to be more important than speed, while others focused mostly on the speed of performance. So the results for speed and precision were different for these two categories of subjects. The other factor that influenced the strategies of task performance was the specific organization of the task elements. As can be seen, the distance between targets 1 and 2 was the same as distance between targets 3 and 4, but they were shorter than the distance between targets 2 and 3. This fact affected the subjects' strategies of attention concentration. Depending on the distance between targets and their quantity, subjects utilized distributed or switching attention strategies. Consequently, the first and the third actions have been considered as the primary ones, and the second action with the larger inter-target "A" has been viewed as auxiliary. This fact influenced strategies of attention concentration, and selection of various strategies of attention concentration in turn influenced the speed and precision of each action's performance.

Function block 8 (assessment of task difficulty) explains how subjects achieved the required precision. Hitting the 50 mm wide target is seen as a relatively easy task, because such targets are subjectively wide enough. The subjects feel that they can manage the speed and precision. This feeling influences motivational block 6 and emotional-evaluative block 7. The significance of speed requirements increases along with precision significance. The subjects become motivated to preserve precision and speed. This in turn influences block 11 *subjective standard of success*. For example, in series 1, when subjects had two targets (width =50 mm) located 60 mm from each other, approximately 80% of the hits were placed in the middle of the targets within a 35 mm range. This means that the subjective standard of success was much narrower than the objectively given standard. The hitting area of the targets and width of the targets is not the same. In series where subjects hit four targets subjects also altered their strategies. The target zone was expanded but remained narrower than the width of the targets.

Let us now consider subjects' strategies when the narrow targets were used. In this case the subjects utilized the entire target area and some of them even expanded the hitting area. As a result, the quantity of errors sharply increased, or the speed of performance decreased.

It has been discovered that when subjects start to hit four wide targets instead of two targets, the time of performance increased by 61%. At the same time, when subjects hit four narrow targets instead of two narrow targets, the performance time of hitting four narrow targets increased only by 14%. Therefore, transfer from two wide targets to four wide targets leads to a more significant slowdown in the pace of performance compared to transferring from two narrow targets to four narrow targets. This factor can be explained by analyzing the relationship between conscious and unconscious or automatic control of actions regulation. When subjects transfer from two narrow targets to four narrow targets, a transition is observed from a simpler level of conscious self-regulation to a more complicated conscious level of self-regulation. When subjects transfer from two broad targets to four broad targets, there is a transition from an automatic level of self-regulation to the conscious level of self-regulation. The transition from an automatic to conscious level of action regulation has a more significant influence on performance time. The action performance time increases gradually within one level of self-regulation, but increases sharply when self-regulation changes from one level to another. This study demonstrated that the grater is the task difficulty, the slower is the pace of task performance. The detailed discussion of this experiment can be found in Bedny, Karwowski, and I. Bedny (2015) and in Bedny, G. Z., and Karwowski, W. (Eds.), (2013).

Thus, the positioning actions cannot be seen as independent ones in the context of the entire activity. Furthermore, it has been discovered that the distance between the targets influences the activity of the high-precision actions, but is not significant for the low precision actions. Positioning actions cannot be considered as isolated ones when they are a part of the task performance. Fitts described only one situation when a subject hits two targets with the maximum speed. According to Fitts's Law a subject can be compared to an informational channel with limited capacity for information processing. However, an individual is not a channel for information processing but rather a subject that actively interacts with the objective world and continually develops a variety of strategies based on self-regulation mechanisms. Emotionally-motivational mechanisms also play a significant role in performance of the positioning actions. Discovering strategies of performance of the positioning actions based on an analysis of activity self-regulation allows us to discover the relationship between speed and precision of such performance.

1.4   INTRODUCTION TO MORPHOLOGICAL ANALYSIS

In SSAT, morphological analysis follows qualitative analysis as the next stage of task analysis. It has been developed independently, and differs from the method of morphological analysis existing outside of SSAT. The purpose of morphological analysis is to discover and describe the structural relationship between elements of complex systems. Morphological analysis is the basis for further application of the quantitative methods of task analysis. In SSAT, such analysis precedes quantitative evaluation of task performance. Through analysis of the relationship between the structure of activity during task performance and the configuration of equipment or interface, it is possible to improve efficiency of design of the considered equipment or interface. Morphological analysis that includes algorithmic and time structure description of activity is the basis for quantitative assessment of the efficiency of task performance. We will consider this method in other chapters. Here we discuss a totally new method of eye movement data interpretation that should be conducted during the morphological analysis of activity.

The detailed description of this method can be found in Bedny, Karwowski, and I. Bedny (2015).

Currently, specialists can see significant progress in the development of technical means for recording eye movements. However, the interpretation of eye movement and its application in applied studies has not changed in a long time. There are many difficulties with the interpretation of eye movement data in applied studies due to the methodological problems researchers experience when trying to interpret eye movements. In cognitive psychology, the method used by Yarbus' (1965, 1969) is limited to analysis of perceptual processes. This traditional method is reduced to analysis

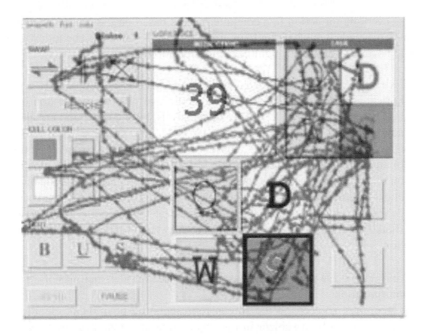

FIGURE 1.4   Cumulative scan path for one completed task.

of the cumulative scan path by using "the point of regard" data. For example, the cumulated scan path is plotted on the interface as demonstrated in the Figure 1.4.

Professionals utilize such data as the length of the scan path (pixels), the cumulative dwell time or average fixation time, and the average movement time. Such parameters of eye movement are useful for analysis of some aspects of task performance. However, this data is not sufficient for task analysis from the SSAT perspective. SSAT offers a totally new method of eye movement interpretation. Eye movement strategies can vary depending on the task goal, the subjective significance of information, motivation, and specificity of motor actions and their correlation with cognitive actions.

We have developed the new method of extracting cognitive and behavioral actions based on an analysis of the eye movement and external motor activity of the subject. As an illustration, we consider the possibility of separating cognitive and behavioral action in the above considered computer-based task in correspondence with parts of the presented cumulative scan path. Eye movement and gaze time in respected areas was the main criterion for extraction of cognitive actions in the visually based task. Eye movement registration was combined with the eye cursor movement registration. Thus, we do not use cumulative scan path for the eye movement interpretation. We divide the cumulative scan path of the eye movement into segments that can integrate only several cognitive actions. For this purpose, we selected the scan path between two mouse clicks. The analysis of dwell time that is associated with the particular area on the screen and its corresponding clicks gives us an ability to extract corresponding cognitive actions between clicks. The mouse clicks have been utilized because every motor action in a computer task is based on the preceding cognitive actions. This made it much easier to interpret the eye movements. The following data allowed us to determine the cognitive actions performed by the user:

1. Extraction of actions: a) each saccade and gaze is considered an action b) actions are classified based on the cognitive process that is dominant in a particular moment c) during this analysis, both correct and erroneous actions should be extracted.
2. Classification of actions: a) analysis of logical organization of actions b) analysis of the purpose of action c) relation of gaze to visible elements on the screen d) purpose of the following action and particularly the mouse clicks e) duration of the gazes and their qualitative analysis f) analysis of debriefing of the subjects and comparison of their reports g) all actions should be described according to the SSAT standardized action classification system.

As an example, we present actions extracted based on such methods of eye movement interpretation.

The goal area showed on the screen represents the position of elements that should be achieved by a subject in the object area as a result of the task performance. The object area is located in the button middle position on the screen. As can be seen, the initial positions of the letters do not correspond to their positions in the goal area. The tool area is on the left. Subjects had to move letters in the object area to match their positions with the goal area using the tool. This is the problem-solving task and the solution should be discovered. Various versions of the task have been presented. As can be seen in Table 1.1, a subject activated element D. The following sequence of actions has been identified. The eyes moved from the start position to the position G_Q (goal area element Q) with dwell time of 180 msec and total time of 330 msec. At this stage the subject only wanted to identify the task goal. Therefore, this is a perceptual action. The next eye movement is from G_Q to O_Q (object is element Q, duration of the dwell time is 220 msec and total time of this action is 370 msec). At the next step eyes move from element O_Q to T_{CB} (tool element, color blue). This move also identifies a perceptual action. At the next stage eyes shift from the tool are to the goal area again (from T_{CB} to G_S— goal area, letter S). The purpose of the movement is not only to receive information (perception). The subject is trying to discover the relationship between elements of the situation, which is not a perceptual action. This is a thinking action performed based on perceptual data. The duration of this eye movement increased as a result of it. The eye movement from G_S to element G_D is also a thinking action because a subject is working on discovering the relationship between elements of the situation. Let us consider the last action. Before activating element D in the object area, the subject should perform a decision-making action to choose the position in the tool group. Hence, the mental action before the click of the *vertical position* tool can be classified as a decision-making action that is based on visual information. It is worth noting that the longer the dwell time is, the higher the probability is that the thinking process is involved in such a complex action. Experts can define

TABLE 1.1

Action classification table (fragment before the first click).

1		2	3	4		5	6	7
Eye Move And Final Position		Activity between successive mouse events (clicks)	Mouse events	Time (ms)		Total Action Time (a + b)	Classification of cognitive actions	Scan path generated/duration
From	To	Mental/motor Actions Involved		a. Approximate Eye Movement time to reqd. position (Ms)	b. Approximate Dwell time at position (Ms)			
Start	G_Q	Goal acceptance and formation,		150	180	330	Simultaneous perceptual actions	
G_Q	O_q	comparison of object and goal areas for creation of		150	220	370	Simultaneous perceptual actions	
O_q	T_{CB}	subjective model of situation relevant to		180	150	330	Simultaneous perceptual actions	
T_{CB}	G_S	accepted goal: Formation sub-goal (selection of		180	220	400	Explorative-thinking action based on visual information	Image 1
G_S	G_D	element in object area O_D for subsequent task		150	190	340	Explorative-thinking action based on visual information	
G_D	O_q	execution. (includes simultaneous perceptual actions with combination		210	220	430	Explorative-thinking action based on visual information	
O_q	O_w	with explorative thinking actions).		150	330	480	Explorative-thinking action based on visual information	

(Continued)

TABLE 1.1

Action classification table (fragment before the first click). (Continued)

Eye Move And Final Position		Activity between successive mouse events (clicks) Mental/motor Actions Involved	Mouse events	Time (ms)		Total Action Time (a + b)	Classification of cognitive actions	Scan path generated/duration
From	To			a. Approximate Eye Movementime to reqd. position (Ms)	b. Approximate Dwell time at position (Ms)			
O_W	O_D			150	190	340	Explorative-thinking action based on visual information	
O_D	Near G_S	Formation program of performance according to developed mental model and accepted goal.		210	190	400	Explorative-thinking action based on visual information	
Near G_S	O_D	Decision-making about possible motor actions. Execution of action.	Click element D in object area.	210	630	840	Decision-making action at sensory perceptual level with simultaneous motor action "$MB + G5 + AP1$".	

wduration of a perceptual action and then subtract this time from the total performance time of the complex action (Bedny, 2015).

This is a brief description of the totally new method of eye movement interpretation. Based on this data we can determine the logical organization of cognitive and behavioral actions (conduct algorithmic description of activity during task performance). The detailed description of this method can be found in Bedny and Karwowski work (2007).

1.5   JOB ANALYSIS, TASK ANALYSIS, AND DESIGN

The traditional job analysis utilizes qualitative methods such as observation, interview, and questionnaires. These methods have limited application in design. Design is a basic concept in engineering and ergonomics, and is one of the most important areas in SSAT. When the design process is analyzed there is often no existing equipment, software, or tasks to look at. The main purpose of design is creation of new equipment, interfaces, and methods of task performance. Design can be defined as "creation and description of ideal images or models of an artificial object in accordance with previously set properties and characteristics with the ultimate purpose of materializing this object" (Neumin, 1984, p.145). According to such (1990, p.40), "the output of the design process is information in the form of drawings, specifications, tolerances, and relevant data required to create the physical entity." The author outlined two main design stages: a creative stage, which includes task definition and development of ideas for the possible solutions, and the formalized methods and/or analytical processes. At the second stage the development of design models and evaluation of the obtained data are the central part of the design process. Observation, questioning, etc., are usually utilized after the prototype of models of equipment or interface are developed. This is why analytical procedures that include quantitative methods play such an important role in SSAT. Various cognitive models are not design models. They are the information processing models that can be used as tools for creation of design models. An example of design models in engineering are drawings and calculations.

We have already briefly discussed the concept of task in general psychology and explained its shortcomings in application to the study of human performance. Ergonomic literature does not offer a clear definition of task either.

Let us consider how the concept of task is viewed in SSAT. Here, each task is regarded as a situation-bounded activity that is directed toward achieving the goal of the task under the given conditions. In engineering equipment, means of work and tools are defined when a subject operates with machines or equipment. For example, in order to cut a part, a worker has to install it into a specific equipment and use certain tools. In manual work workers manipulate with hand tools directly. Means of work is a general term that identifies a combination of physical equipment and tools. Computer is not a tool but means of work that make various artificial tools and objects that can be modified by a user available. Such concepts as material and mental actions, material and mental objects, and tools are important concepts in task analysis from the SSAT perspective. The structure of activity during task performance is determined by the spatially-temporal and logical organization of elements of activity.

The stage of adequate subjective acceptance of a goal and a task as a whole is not only an important step in task analysis, but also vital when designing psychological experiments. There are task requirements and associated with them is the task goal. There are also conditions in which requirements are presented. The continuum of tasks vary from skill-based tasks on one side to problem solving tasks on the other side. Rule based tasks includes multiple decisions and can be described by a human algorithm. There are deterministic and probabilistic-algorithmic tasks. Such tasks can often be very complex. We will consider such tasks in the following chapters. Purely creative tasks are heuristic tasks-problems that cannot be designed.

Software engineers analyze the tasks primarily from the standpoint of the technological or control analysis point of view. Ergonomists and psychologists conduct task analysis from cognitive psychology or activity theory perspectives. The structure of activity during task performance is the critical factor in task analysis. Task analysis in the field of HCI can be defined as the comparison of activity structure with the structure of the interface. It is important to distinguish the goal of the task from the goal of the system. One common drawback of cognitive approach in task analysis is the disregard for emotionally evaluative and motivational aspects. These aspects of task analysis in cognitive psychology are limited to studying performance in stressful conditions. This approach does not pay sufficient attention to positive emotions in task performance, to the relationship between emotions and motivation, to the significance of the task for a subject, and so on. The concept of affective design emerges, and the pleasure-based task analysis for the computer-based tasks is suggested (Helander, 2001). SSAT always views emotionally motivational aspects of task analysis as being important. These aspects should not be separated from cognition. Games have a significant role in the study of HCI in the nonproduction environment. The purpose of play is not to achieve a useful result but rather the activity process itself. However, this fact does not eliminate goal formation and motivational aspects in this activity because a subject performs goal-directed actions during the game. Moreover, imaginative aspects of the game become the critical factor. In a game, the same as in the work environment, the task can be formulated in advance by others or formulated by a subject independently. Analysis of the games demonstrates that goal is one of the central concepts for computer-based tasks in a nonproduction environment. We cannot agree with the opinion of some ergonomists who try to eliminate this concept when analyzing such tasks. Sometimes, the goal of the game is not precisely defined at the beginning, or is presented in a very general form. Only at the final stage does the goal become specific and clear. Therefore, the goal of a task is not the end state of the system that the human or machine wishes to achieve as it is stated by Preece et al. (1994, p. 411). The goal of the system and the human is not the same. Karat et al., (2004, p. 587) mixes the goal of the task with motive, and insists that in the nonproduction environment and particularly in games, task and goal do not exist. However, as has been demonstrated earlier, a game does not exist without a goal. In contrast to cognitive psychology, in SSAT any task has its desired intermediate and final goals and motivational forces. Human information processing inherently interacts with emotionally motivational aspects of activity that has a goal directed character. From the perspective of the activity self-regulation, the more complex and significant the task is for a person, the more mental effort a person will devote to reach the task goal. This is also true for the pleasure-based tasks. Further discussion of this subject can be found in Bedny (2015) and Bedny and Karwowski, (2011c).

In the following chapters we will consider morphological analysis of tasks with a complex probabilistic structure and some quantitative methods of their evaluation.

CONCLUSION

The purpose of this article is to give a basic introduction to applied, and specifically, systemic-structural activity theory. It offers guidelines to specialists in the field and provides information on where to find detailed data presented in our previous publications and in the chapters of this book. General activity theory is an important predecessor of applied and systemic-structural activity theory. Chronological development of activity theory can be seen as general activity theory, applied activity theory, and systemic-structural activity theory. Rubinshtein is the founder of general activity theory. Applied activity theory is not a unitary field. The main contributors in this area are Galactionov, Kotik, Konopkin, Ponomarenko, Zarakovsky, and V. Zinchenko. The founder of systemic-structural activity theory is Gregory Bedny. This field has been further developed by him and his colleagues W. Karwowski and I. Bedny, as well as others. New principles of design have been introduced in this field. Analytically design models of activity during task performance have been developed. These models of activity describe the structure of activity and compare it with configuration of equipment or interface. Qualitative, morphological, and quantitative methods have been created for this purpose.

REFERENCES

Barnes, P. M. (1980). *Motion and Time Study Design and Measurement of Work*. New York: John Wiley & Sons.

Bedny, G. Z. (1979). *Psycho-Physiological Aspects of a Time Study*. Moscow: Economics Publishers.

Bedny, G. Z. (1981). *The Psychological Aspects of a Timed Study During Vocational Training*. Moscow: Higher Education Publisher.

Bedny, G. Z. (1987). *The Psychological Foundations of Analyzing and Designing Work Processes*. Kiev: Higher Education Publishers.

Bedny, G., Meister, D. (1997). *The Russian Theory of Activity: Current Application to Design and Learning*. Mahwah, New Jersey: Lawrence Erlbaum Associates, Publishers.

Bedny, G.Z., Karwowski, W. (2007). *A Systemic-Structural Theory of Activity. Application to Human Performance and Work Design*. Boca Raton, FL: CRC, Taylor and Francis.

Bedny, G. Z. (2015). *Application of Systemic-Structural Activity Theory to Design and Training*. Boca Raton, FL: CRC, Taylor and Francis.

Bedny, G. Z., Karwowski, W., I. Bedny. (2015). *Applying Systemic-Structural Activity Theory to Design of Human-Computer Interaction Systems*. Boca Raton, FL: CRC, Taylor and Francis.

Bedny, G. Z., Bedny, I. S. (2018). *Work Activity Studies Within the Framework of Ergonomics, Psychology, and Economics*. CRC Press, FL: Taylor and Francis Group.

Bedny, G. Z., Karwowski, W. (Eds.). (2011a). *Human-computer interaction and operator's performance. Optimization work design with activity theory*. Boca Raton, FL: CRC, Taylor and Francis.

Bedny, G.Z. (Ed.). (2004). Special Issue. *Theoretical Issues in Ergonomics Science*, V. 5 # 4.

Bedny, G., Karwowski, W. Bedny, M. (2001). The principle of unity of cognition and behavior: Implications of activity theory for the study of human work. *International Journal of Cognitive Ergonomics*. V. 5, # 4, pp. 401–420.

Bedny, G. Z., Karwowski, W., & Voskoboynikov, F. (2011b). The relationship between external and internal aspects in activity theory and its importance in the study of human work. In G. Z. Bedny and W. Karwowski (Eds.) *Human-computer interaction and operators' performance. Optimization of work design with activity theory.* (pp. 31–62). Boca Raton, FL: Taylor and Francis, CRC Press.

Bedny, G. Z., Karwowski, W. (2011c). Task concept in production and nonproduction environments. In G. Z. Bedny and W. Karwowski (Eds.) *Human-computer interaction and operators' performance. Optimization of work design with activity theory.* (pp. 89–116). Boca Raton, FL: Taylor and Francis, CRC Press.

Bedny, G. Z. & Karwowski, W. (2013). Analysis of strategies employed during upper extremity positioning actions. *Theoretical Issues in Ergonomics Science.* V. 14, N. 2, pp. 174–175.

Bedny, G. Z., Harris, S. (2013). Safety and reliability analysis methods based on systemic-structural activity theory. *Journal of Risk and Reliability.* Sage Publisher, V. 227, Number 5, pp. 549–556.

Bedny, I. S. (2004). General characteristics of human reliability in in system of human and computer. *Science and Education*, Ukraine, Odessa: #8-9, pp. 58–61.

Bedny, I. S. (2006). On systemic-structural analysis of reliability of computer based tasks. *Science and Education*, Ukraine, Odessa: 1-2, #7-8, pp. 58–60.

Chebykin, O. Y., Bedny, G. Z. and Karwowski, W. (Eds.). (2008). *Ergonomics and psychology. Developments in Theory and Practice.* CRC Press, FL. Taylor and Francis.

Endsley, M. R. (2000). Theoretical underpinnings of situation awareness: a critical review, in M. R. Endsley, D. J. M. Garland (Eds.). *Situation awareness analysis and measurement*, pp. 3–32.

Fitts, P. M. (1954). The information capacity of the human motor system in controlling the amplitude of movement. *Journal of Experimental Psychology, 47*, 381–391.

Galactionov, A. I. (1978). *The Fundamentals of Engineering Psychological Design of Automatic Control Systems of Technological Processes.* Moscow: Energy Publishers.

Gordeeva, N. D., & Zinchenko, V. P. (1982). *Functional Structure of Action.* Moscow: Moscow University Publishers.

Helander, M.G. (2001). Theories and methods in affective human factor design, in G. D'Ydealle and J. Van Rensberggen (Eds.), *Perception and Cognition*: Elsevier Science Publishers, pp. 37–50.

Karat, J., Karat, C-M, and Vergo, J. (2004). Experiences people value: The new frontier for task analysis. In D. Diaper and N. Stanton (Eds.). *The Handbook of Task Analysis for Human-Computer Interaction*, pp. 585–602. Mahwah, NJ: Lawrence Erlbaum Associates, Publishers.

Konopkin, O. A. (1980). *Psychological Mechanisms of Regulation of Activity.* Moscow: Science.

Kotik, M. A. (1974) *Self-Regulation and Reliability of Operator.* Tallin: Valgus.

Kotik, M. A. (1978). *Textbook of Engineering Psychology.* Tallin: Valgus.

Kotik, M. A. (1987). *Psychology of Safety. Tallin*, Estonia: Valgus.

Kotik, M. A., Yemelyanov, A.M. (1985). *Errors of Control Psychological Causes. Method of Automated Analysis.* Tallin: Valgus.

Kotik, M. A., Yemelyanov, A.M. (1993). *Causes of Human-Operator's Errors when Controlling Transportation Systems.* Tallin: Valgus.

Lee, T. W., Locke, E. A., & Latham, G. P. (1989). Goal setting, theory and job performance, in A. Pervin (Ed.), *Goal Concepts in Personality and Social Psychology* (pp. 291–326). Hillsdale, NJ: Lawrence Erlbaum Associates.

Leont'ev, A. N. (1978). *Activity, Consciousness, and Personality.* Englewood Glifts: Prentice Hall.

Leont'ev, A. N. (1977). *Activity, Consciousness, Personality.* Moscow: Political Publishers.

Locke, E. A. (1994). The emperor is naked, in R. Lord, P. E., Levy (Eds.), *Applied psychology: An international review,* pp. 367–370.

Luria, A. R. (1970). *Traumatic Aphasia.* The Hague, the Netherlands: Mouton.

Neumin, Y. G. (1984) *Models in Science and Technic.* Leningrad: Science Publishers.

Pervin, L. A. (Ed.), 1989, *Goal Concepts in Personality and Social Psychology.* Mahwah, New Jersey: Lawrence Erlbaum Associates, Publishers.

Preece, J. Rogers, Y. Sharp, H., Benyon, D., Holland, S., Carey, T. (1994) *Human-Computer Interaction.* Addison-Wesley.

Ponomarenko, V. A., & Zavalova, N. D. (1981). Study of psychic image as regulator of operator actions. In B. F. Lomov, V. F. Venda (Eds.). *Methodology of Engineering Psychology and Psychology of Work of Management.* pp. 30–41. Moscow: Science.

Suh, N. P. 1990, *The Principles of Design.* (New York, Oxford: Oxford University Press).

Rubinshtein, S.L. (1954) The problem of psychology in Marx'x works. Soviet psychotechnics. 1: 14.

Rubinshtein, S.L. (1959). *Principles and Directions of Developing Psychology.* Moscow: Academic Science.

Vygotsky, L.S. (1962). *Thought and Language.* Cambridge: MIT Press.

Vygotsky, L.S. (1971). *The Psychology of Arts.* Cambridge: MIT Press.

Vygotsky, L.S. (1978). *Mind in Society. The Development of Higher Psychological Processes.* Cambridge, MA: Harvard University Press.

Wertsch, V. (Ed.). (1981). *The Concept of Activity in Soviet Psychology.* New York: M. E. Sharpe, Inc., Armonk. pp. 383–433.

Yarbus A. L. (1965). *The Role of Eye Movements in the Visual Process.* Moscow: Science Publishers.

Yarbus, A. L. (1969). *Eye Movement and Vision.* New York: Plenum.

Zarakovsky, G. M., Korolev, B. A., Medvedev, V.I., & Shlaen, P. Y. (1974). *Introduction to Ergonomics.* Moscow: Soviet Radio.

Zarakovsky, G. M., & Pavlov, V.V. (1987). *Laws of Functioning Man-Machine Systems.* Moscow: Soviet Radio.

Zinchenko, V. P., & Vergiles, N.Y. (1969). *Creation of Visual Image.* Moscow: Moscow University.

within given constraints. Consequently, an operator is given discretion when making task-performance decisions (Vicente, p. 69). In fact, this approach contradicts design principles (Bedny, 2015).

Assessing the probability of decision-making outcomes involved in task performance is an important stage of task analysis that makes it possible to determine the probabilistic structure of the performed task. Knowing the decision-making steps and associated task performance steps enables algorithmic description of the task performance.

SSAT offers the morphological analysis of activity, which includes its algorithmic and time structure description (Bedny, Karwowski, 2007; Bedny, Karwowski, I. Bedny, 2015). The algorithm describes the logical organization of cognitive and behavioral actions during task performance and is called *human algorithm*. Decision-makings are the main factors that define the probabilistic structure of activity in the human algorithm. More specifically, the logical and probabilistic structure of activity depends on the probabilities of outcomes of such decisions. There is a need to determine the probability of the events that influence decision-making outcomes. This can be done by subject matter experts (SMEs). In this chapter, we consider some examples of evaluating the probability of events by an SME. Some examples of assessing the probability of decision-making outcomes by specialists who conduct the task analysis are also considered.

There are two basic methods of determining probability of events: the objective method of evaluation and the subjective method of interpretation of the probability of events (Savage, 1954). In this paper, we give a brief analysis of these two methods as well as describe the method of subjective evaluation of events for rule-based tasks developed in SSAT. As we show below, this method should be distinguished from determining probability of events for decision-making tasks considered not only in ergonomics but also in economics.

2.2   SYSTEMIC-STRUCTURAL ACTIVITY THEORY AS AN ANALYTICAL FRAMEWORK IN ERGONOMICS AND WORK PSYCHOLOGY

Systemic-structural activity theory (SSAT) was founded by Bedny (1987), Bedny, Meister, (1997), and further developed in the US with his colleagues and published in a number of original books such as (G. Bedny, Karwowski, 2007; I. Bedny, 2015; G. Bedny, Karwowski, I. Bedny, 2015, G. Bedny, I. Bedny, 2018). A collection of articles in this field were presented in Bedny, G. Karwowski, Eds. (2011). There are also multiple articles published in this field. A brief overview of SSAT is presented in this section. A cognitive and analytically oriented viewpoint on ergonomic design requires a clearly developed and standardized terminology for human activity description. Many misunderstandings among practitioners and researchers in various areas of specialization derive from terminology issues.

Currently, the main approach to the design issues in ergonomics is based on cognitive psychology. The methods of study in this field include task analysis and associated experiments. A lot of practitioners and researchers point out that existing methods of task analysis are ambiguous and don't have a clear theoretical background. Miller (1962) has argued that task analysis is an art, and the value of a particular study depends on the skill of the task analyst. A large number of unrelated limited techniques, which do not have a unified theoretical background, are used in task analysis. Meister (1999) pointed out that contemporary research in human factors and ergonomics ignores the needs for a much more comprehensive human factors ergonomics (HFE) contribution to design. Similar limitations in task analysis and associated experiments have been discussed by Schraagen, et al., (2000). Usually, professionals accept the experiment as the single most powerful research instrument available in HFE. However, experiments are often quite artificial and do not have any external validity. This approach is limited by focusing on already existing tasks, which makes such task analysis inadequate for design purposes. The main purpose of design is the development of completely new systems. Only analytical methods can be utilized in the design process. Moreover, redesign also requires analytical methods. For example, in engineering where the concept of design originated, analytical methods are always utilized first.

In engineering design, experiments are also utilized but they are combined with analytical methods of study. Objects of design in ergonomics and in engineering are often the same. However, the specifics of ergonomic design is that design should take into account the capabilities and limitations of a human being. Ergonomic professionals often use the term "design" inaccurately and even incorrectly. Design should be considered the creation and description of ideal models of a designed object, in accordance with the previously set properties and characteristics, for further creation of this object. For example, creation of manufacturing equipment starts with the qualitative stage, which is followed by the analytical stage that includes development of drawings and accompanying calculations. Only after that, physical models or prototypes of designed objects are created. At the final stage, prototypes or designed objects can be experimentally evaluated. The analytical stage is also used when equipment is redesigned. Thanks to analytical stages, experiments with physical models or prototypes have external validity. In ergonomics, analytical stages are not used outside of SSAT, and utilized physical models do not provide external validity of experiments.

SSAT does not reject cognitive psychology. It uses cognitive psychology in its theoretical developments and as a possible stage of analysis. However, due to the new information obtained from SSAT theoretical data, it becomes possible to develop and utilize analytical procedures in the ergonomic design.

Below we present in abbreviated manner the concept of activity from SSAT perspective, and how it can be used in the ergonomic design process. Contrary to behavioral or cognitive psychology, SSAT views activity as a structurally organized self-regulated system rather than as an aggregation of responses to various stimuli, or a linear sequence of information stages. Human activity is a goal-directed rather than a homeostatic self-regulative system. The system can reformulate the goal during task performance. It integrates cognitive, behavioral, and motivational subsystems that influence each other. The goal of activity is a conscious image and logical representation of the desired future result that is connected with motives. It creates the vector "motives → goal" and gives activity its goal-directed character. This is a totally different interpretation of the concept of goal from its understanding in cognitive psychology and the theory of motivation. In work motivation, goal integrates cognitive and motivational (energetic) components. The goal can be either intensive or not intensive. According to this approach, the more intensive the goal is, the more it pulls activity. According to SSAT, motives can push activity to reach its goal. Moreover, human activity is poly-motivated. Therefore, due to such misinterpretation, there is the possibility of the same task having multiple goals. However, according to SSAT this is impossible. In task analysis, we have to distinguish between the overall goal of the task and intermediate goals of subtasks, or separate actions that are basic constituent elements of the activity.

Activity during task performance can be described as a sequence of cognitive and behavioral actions. In SSAT, actions are described in a standardized manner. The terms activity and actions are often considered interchangeable, but in activity theory, these terms have very different meanings. Classification of cognitive and behavioral actions is described in greater detail in G. Bedny, Karwowski, I. Bedny (2015) and G. Bedny (2015).

Actions are always involved in manipulation of ideal or material objects. Ideal objects such as images, concepts, and propositions are transformed according to the goal of the activity and actions. Actions are means of forming images and concepts, and transforming external material objects. Images and various concepts are main components of our knowledge. For example, actualization of images or information in memory is a result of conscious mnemonic actions or unconscious mnemonic operations. More complex processes of recalling information can involve thinking actions. Thus, declarative and procedural knowledge are the result of human actions. Procedural knowledge that is needed to achieve a conscious goal is not sufficient for a subject to perform procedures. A subject may know the procedures but does not have the required skills to perform them. As can be seen in these examples, cognitive and behavioral actions are key concepts for psychology and ergonomics that are theoretically and experimentally grounded.

Knowledge is formed through actions and actions are basic units of activity analysis.

Concepts of cognitive and behavioral actions have fundamental meaning for the morphological analysis of activity, which is the important stage of analysis in design that allows introducing quantitative methods in ergonomics. Below we briefly consider morphological analysis of human activity using the hypothetical examples to illustrate the method. We have selected the fragment of the simplest version of this operation studied in laboratory conditions utilizing the physical model of production operation. This model contained a pin board with six holes. There was a box behind the pin board that contained six pins. The subjects had to move two hands over the pin board, and grasp one pin by each hand. After that, subjects moved both hands with pins and installs each pin into the hole simultaneously and releases them. The subjects repeated these movements three times to complete the task. This task included only motor actions that were performed in the same sequence. Let us describe this task algorithmically (the description is given in a simplified manner to illustrate the method and terminology; see Table 2.1).

We can also describe this algorithm in the following way:

$$O^\varepsilon_1 O^\varepsilon_2 O^\varepsilon_3 O^\varepsilon_4 O^\varepsilon_5 O^\varepsilon_6$$

TABLE 2.1

Algorithmic description of hypothetical task that includes only behavioral components.

Member of the algorithm	Description of a member of the algorithm	Description of a member of the algorithm
O^ε_1	Move left hand to the box and grasp a pin	Move right hand to the box and grasp a pin
O^ε_2	Move left hand with a pin to the pin board and put a pin in a hole	Move right hand with a pin to the pin board and put a pin in a hole
O^ε_3	Move left hand to the box and grasp a pin	Move right hand to the box and grasp a pin
O^ε_4	Move left hand with a pin to the pin board and put a pin in a hole	Move right hand with a pin to the pin board and put a pin in a hole
O^ε_5	Move left hand to the box and grasp a pin	Move right hand to the box and grasp a pin
O^ε_6	Move left hand with a pin to the pin board and put a pin in a hole	Move right hand with a pin to the pin board and put a pin in a hole

This is an algorithm described as a formula. Usually each member of an algorithm contains from one to four actions (capacity of working memory). O^ε means that this member of the algorithm depicts only motor actions. In SSAT this is a standardized method of describing motor components of activity. The formula demonstrates that there are no independent cognitive components. The members of the algorithm always follow each other in the same sequence, which means that this is not a variable task. In the left column, we utilize psychological units of analysis because classification and description is given in a standardized manner according to psychological criteria. The probability of the appearance of these members of the algorithm is equal to one. In the second and the third columns we utilize technological units of analysis, described by common language or technological terminology. The combinations of psychological and technological units of analysis allows interpretation of what the operator is performing based on a presented description and without observation.

Let us now consider the simplest task that contains cognitive components, which makes it variable with probabilistic structure. For example, if the red light is turned on, press the red button; if the green light comes on, press the green button. The red light is turned on with probability $P_1 = 0.2$ and the green light is turn on with probability $P_2 = 0.8$. We can describe this symbolically as follows (see Table 2.2).

We can also describe this algorithm as a formula

$$\left\{ O^\alpha_1 \; l_1 \uparrow \left(\overset{(1-2)}{\downarrow} O^\varepsilon_2 \omega \overset{1(1)}{\uparrow} \downarrow O^\varepsilon_3 \right) \right\} \overset{1(2)}{\downarrow} \text{- task is completed}$$

TABLE 2.2

Algorithmic description of task that includes behavioral and cognitive components.

Member of the algorithm	Description of a member of the algorithm	Description of actions included in each member of the algorithm
$O^{\alpha}{}_1$	Receiving information from red or green bulb	Simultaneous perceptual action
$l_1 \uparrow^{1(1-2)}$	If l_1 has the first outcome go to $O^{\varepsilon}{}_2$. If l_1 has the second outcome go to $O^{\varepsilon}{}_3$.	Decision making action with two outcomes. The first outcome has probability $P_1 = 0.2$. The second one has probability $P_2 = 0.8$
$\downarrow^{1(1)} O^{\varepsilon}{}_2$	Press the red button	Motor action
$\omega\uparrow$	Always false logical condition	
$\downarrow^{1(2)} O^{\varepsilon}{}_3$	Press the green button	Motor action
Task is completed		

There are four members in this algorithm. Each member of the algorithm includes only one action. Actions have a standardized description in SSAT. This is a human algorithm because the main units of analysis are human actions. In the algorithmic description, $O^{\alpha}{}_1$; $O^{\varepsilon}{}_2$; $O^{\varepsilon}{}_3$ are called operators and l_1 is called a logical condition (decision-making action). In our example, the logical condition has only two outputs with different probabilities.

The value of P_1 (probability of output $l_1 \overset{1(1)}{\uparrow} \overset{1(1)}{\downarrow} O^{\varepsilon}{}_2$) is 0.2 and P_2 is equal to 0.8. Hence, the probability of $O^{\varepsilon}{}_2$ is 0.2 and the probability of $O^{\varepsilon}{}_3$ is 0.8. After $O^{\alpha}{}_1$ logical condition l_1 is performed, and then $O^{\varepsilon}{}_2$ or $O^{\varepsilon}{}_3$ follow it. These are examples of transitions from one member of the algorithm to another. Suppose that this task is performed 100 times during the shift. Then, $O^{\varepsilon}{}_2$ is performed 20 times and $O^{\varepsilon}{}_3$ is performed 80 times. Thus, the relative frequency of $O^{\varepsilon}{}_2$ is 0.2 and $O^{\varepsilon}{}_3$ is 0.8. In our formula, we use the symbol $\omega\uparrow$, which is an always false logical condition. Such a logical condition does not describe elements of activity. In our example, the false logical condition means that after performance of $O^{\varepsilon}{}_2$, the task is completed and $O^{\varepsilon}{}_3$ is not performed.

As can be seen, the described activity has a logical and probabilistic structure. In an example to come, we have a much more complex task. However, the utilized terminology is the same.

Presented material helps readers to understand further discussion even if they have a limited background in SSAT.

2.3 A BRIEF OVERVIEW OF EVALUATION OF THE PROBABILITY OF VARIOUS EVENTS

There are different approaches to the assessment and interpretation of the probability of events. The most widely used in mathematics and statistics is the concept of the frequency interpretation of events. The other main concept is the relative frequency. The latter refers to the ratio of the number of trials accompanied by the occurrence of events to the total number of trials. In statistical terms, the probability of the occurrence of an event can be interpreted as the relative frequency of the event. The value, around which the frequency of occurrence of some event fluctuates with an increasing of number of trials to infinity, is called the probability of the event. However, we can observe the frequency of occurrence of the events using a limited number of trials. In spite of the difficulties in calculating the frequency of such events, the frequency interpretation of the probability method has an important role in statistics, psychology, economics, and other areas of science. Probability that is based on the evaluation of frequency of events is also called objective probabilities.

Another approach is called the subjective method of interpretation of the probability of events (Savage, L. J., 1954). According to this approach, the probability can be considered a belief of the subject, who can rationally estimate the probability of events. Probability obtained with this method of assessment is called subjective probability. This approach can also be related to the method of subjective evaluation of probability, which is called the logical method of interpretation of probability (Kozelecki, 1979). This approach evaluates the probability of correctness of judgment about occurrence of the events, which can be either true or false. Since the subject does not always possess all the necessary information about considered events, one could only assume that the judgment is likely to be true. To determine the probability that the judgment is correct is possible based on comparison with other judgments and the drawing of correct conclusions or inferences from observation or facts. Such methods of assessment of subjective probability are closely related and don't have a definite borderline. This can be explained by the fact that the degrees of certainty or degrees of belief of the subject (subjective interpretation) affect the rational judgments about the likelihood of events (logical interpretation). Subjective probability is an important concept in psychology. Specifically, it is the critical factor in evaluating the probability of occurrence of risky events. Methods for assessing subjective probabilities are based on how a subject forms its judgments about the probability of events. However, the subjective assessment of the probability may be inaccurate. For example, the subject may not be competent enough to assess the probability. Anticipating various events, subjects use not only the laws of theory of probability but also utilize some heuristic principles. The latter may result in errors in assessing the probability of events, but such strategies are simple and do not require significant mental efforts.

One heuristic strategy that reduces mental effort is called representativeness (Tversky, Kahneman, 1974). This strategy is based on the degree of similarity between the sample and the population. The sample, based on which assessment of probabilistic characteristics of population is performed, has to reflect probabilistic features of real events. However, the strategy of representativeness does not guarantee correctness of evaluation of probability of considered events. Another strategy that increases the accuracy of the estimate of probability is the principle of availability. One of the factors that can reduce the accuracy of the estimation of the probability is that familiar events that are encountered by a subject more frequently can be evaluated differently in comparison to those that are seldom encountered. For instance, a boxing fan who knows that his country often wins Olympic medals in boxing can decide that men in his country are usually stronger than men in other countries. However, in reality, boxing might be very popular in his country and therefore there are many boxing clubs with good coaches in this country. Examples of the importance of the principle of representativeness have been shown by Tversky and Kahneman (1973) in various studies. It has also been demonstrated that properties of long-term and short-term memory are important for the evaluation of subjective probability of events. Despite the fact that in some cases subjective assessment of probability of events is not entirely accurate, and in some cases even erroneous, correctly selected experts can assess the probability of events with high levels of accuracy. We believe that in the specific production environment, properly chosen experts can estimate with high accuracy the probability of certain events. In some cases, two to three experts can be selected to estimate the probability of events. With significant discrepancies in evaluation of data, it's necessary to ascertain the reasons for their differences and repeat the assessment. Analysis of field data has allowed us to enhance the accuracy of estimating the probability of the decision-making outcomes. Below, we describe a method developed by us of obtaining the numeric value of probability of the decision-making outcomes in a production environment with high precision.

2.4   SUBJECTIVE ASSESSMENT OF THE PROBABILITY OF EVENTS IN ANALYSIS OF COMPUTERIZED TASKS

In this section, we describe the basic principles of evaluation of decision-making outcomes developed in SSAT. As we've discussed above, performance of rule-based tasks requires making various decisions. Some of them are carried out based on the assessment of perceptual information, while others involve extracting information from memory and using thinking operations. Simple decisions have only two outcomes with the probability of 0 or 1. Additionally, probabilistic algorithmic tasks include decisions with multiple outcomes and their probability varies from 0 to 1. In algorithmic description, such decisions are called logical conditions and they're depicted by the symbol l. Decision-making or logical conditions are major factors of task complexity. The probability of the outcomes of certain logical conditions or decision-making determines the probability of performing the next steps of a task that follow such outcomes.

Probabilities of some decision-making (logical conditions) outcomes depend on the preceding external events that in turn can appear with certain probabilities. For instance, suppose there are two mutually exclusive events. One of them has a probability of 0.8 and the other one has a probability 0.2. If event one occurs, then the corresponding decision-making outcome should have the same probability unless the operator makes a mistake. The same is true for the second outcome. Therefore, the probabilities of some decision-making outcomes depend on the preceding events. The probabilities of these outcomes determine the probabilities of the following members of the algorithm of the task performance, until the next decision that can change the probabilities of the next members of the algorithm (steps of task performance). Hence, the evaluation of the probability of events is the critically important stage of the analysis of tasks with a complex probabilistic structure.

The outcomes of certain decisions depend on the specifics of a situation in which such decisions are made. Thus, we have to evaluate the probability of events in specific situations. Experimental evaluation of such probabilities is often complex, time consuming, and sometimes even impossible. In these cases, researchers utilize subjective measurements. There are several good publications that compare objective versus subjective methods of measurement (Hennsy, 1990; Muclker, 1992, and so on). These studies demonstrate that in some situations subjective measurements arc more advantageous than the objective ones. Below we consider subjective measurement procedures developed in the framework of SSAT.

In the following sections, we consider original methods of developing event trees that facilitate determining the probability of the external events that define the probability of decision-making outcomes. In our study, subject matter experts (SMEs) estimated the probability of events that influence decision-making outcomes.

We present the procedures newly developed by us, and demonstrate how they are utilized for reliability assessment. Our methods involve utilizing special scales for evaluation of the subjective probability of various decision-making or logical conditions. This method also involves a new event tree technique. Probabilistic event trees reduce the possible range of the probabilistic evaluation of the following outcomes depending on the probability of preceding outcomes. This method also helps us to transfer qualitative data about outcomes of various decisions into the quantitative one, as well as to transform verbal subjective judgment of experts into quantitative data. Of course, such quantitative data is obtained with some approximation. Nevertheless, the considered method has precision that's sufficient for practical application.

Figure 2.1 below illustrates the scale of the subjective probability of events that has been utilized for evaluating the probability of decision-making outcomes and of the following corresponding members of the algorithms. Here we consider this scale in details.

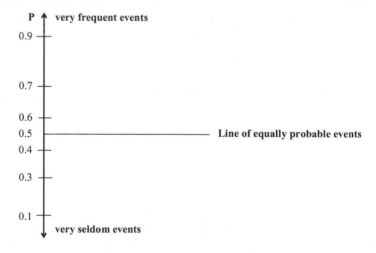

FIGURE 2.1   The scale of the subjective probability of events.

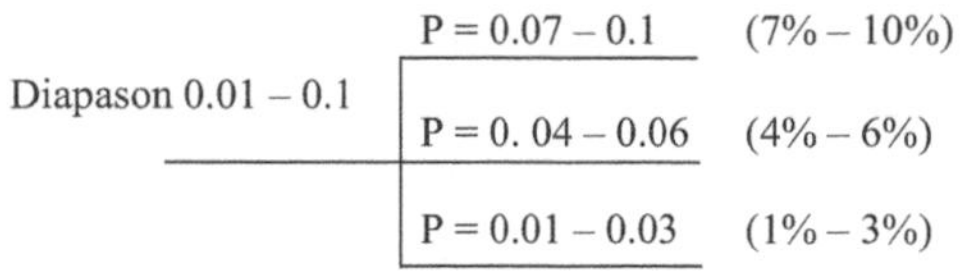

$$P = 0.07 - 0.1 \quad (7\% - 10\%)$$

Diapason 0.01 – 0.1

$$P = 0.\,04 - 0.06 \quad (4\% - 6\%)$$

$$P = 0.01 - 0.03 \quad (1\% - 3\%)$$

FIGURE 2.2   The extremely low scale diapason that covers very seldom events with P from 0.01 to 0.1.

The scale of the subjective probability of events has a line of equally probable events that reflects the numerical value $P = 0.5$. Events with probabilities greater than $P = 0.5$ are above this line, and events with probabilities less than $P = 0.5$ are below it. There are two extreme scale ranges of values: the extremely high scale diapason corresponds to events with a probability from 0.9 to 0.99, and the extremely low scale diapason reflects probabilities from 0.1 to 0.01. The probability values on the vertical scale are divided into different ranges or diapasons of probability. In our scale, eight diapasons are suggested. Four of them are above and four are below the equally probable line.

The main characteristic of each diapason is its position in relation to the equal probability line. The extremely high scale diapason ($P = 0.9 - 0.99$) and the extremely low diapason ($P = 0.1 - 0.01$) are considered additional reference points. When analyzing the events that might occur with different probabilities inside each diapason, it's necessary to determine whether the event probability is closer to the top or bottom border of each considered diapason. Based on this method, we can determine the possible probability of considered events with sufficient accuracy. Let us consider the two extreme diapasons. Figure 2.2 depicts the extremely low scale diapason with probability $P = 0.01 - 0.1$ (very seldom events), and Figure 2.3 reflects the extremely high scale diapason with probability $P = 0.9 - 0.99$ (very frequent events). The diapason with very seldom events can be further divided into three areas with the following probabilities: a) $P = 0.01 - 0.03$ b) $P = 0.04 - 0.06$ c) $0.07 - 0.1$. These diapasons are presented graphically on Figure 2.2. If we consider the events that fall into one of these diapasons or into a considered range of probabilities, then based on qualitative analysis we can find out in which specific area of this diapason their probability belongs.

The extremely high diapason for very frequent events can be presented similarly (see Figure 2.3).

Thus, the subjective assessment of the probability of events consists of a few stages. First, using the scale, the SME determines in what diapason a considered event falls, and then defines if this event is close to the top, middle, or bottom of this diapason. Based on this, the probability of the events can be determined within the considered range of probabilities. There are other types of diapasons such as a

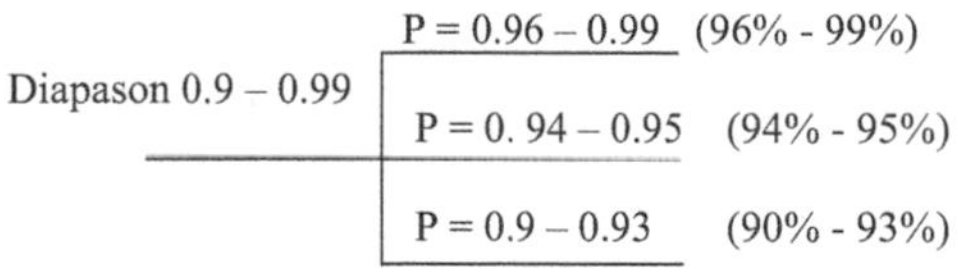

$$P = 0.96 - 0.99 \quad (96\% - 99\%)$$

Diapason 0.9 – 0.99

$$P = 0.\,94 - 0.95 \quad (94\% - 95\%)$$

$$P = 0.9 - 0.93 \quad (90\% - 93\%)$$

FIGURE 2.3   The extremely high scale diapason that covers very frequent events with P from 0.9 to 0.99.

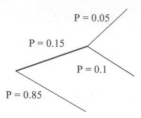

FIGURE 2.4   The hypothetical event tree model.

diapason with probability P = 0.51 − 0.6 that we designate as *diapason for events with just noticeably higher than equal probability of occurrence.* A diapason with P = 0.4 − 4.99 is designated as *diapason for events with just noticeable lower than equal probability of occurrence.* We also distinguish such diapasons as: *noticeably more frequent events* with probability P = 0.60 − 0.69; *rather frequent events* with probability P = 0.7 − 0.89; *noticeably more seldom events* with probability P = 0.4 − 0.31; *rather seldom events* with probability P = 0.3 − 0.1 and *very seldom events* with probability P = 0.01 − 0.1 (Figure 2.2).

This method allows determining the probability of considered events with adequate precision. We would like to point out that the line of equally probable events and the diapasons used as reference points should be constant (extremely high diapason, extremely low diapason). Other diapasons may vary in size depending on the specifics of the task being analyzed.

As mentioned above, the suggested method also utilizes event tree modeling. However, this method is used with some modifications. Below we demonstrate an example of a hypothetical probabilistic event tree modeling built utilizing our method (see Figure 2.4).

The first two branches in this example have probabilities P = 0.15 and P = 0.85. The second two branches have probability P = 0.05 and probability P = 0.1. As can be seen, these last two branches have a combined probability P = 0.15. The probability of the last two branches together is the same as the probability of a branch that is designated by the bold line. The suggested method of developing event tree models helps SMEs determining the probability of events with high accuracy.

Let us to compare this event tree model with the event tree model developed utilizing the traditional method depicted in Figure 2.5. This model was used in the

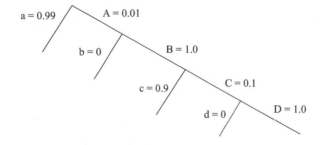

FIGURE 2.5   Example of human reliability analysis model.

human reliability assessment study (Kirwan, 1994). As can be seen in this model, the probability of each following branch does not depend on the probability of preceding branches, and each pair of branches has the accumulated probability of 1.

2.5 EXAMPLES OF THE APPLICATION OF THE NEW METHOD

A task performed in the inventory processing system has been selected for our study. The inventory process includes receiving items, putting them away, storing items, picking the items from the shelves, and moving them through work-in-process (WIP) while simultaneously tracking their movements as well as recording all of these events in the database. In this study, we considered the receiving task. This is a computerized task, in which an operator performs manual work and utilizes the software. The task analysis for this task has been performed within the framework of systemic-structural activity theory (SSAT). This approach includes qualitative analysis, algorithmic description of work activity, time structure analysis, and quantitative evaluation of task performance (complexity, reliability evaluation). Each of the listed stages often requires reconsidering prior stages of analysis. In this chapter, we only consider the method of evaluation of the probabilistic structure of activity during task performance.

The considered task includes three versions of task performance that occur with various probabilities. Each version of task includes a number of different decisions and has a complex probabilistic structure. We do not consider all stages of analysis but just briefly describe some of them for further demonstration of some principles of probability evaluation of decision-making outcomes, and the evaluation of probability of the following members of the human algorithm.

SSAT utilizes various qualitative methods of task analysis and here we utilize the simplest objectively logical analysis. It consists of a short verbal description of task performance, analysis of related tasks, description of technological process, and work conditions. Discussion with workers or supervisors, observation, review of documentation, analysis of similar work processes, possible prototypes of equipment, and so on, can be used at this stage of analysis.

We describe here in an abbreviated manner the computerized inventory-receiving task. It consists of the setup task and the main task that includes three versions of tasks with different probabilities (Bedny, I. Bedny, 2018). Thus, the task as a whole has a complex logical and probabilistic structure. The setup task involves choosing the right menu option, opening the box with items, taking the purchase order (PO) out of the box, and keying or scanning the PO number. The first version of the main part of the task included getting the item out of the box, matching the item number on the physical item with the item number in the list of items on the computer screen, activating the screen with the information for the item at hand, as well as comparing received quantity with the ordered quantity. Figure 2.6 depicts probabilities of the decision-making outcomes for item quantity evaluation, which is the first version of this task. If the ordered and received quantity of items is the same ($P = 0.9$), an operator continues to receive the item (version 2 and 3 of the task). If the quantity is less (partial shipment), the item is processed further (quantity of items is accepted) with $P = 0.03$. Otherwise, the item is rejected and put aside with $P = 0.07$ (version 1 is completed).

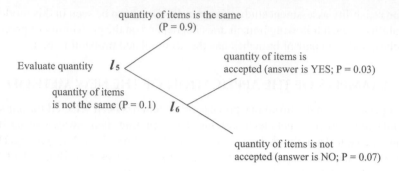

FIGURE 2.6   Event tree of events that influence decisions l_5 and l_6 when item quantity is evaluated.

Version 2 of the task includes version 1 with the outcome of the quantity of the item being accepted and the next stage of the process when the price of the item is compared to the order price. If the item price is the same as the order price, the item is accepted. If the price is not the same, the item can be still accepted or rejected according to certain rules. Rejection of items according to their price is the final step in version 2 of the task. Version 3 includes Versions 1 and 2 as described above, with the outcome of item quantity and price being accepted. It also has an additional stage of the task when the location of the item in the warehouse is determined, the label is placed on the item for the put away process, and the item is placed in the corresponding tote. Thus, only the items that are processed through the third version of the task are eventually delivered to the warehouse. Performance of versions 1 and 2 leads to the rejection of the items.

Each of the versions of the task is performed with a different probability. Therefore, it's necessary to determine the probability of emergence of each version in the context of the main task. In order to evaluate the probabilistic structure of the task, we need to determine the probabilities of the following events: the item is rejected due to the unacceptable quantity; the quantity of the items is acceptable but the price is unacceptable; and both quantity and price are accepted and the item is received and sent to the warehouse.

In further discussion, we do not describe each version of the task separately. Only the consolidated version of the task is presented in this chapter. This version of the task facilitates determining probabilities of all members of the algorithm of task performance during the shift. It demonstrates the combination of all three versions of the task. In reality, the operator performs version 1, version 2, or version 3 depending on the received and ordered quantity and price of the items. The consolidated version of the task depicts the average complexity, probability, strategy, and performance time of the task during the shift. The probabilities listed in this table are unconditional probabilities of the individual members of the human algorithm. They depict the absolute probability of occurrence of these events. When we describe the individual tasks, the probabilities of the members of the algorithm are conditional probabilities within each task when it occurs.

The development of various event trees helped SMEs determine the probabilities of these events. Probabilities of events for the event tree that is depicted below have been determined based on the subjective scale of events and the notion of scale dia-pasons. The event-tree presented below includes two decision-making actions that are designated as logical conditions l_5 and l_6 in the algorithmic descriptions of activity during task performance (see Figure 2.6).

The first decision l_5 has two outcomes. One of them reflects the situation when the item quantity and ordered quantity are equal. The extremely high scale diapason covers very frequent events with P from 0.9 to 0.99. SMEs related the event when ordered and received quantity of the item is the same to the low border of the high diapason (P = 0.9 – 0.93; see Figure 2.3). Therefore, SMEs assign probability P = 0.9 to this event. This probability falls into the extremely high scale diapason. The second outcome corresponds to the situation when the item quantity is not the same. Hence, the probability of the opposite outcome is 0.1, which belongs to the low scale diapason (see Figure 2.2). The extremely low scale diapason covers very seldom events with P from 0.01 to 0.1.

Let us consider the second decision or logical condition l_6. The probability of log-ical condition l_6 is P = 0.1. Therefore, in order to determine outcomes of this logical condition, SMEs utilized the extremely low scale diapason depicted in Figure 2.2. SMEs evaluated that items rejection when the quantity does not match the ordered quantity is a highly probable event inside this diapason, assigning to it the probabil-ity P = 0.07. The opposite outcome should be P = 0.03. The event tree in Figure 2.6 demonstrates that the probability of an event in which items are accepted based on quantity criterion is P = 0.93 and rejected according to this criterion with P = 0.07.

Figure 2.7 depicts the event tree for the price evaluation. It shows that the received price matches the ordered price with the probability P = 0.8. When the price is dif-ferent it can be less (P = 0.1) or greater than the purchase order price. If the price is

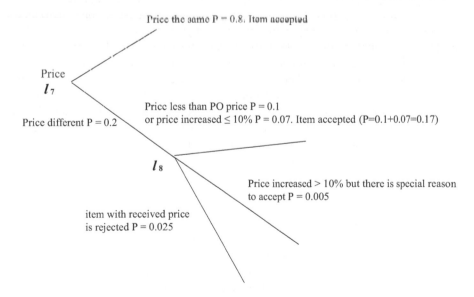

FIGURE 2.7 Evaluation of probability of outcomes of logical conditions and associated members of the algorithm during item price estimation based on the event tree analysis.

the same or less the item is accepted according to the price criteria. It's also accepted if the price is greater but the price increase is less or equal to 10% (P = 0.07). Thus, the combined probability is P = 0.1 + 0.07 = 0.17. If the price increase is greater than 10%, the item is usually rejected. In rare cases, the item can still be accepted for business reasons (P = 0.005). Therefore, the items are rejected according to price criteria with a probability P = 0.025 and accepted with probability P = 0.8 + 0.1 + 0.07 + 0.005 = 0.975. As we can see, the considered event tree has three outcomes.

Algorithmic description of the consolidated version of the task (fragment) demonstrates how probabilities of events (as depicted by the event tree) are reflected in the decisions made by the operators (see Table 2.3). This table depicts the consolidated version of the task that accounts for all three versions of the task performance and their probabilities of occurrence. This paper does not include a detailed description of each version of the task. For analysis of all three versions of this task and their complexity evaluation, reference Bedny, I. Bedny (2018). Here we only demonstrate the method of probability evaluation of events associated with considered above decisions and how they influence the probability of the related operators. The following is a brief description of Table 2.3. Table 2.3 shows how a scale of the subjective probability of events in combination with scale diapasons and event-tree modeling facilitates the determination of probability of outcomes of various decisions during task performance. Based on such information it becomes possible to assign probabilities of all associated members of the algorithm of human performance.

Creation of such tables is possible only when there's a standardized terminology for activity description. This terminology has been developed in SSAT (Bedny, Chebykin, 2013). Using the standardized language for describing activity structure is fundamentally important for the development of principles of design in ergonomics and engineering psychology, where analytical models plays a leading role.

Algorithmic description of task performance enables the description of extremely variable human activity. Along with temporal analysis of activity, this table presents morphological analysis of activity during task performance. Members of the human algorithm include operators and logical conditions. Operators integrate several cognitive or behavioral actions that transform objects, information, and energy. The number of possible actions in one member of the algorithm are restricted by the capacity of working memory. In SSAT, a motor action is considered as a group of motor motions that are integrated by a goal of this action. Similarly, cognitive actions integrate several mental operations. There are such cognitive actions as sensory actions, simultaneous and successive perceptual actions, imaginative, mnemonic, thinking, decision-making actions, and so on. For instance, logical conditions can be considered decision-making actions. Members of the algorithm in Table 2.1 are described using technological and psychological units of analysis, which facilitates the clear understanding of what an operator does during task performance even without directly observing it. Description of each member of the algorithm is performed with various levels of decomposition.

The standardized symbols help to understand the logic of the transition from one member of an algorithm to another. For example, O^{ε}_{15} includes several executive or motor components of activity, while $O^{\alpha\mu}_{14}$ includes perceptual actions with mnemonic operations, and l_5 with the corresponding arrow depicts a decision-making

TABLE 2.3

Algorithmic description of activity and its time structure during computerized task performance (consolidated version of the task, fragment).

Members of Algorithm (psychological units of analysis)	Description of Elements of task (Technological units of analysis)	Description of Elements of activity (Psychological Units of Analysis)	Time sec
O^α_1	Check for presence of inventory receiving screen	Simultaneous perceptual action (ET + EF)	$0.42 + 0.3 = 0.72$
$\overset{1}{\downarrow} O^\epsilon_2$	Type 1 and press ENTER to choose ADD INVENTORY RECEIVING screen	(R50B + AP1) + (R30B + AP1)	$1.68 \times 1.2 = 2.01$
O^ϵ_{13}	Press ENTER to go to the screen with detailed item information	Motor action (R26B + AP1)	$0.76 \times 1.2 = 0.9$
$O^{\alpha\mu}_{14}$	Compare received quantity with PO (purchase order) quantity	Combination of two simultaneous perceptual actions $- 2 \times$ (ET + EF) with simultaneously performed mnemonic operation (MO)	$(0.42 + 0.4) \times 2$ $= 1.64$
$\overset{5}{l_5 \uparrow}$	If received quantity and order quantity are the same, go to O^ϵ_{22} (P = 0.9). If received quantity is greater or less than order quantity, go to O^ϵ_{15} (P = 0.1)	Decision-making action performed based on visual information	0.4
O^ϵ_{15}	Type received quantity and press ENTER to get a question at the bottom of the screen (P = 0.1)	Motor action (R20B + AP1) + (R12B + AP1) (example with two digits number)	$(0.8 \times 1.2) \times 0.1$ $= 0.096$
O^α_{16}	Read the statement: THE RECIVED QUANTITY AND ORDER QUANTITY DO NOT MATCH. DO YOU ACCEPT? (YES/NO). (P = 0.1). Scan and read about four words.	Successive perceptual action. ET + $4 \times EF$	$(0.42 + 4 \times 0.18)$ $= 1.14 \times 0.1$ $= 0.11$
$*O^{\mu th}_{17}$	Recall instructions and perform required calculations and estimate (P = 0.1)	Combination of mnemonic action (retrieve simple information from memory) and logical thinking action	$\approx (1.2 + 3) \times 0.1$ $= 0.42$

(*Continued*)

TABLE 2.3

Algorithmic description of activity and its time structure during computerized task performance (consolidated version of the task, fragment). (Continued)

Members of Algorithm (psychological units of analysis)	Description of Elements of task (Technological units of analysis)	Description of Elements of activity (Psychological Units of Analysis)	Time sec
$l_6 \overset{6}{\uparrow}$	If quantity is not accepted (computer defaults to N), go to O^ε_{18}. (P = 0.07). Otherwise, go to O^ε_{21} (P = 0.03)	Decision-making action that includes simple syllogistic conclusion	$1.5 \times 0.1 = 0.15$
O^ε_{18}	Press ENTER (default is confirmed and after performing O^ε_{19} start working with a new item). (P = 0.07)	Motor action (R26B + AP1)	$(0.7 \times 1.2) \times 0.07 = 0.058$
O^ε_{19}	Put rejected item in the Put-Aside Area. Return to the base unit. (P = 0.07) O^ε_{19} includes: 1) the left hand grasps the item; 2) worker turns her/his body and makes approximately 6 steps to the put aside area; moves left hand and releases the item; 3) turns and makes approximately 4 steps back to the base unit.	Motor actions involving hand and leg movements: (R6B + G1B) + (TBC1 + 6WP + TBC2) + (R40B + RL1) + (TBC1 + 4WP + TBC2)	$(10.4 \times 1.2) \times 0.07 = 0.87$
O^α_{20}	Check if there are other items in the box to receive. (P = 0.07)	Simultaneous perceptual action ET + EF	$(0.42 + 0.3) \times 0.07 = 0.05$
$\downarrow O^\varepsilon_{21} \overset{6}{}$	Change "N" to "Y", quantity is accepted. (P = 0.03)	One motor action: (R60B + AP1)	$(1.05 \times 1.2) \times 0.03 = 0.038$
$\downarrow O^r_{22} \overset{5}{}$	Press ENTER and go to O^α_{23}. (O^ε_{22} has probability P = 0.9 + 0.03 = 0.93)	One motor action (R40 B + AP1)	$0.84 \times 1.2 \times 0.93 = 0.94$
$^{**}O^\alpha_{23}$	Compare PO price of the item with received price. (P = 0.93)	Combination of two simultaneous perceptual actions including one perceptual action (double check) $- 2 \times (ET + EF)$ + (ET + EF)	$2 \times (0.42 + 0.4)$ $+ (0.42 + 0.3)$ $\times 0.93 = 2.19$

TABLE 2.3

Algorithmic description of activity and its time structure during computerized task performance (consolidated version of the task, fragment). (Continued)

Members of Algorithm (psychological units of analysis)	Description of Elements of task (Technological units of analysis)	Description of Elements of activity (Psychological Units of Analysis)	Time sec
$l_7 \uparrow^{7}$	If the price on the screen and on the packing slip are different, go to O^{ε}_{24} (P = 0.2). If price is the same go to O^{ε}_{30} (P = 0.8).	Simple decision-making action	$0.4 \times 0.93 = 0.37$
O^{ε}_{24}	Key in the new price and hit ENTER (P = 0.2)	Four motor actions and mnemonic operations $(R20B + AP1) + 2 \times (R6B + AP1) + (R12B + AP1)$	$(2.15 \times 1.2) \times 0.2 \times 0.93 = 0.48$
O^{α}_{25}	Look at information on the screen (Cursor moves to the next field if price is accepted. Otherwise there is a message on the screen "Price increase is > 10%". Do you wish to proceed?) P = 0.2. Scan and read about four words	Successive perceptual action. $ET + 4 \times EF$	$(0.42 + 4 \times 0.18) \times 0.2 \times 0.93 = 0.22$
$l_8 \uparrow^{8(1-3)}$	If deference > 10% go to O^{ε}_{26}. However, if there is a special reason to accept the item go to O^{ε}_{29}. (Probability of O^{ε}_{26} and O^{ε}_{29} together is 0.03). If there was no massage and cursor moved to the next field go to O^{ε}_{30}. Probability of O^{ε}_{30} is 0.17 (Price is lower P = 0.1 or price is greater but increase is ≤ 10% P = 0.07).	Decision-making action from three alternatives that include simple syllogistic conclusion and requires actualization of information in memory.	$\approx (2 + 1.5) \times 0.2 \times 0.93 = 0.65$
$\downarrow O^{\varepsilon}_{26}{}^{8(1)}$	Hit ENTER to accept system default (N). (P = 0.025) **Members of algorithm $O^{\varepsilon}_{27} - l_8$ are performed in version 2 of the task with probability P = 0.03**	Motor action (R26B + AP1)	$(0.7 \times 1.2) \times 0.025 \times 0.93 = 0.016$

(Continued)

TABLE 2.3

Algorithmic description of activity and its time structure during computerized task performance (consolidated version of the task, fragment). (Continued)

Members of Algorithm (psychological units of analysis)	Description of Elements of task (Technological units of analysis)	Description of Elements of activity (Psychological Units of Analysis)	Time sec
$O^\varepsilon{}_{27}$	Put rejected item in the Put-Aside Area. Return to the base unit. $O^\varepsilon{}_{27}$ includes: 1) the left hand grasps the item; 2) an operator turns and takes approximately 6 steps to the put aside area; moves the left hand and releases item; 3) an operator turns and takes approximately 4 steps back to the base unit (P = 0.025).	One motor action involving a hand and leg movements: (R6B + G1B) + (TBC1 + 6WP + TBC2) + (R40B + RL1) + (TBC1 + 4WP + TBC2)	$(10.4 \times 1.2) \times 0.25 \times 0.93 = 0.29$
$^{***}O^\alpha{}_{28}$	Check to see if there are other items in the box to receive. (P = 0.025)	Simultaneous perceptual action	$0.3 \times 0.025 \times 0.93 = 0.001$
$l_9 \stackrel{9(1-2)}{\uparrow}$	If there are no more items in the box, go to $O^\varepsilon{}_4$, otherwise press F3 to return to previous screen and go to $O^\varepsilon{}_{11}$	Simple decision-making	$0.3 \times 0.025 \times 0.93 = 0.001$
$\stackrel{8(2)}{\downarrow} O^\varepsilon{}_{29}$	Type "Y". (P = 0.005)	Motor action: (R30B + AP1)	$0.74 \times 1.2 \times 0.005 \times 0.93 = 0.004$
$\stackrel{8(3)7}{\downarrow\downarrow} O^\varepsilon{}_{30}$	Press ENTER to go to the Completion Flag (P = 0,975)	One motor action (R25B + AP1)	$0.69 \times 1.2 \times 0.9 = 0.75$
$\stackrel{\omega 1}{\downarrow} O^\alpha{}_{49}$	Check if there are other items in the box to receive (This is the beginning of processing another item).	Simultaneous perceptual action	— — — — —
l_{15}	If there are no more items in the box, go to $O^\varepsilon{}_4$, otherwise go to $O^\varepsilon{}_{11}$ (This decision-making is involved in processing another item)	Simple decision-making	— — — — —
Total performance time of the task			**36.8099**

action. The associated arrows demonstrate possible transitions from logical conditions to the following corresponding member of the algorithm based on the outcome of the decision-making action. Detailed psychological description of members of the algorithm using psychological units of analysis (in terms of actions and their operations), is presented in column three. The first member of the algorithm is O^α_1 and the last one is O^α_{49}. The last logical condition (decision making) is l_{15}. Hence, the entire task has 49 members of the algorithm that are involved in receiving information, comparing it with information in memory, executing manual components of work, and so on. This task also has 15 decision-makings or logical conditions. The quantity of decisions shows that this task is extremely variable, which therefore significantly increases complexity of the task.

Let us consider how event tree data (see Figure 2.6) is used in the algorithmic task description. This event tree includes decisions l_5 and l_6 that are involved in evaluating the item quantity. Decision l_5 has two outcomes. According to this event tree, one outcome has the probability of $P = 0.9$ and the other one has the probability of $P = 0.1$. As a result, the probability of all members of the algorithm that directly follow l_5 starting from O^ε_{15} and up to $O^{\mu th}_{17}$ have the probability $P = 0.1$. A member of the algorithm O^ε_{22} has the probability $P = 0.9 + 0.03 = 0.93$. Probability 0.9 is the outcome of the logical condition l_5. A decision described as logical condition l_6 introduces new probabilities in the structure of the task. As can be seen, the item is rejected based on its quantity with probability $P = 0.07$, and accepted with probability $P = 0.03$ when the ordered and received quantities are not the same. These probabilities of outcomes influence the probability of the following members of the algorithm. Members of the algorithm from O^ε_{18} up to O^α_{20} have a probability $P = 0.07$. A member of the algorithm O^ε_{21} has the probability $P = 0.03$ (see upside down arrow from decision l_6). Finally, O^ε_{22} has an accumulated probability. One probability is 0.9 (see upside down arrow from l_5) and another probability comes from O^ε_{21}. Hence, the accumulated probability of O^ε_{22} is $P = 0.9 + 0.03 = 0.93$.

Let's now consider members of the algorithm associated with the evaluation of the price of the item. l_7 depicts the decision-making at the first stage of the price evaluation. If the price of the received item on the packing slip matches the PO price, which happens with probability $P = 0.8$, the next member of the algorithm to perform is O^ε_{30}. Otherwise O^ε_{24} is performed, which means that the operator keys in the new price and hits ENTER (see Table 2.3). Logical condition l_8 describes how the price difference is evaluated. If the price has decreased or the increase is less or equal than 10% ($P = 0.17$) then the item is still accepted. Even if the price increase is over 10%, the items might still be accepted in rare cases for special business reasons ($P = 0.005$). Therefore, the following member of the algorithm O^ε_{29} is performed with probability 0.005. An item is rejected based on the price criteria with probability $P = 0.025$ and the following member of the algorithm O^α_{28} has the same probability. Therefore, O^ε_{30} has an accumulated probability of 0.975 ($P = 0.8 + 0.17 + 0.005$). As can be seen, the logical conditions are the points where the probabilities of the following members of the algorithm change.

In this paper we do not consider the temporal characteristics of activity. Here we only demonstrate how to analyze the probabilistic structure of activity during task performance.

We briefly consider temporal characteristics of two members of the algorithm in order to demonstrate some basic ideas of the temporal task analysis. The performance times are depicted in Table 2.3 in the first column from the right. Temporal data for cognitive components of activity has been obtained from various handbooks, and the MTM-1 system has been utilized to determine performance time of the motor components. We have also performed chronometrical analysis when it was necessary (Bedny, 2015); Myasnikov and Pertrov (Eds.; 1976), Lomov (Ed.; 1986), and so on. Let us consider as an example logical condition l_5. Its performance time is equal to 0.4 sec. O^{ε}_{15} is the motor component of activity. According to the MTM-1 system, performance time of this member of the algorithm is 0.8 sec. The coefficient of pace (1.1 or 1.2) should be applied in this case based on SSAT rules. The probability of occurrence of O^{ε}_{15} in the task performance is equal to P = 0.1 (see column 2 from the right, Table 2.3). Thus, the performance time of O^{ε}_{15} is calculated as follows:

$$t = (0.8 \times 1.2) \times 1.1 = 0.096,$$

where 0.8 is performance time of this motor component of activity according to MTM-1; 1.2 – coefficient of pace according to SSAT rules; 0.1 – probability of occurrence of O^{ε}_{15} in the structure of activity. Performance time of other members of the algorithm have similarly been determined.

CONCLUSION

There are varieties of tasks that involve decision-making processes. According to Rasmussen terminology, specialists concentrate their attention on the decision-making tasks with knowledge-based behavior at their core. The main goal of these tasks is the selection of one choice from a number of choices, and the key factors are the amount of available information, the uncertainty of consequences of selected actions, time limit, possible risk, and so on.

However, rule-based tasks, the main purpose of which is achieving the assigned production goal according to prescribed instructions, dominate in production environments. Achieving the goal of a task is possible when an operator makes a number of necessary decisions that can have a number of different outcomes with various probabilities. These decisions are critical points of performing such tasks. In most cases, heuristic strategies cannot be used in these cases because operators follow certain business rules. Application of these rules can sometimes be a challenge for operators, especially in non-standard situations, or if an operator is fatigued, under stress, or affected by task monotony, and so on. Hence, there are decision-making tasks that are based on knowledge-based behavior and decision-making tasks that are based on rule-based and/or skill-based behavior. SSAT distinguishes between two types of rule-based tasks: deterministic-algorithmic tasks and probabilistic-algorithmic tasks. Probabilistic-algorithmic tasks can be rather complex even under normal circumstances and presently are not paid sufficient attention by the researchers. Human activity is a self-regulated system and as a result, variability of human performance is its basic characteristic. If this variability is in the range of acceptable tolerance, it should be considered the same activity. Another factor of human performance variability is that it depends on decisions that are involved in the task performance. Analysis and description of various decisions that are involved in the task performance is a challenging scientific undertaking. An important stage of the design of such tasks is the descriptions of their probabilistic structures that can be based on objective and/or subjective evaluation of decision-making outcomes. Objective methods are often not available in task analysis. So, the subjective principles of probability evaluation of decisions-making outcomes should be utilized instead. In order to improve the accuracy of the subjective evaluation, we have developed a scale of the subjective evaluation of the probability of events, along with the notion of scale diapasons in addition to the event tree modeling method. Combining this method with morphological analysis of activity (algorithmic and time structure analysis), allows us to determine the probabilistic structure of a task and therefore evaluate the time of task performance, its complexity, and its reliability with high precision.

In this chapter we have demonstrated principles of evaluation of the probabilistic structure of the complex rule-based task that includes various versions occurring with various probabilities. Here we described the analysis of a probabilistic structure of a consolidated version of a task before improvement. There is also a need to analyze the consolidated version of the task after improvement, and to evaluate the probabilistic structure of different versions of the task before and after improvement. These topics are covered in our publications.

REFERENCES

Bedny, G. Z. (1987). *The psychological foundations of analyzing and design work processes.* Kiev, Ukraine: Higher Educational Publishers.

Bedny, G. Z., Meister, D. (1997). *The Russian theory of activity. Current applications to design and learning.* Lawrence Erlbaum Associates, Publishers: Mahwah, New Jersey.

Bedny, G.Z., Karwowski, W. (2011). (Eds.). Human-computer interaction and operators' performance. *Optimizing work design with activity theory.* CRC Press, Taylor and Francis Group.

Bedny, G. Z. (2015). *Application of systemic-structural activity theory to design and training.* CRC Press, FL: Taylor and Francis Group.

Bedny, G. Z., Bedny, I. S. (2018). *Work Activity Studies Within the Framework of Ergonomics, Psychology, and Economics,* CRC Press, FL: Taylor and Francis Group.

Bedny, G. Z., Karwowski, W., Bedny, I. S. (2015). *Applying systemic-structural activity theory to design of human-computer interaction system.* CRC Press, FL: Taylor and Francis Group.

Bedny, G. Z., Karwowski, W. (2007). *A systemic-structural theory of activity. Application to human performance and work design.* CRC Press, FL: Taylor and Francis Group.

Bedny, G. Z., Chebykin, o. Ya. (2013). Application the basic terminology in activity theory, *IIE Transactions on Occupational Ergonomics and Human Factors, 1*(1), 82–92.

Heukelom, F. (2014). *Behavioral Economics A history.* Cambridge, Cambridge University Press.

Henessy, R. T. (1990). Practical human performance testing and evaluation. In H. R. Booher (Ed.), MANPRINT: *An approach to system integration* (pp. 433–479). New York: Van Nostrand Reinhold.

Meister, D. (1999). *The history of human factors and ergonomics.* Lawrence Erlbaum Associates, Publishers: Mahwah, New Jersey.

Miller, R. B. (1962). Task description and analysis. In R. M. Gagne (Ed.), *Psychological principles in systems development.* New York: Holt, Rinehart, & Winston.

Muckler, F.A. (1992). Selecting performance measures: "Objective" versus "subjective" measurement. *Human Factors, 34*(4), 441–455.

Kirwan, B. (1994). *A guide to practical human reliability assessment.* Taylor and Francis.

Lomov, B. F. (Ed.) (1986). *Foundation of engineering psychology.* Moscow: Russia: Higher Education Publishers.

Myasnikov, V. A. and Petrov, V. P. (Eds.) (1976). *Aircraft digital monitoring and control systems.* Leningrad: Russia: Manufacturing Publishers.

Schraagen, J. M., Chipman, S. F., Shalin, V. L. (Eds.) (2000). *Cognitive task analysis.* Lawrence Erlbaum Associates, Publishers: Mahwah, New Jersey.

Tversky, A., Kahneman, D. (1973). Availability: a heuristic for judging frequency and probability. *Cognitive Psychology, 5,* pp. 207–232.

Payne, J. W. (1982). Contingent decision behavior. *Psychology Bulletin, 92,* 382–402.

Rasmussen, J. (1983). Skills, rules, knowledge: Signals, signals, and symbols and other distinctions in human performance models. *IEEE Transactions on Systems Man, and Cybernetics, 13*(3), 257–267.

Savage, L.J. (1954). *The foundation of statistics.* Waley: New York.

Simon, H. A. (1957). *Models of man, social and rational – mathematical essays on rational human behavior in a social setting.* New York: John Wiley & Sons.

Vicente, K. J. (1999). *Cognitive task analysis: Toward safe, productive, and healthy computer based work.* Mahwah, NJ: Lawrence Erlbaum Associates, Publishers.

3 Studying Human-Autonomous Technology Interaction Through the Lens of Systemic-Structural Activity Theory

Julian P. Vince
Defence Science & Technology Group,
Department of Defence, Australia

Gregory Z. Bedny
Essex County College, New Jersey, US

CONTENTS

3.1 INTRODUCTION

Human interaction with complex automation and autonomous technologies is one of the most pressing domains of ergonomic research. Much research effort is being invested in examining how humans can usefully and safely interact with "autonomous systems," "machine learning," and "artificial intelligence." While there has been much research that seeks to examine these interactions using the metaphor of social interaction, this work has seemed to underemphasize an alternative metaphor proposed by activity theory (AT), and other related theoretical paradigms. This alternative metaphor relates to the development of human tool use as an extension of the body and mind and treats automated and autonomous technologies primarily as tools rather than social actors.

Recent attempts at describing the issues of human-machine interaction (e.g., Chen & Barnes, 2015; Schaeffer, Chen, Szalma & Hancock, 2016) claim that answers to the problems of interacting with complex and opaque automation/autonomous technologies appears to lie in the inherent social quality of these technologies. These descriptions make social psychological concepts such as "teaming" and "trust" central to their understanding of how humans interact with these technologies.

However, the evolution of autonomous technologies from more simple forms of automation makes these more complex forms of automated/autonomous technology possibly not as novel in the control and interaction issues they present as some may argue. These issues may be more tractable when automated and autonomous technologies are examined as tools or artifacts that are essentially extensions or disembodied augmentations to human physical and cognitive capabilities, which in turn help humans transform some aspect of the world.

Theoretical approaches that lend themselves to viewing new technologies as mediational tools and artifacts include distributed cognition (Hollan, Hutchins & Kirsh, 2000) and AT (Leont'ev, 1978, Rubinstein, 1935). The development of AT has involved some distinct interpretations, at least in part due to differing objects of analysis (e.g., cultural-historical activity theory; Engeström, 1987). After some historical descriptions of the general AT treatment of the nature of subject-object relations and the mediational role of tools in those relations, this chapter will focus on the particular lens provided by the "systemic-structural" version of activity theory (SSAT; Bedny, Seglin & Meister, 2000; Bedny, Karwowski & Bedny, 2015) to describe human functional mechanisms used to regulate the use of any tool, even those complex enough to be considered "autonomous."

AT provides a theoretically coherent lens with which to understand the anthropocentric development and employment of technology. AT posits that human agency is mediated by the use of conceptual and physical tools, and this mediation is essential to human individual and collective development. The ability to interact with the world conceptually and physically is based on the capacity to develop and use representational systems and physical artifacts. This is done in order to satisfy human motives and will be considered useful only if in the service of those human ends. AT emphasizes an anthropocentric approach to examining human-technology interaction. This approach places the human not as a separate unit to the technological system, but rather as a core organizing element of the system—helping the human-technology unit realize individual and collective human goals (Nosulenko, Barabanshikov, Brushlinsky & Rabardel, 2005).

The following section discusses the ontological subject-object relations that are fundamental to AT. "Ontology" here refers to the study of what entities exist and their relations. These foundational ontological relations applied in AT speak to why treating interactions with technologies in terms of a social relationship obscures a useful way of considering how human interaction with tools aids the evolution of both tools and human creative capacity and culture.

3.2   MEDIATION IN SUBJECT-OBJECT
AND SUBJECT-SUBJECT RELATIONS

AT posits that human subjects make use of physical and conceptual tools to undertake "creative" activity. Human activity is mediated by physical and conceptual tools to modify an object. An object is the thing or state that is modified by human activity, which is mediated by tools and artifacts. It is important to make clear that in AT, the object that is modified through human activity is therefore not necessarily the tool or artifact that mediates the modification of the object. There is, however, a reciprocity of modification and evolution of both tools and their user through "object-oriented" activity. This reciprocity is no less true for conceptual tools in "subject-oriented" activity, that is, activity between human subjects.

AT provides conceptual tools with which to examine these subject-object and subject-subject relations regarding the use of complex automation and "autonomous" tools. In its general form, AT provides an ontological frame of reference, which describes the relationship between subject and object, and subject and subject, and the mediational role of tools. There have been historical theoretical debates over the relative primacy of the social and the technological in mediating human understanding of the world. This debate provides a useful basis for thinking about the ways in which our technological tools carry social meaning, and how they mediate the understanding we have of the world.

Activity theory ideas were part of a broader movement in scientific and philosophical thought in the Soviet Union, which focused on the importance of language as a mediational tool between consciousness and culture (Vygotsky, 1978). The movement also focused on the concomitant role of physical tools to the same end (Rubinstein, 1935; Leont'ev, 1978). Vygotsky proposed that higher cognitive functions are based on semiotic meaning, which is conveyed through the development and use of conceptual tools of signs (as semiotic referents to objects in the world) and then language (symbolic referents). The genesis for the development of the apparatus of meanings is predicated on engagement with the external social world.

A parallel example of the broader philosophical movement in the Soviet Union of the time also includes the work of literary theorist and philosopher, Mikhail Bakhtin (1982; 1986), whose writings provided a comparable emphasis to that proposed by Vygotsky on transmission of meaning mediated by the use of culturally recognized and refashioned use of language. Bakhtin's concept of the "dialogic" applied to both language and literary texts. Dialogism refers to how language and texts are in dialogue with other historical forms of meaning, communication, and thought. Words, phrases, and texts are laced within their historical context of usage while remaining open to the future changing context of their use. The use of euphemism or the referencing of movie dialogue in everyday conversation, perhaps to help underscore one's point through humor, might be an example of how language used in one context is adapted for use by new users in a novel context—because it helps mediate the transmission of meaning.

Vygotsky's concept of "sociality" and Bakhtin's concept of "dialogicality" both view human "being" as inherently social, but with an associated need to be able to express and interpret meaning to enable social interaction (Stetsenko, 2007). For Vygotsky, the language tools we use carry the intentions and thinking of those who have used them before, and if novel situations arise, we are able to fashion modifications that give novel form to our innovations and thereby influence the culture of language and thought. Similarly, for Bakhtin, the dialogicality of language and thought involves an understanding of the "contraposition of the self and other, which makes the act of making meaning necessarily mediated, and culturally and historically influenced. This conception of mediation is not limited in scope to language tool use in meaning making but is also extensible to physical tools and their development and use. This is because a new advance or modification to a technology or a practice is a response to earlier versions and approaches to that practical question and its evolving context.

While Vygotsky and Bakhtin independently emphasized the role of language as a mediational tool in understanding and contributing to culture, others within Soviet psychology reacted to Vygotsky's emphasis on the relation between language, culture, and consciousness by placing a renewed emphasis on physical engagement with the world. Rubinstein (1935), Leont'ev (1978), and others felt that Vygotsky did not pay adequate attention to the role of tool mediation in physical activity in the development of concepts and meanings. Rubinstein (1957) also argued that Vygotsky's cultural-historical theory did not take enough account of individual psychological characteristics relative to Vygotsky's emphasis on the social environment regarding the development of the mind and consciousness. Rubinstein placed greater emphasis on interaction with and reflection on the objective world, rather than the acceptance and internalization of cultural standards, than did Vygotsky (Bedny & Karwowski, 2007). Rubinstein stated that as a young child becomes able to physically engage with the world, they are able to use the external and physical (the body, then implements) in beginning to make adjustments to the world for themselves and in collaboration with others, as well as then being able to self-regulate their own behavior as well as contribute to the development of culture (Bedny and Karwowski, 2007). This occurs through creative engagement with the external physical world mediated by the use of external tools, which are transformed into an internal signs and artifact that can in turn be mentally manipulated. A reciprocal relationship of change and development results through the creative interaction between the human and the world.

In writing about human creativity, Rubinstein argued that the subjects manifest and create themselves through activity. Rubinstein emphasized the dual interaction or reciprocity involved in object-oriented activity. Each human act changes not only the object, but also the subject as well. An individual actively changes the objective world and culture, and thereby changes their self (Nosulenko, Barabanshikov, Brushlinsky & Rabardel, 2005).

3.3 TYPES OF ACTIVITY

Activity theory proposes two types of activity: object-oriented activity and subject-oriented. Object-oriented activity is performed by a subject using tools on a material object:

$$Subject > Tools > Object$$

What is important to note is that the use of tools to modify an object is undertaken in order to achieve a predefined goal. There are a series of determinate stages through which an individual will pass in order to modify an object: 1) a human motive will create an energetic impetus for activity, 2) the individual will develop and accept a formulated image of the desired outcome or goal, 3) the individual will orient themselves with reference to their work conditions and the goal, 4) the individual will formulate a representation of the progress of the task, 5) there will be an evaluation of the individual's resources in comparison with the requirements of the task such as its difficulty or complexity, and 6) the individual will develop strategies in order to undertake the tasks that will achieve the desired end (Bedny & Karwowski, 2007).

Subject-oriented activity refers to social interaction, and involves using semiotic (syntactic, semantic, pragmatic) tools to modify understanding. Subject-oriented activity may be formulated as follows:

$$Subject <-> Tools <-> Subject$$

Bedny & Karwowski (2007) argue that subject-oriented activity begins with a subject's goals and orientation to the situation, much the same as object-oriented activity. However, subject-oriented activity also entails an understanding of other people and making predictions about their intended activity, evaluating their goals, abilities, and other personal features, and then weighing up strategies and actions in responses to each other.

Social interactions are made up of three types of phenomena: 1) the exchange of information, 2) personal interactions, and 3) mutual understanding. The exchange of information includes both verbal and nonverbal communication. Current conceptualizations of the social nature of autonomous technologies tend to focus on this process of information exchange, and regard this as all that is required to qualify as social interaction. However, personal interactions involve the coordination of actions among individuals' social roles and norms, standards, and values. These considerations can be difficult to make explicit in all but the most straightforward or constrained social situations. It is something with which even humans all struggle with, at least from time to time. Mutual understanding involves the comprehension of another's inner experience and particularly their goals and motives and feelings. These too can be opaque, even to other humans.

The issue of confusing subject-subject relations for subject-object relations is understandable given the mechanisms humans tend to use to navigate interactions with objects, which are hardly unique to human-autonomy interaction. For example, many languages give inanimate objects a gender. However, this subjective attribution is largely arbitrary and doesn't represent the true nature of the object.

In general, object-oriented activity and subject-oriented activity are mostly distinct. However, during job performance they will transform into one another (Bedny and Karwowski, 2007). Bakhtin's emphasis on the interdependence of subject-object and subject-subject relationships through the concept of "inner dialogue" describes situations in which intersubjective interactions are found in subject-object activity (Bedny, 2015). These relationships arise through the observations of others without direct interaction with them, or by using socially developed norms or informal instructions.

Bakhtin's concept of "inner dialogue" informs the social significance of tool use in three chief ways. Firstly, there is the social dialogue created through the outcomes of tool use on the object and what this means to others, who in turn respond. Secondly, there is the dialogue with the tool as a genre of human development and the use of it, which has both historical and immediate dialogical implications in that through the employment of the tool toward an end, one is engaging inter-subjectively with those previously involved in the same activity. In using the tool in a novel way, and particularly when modification to the tool is made or techniques are socially transmitted, one is contributing to the development of human tool use culture. Thirdly, in engaging with the tool there is a form of dialogue that is immediate in that the human is investing certain inputs and receiving certain outputs, and adjusts the use of the tool in response to those outputs toward their human end. The nature of the feedforward and feedback mechanisms and their role in helping guide the human in appropriate tool use is central to SSAT and will be elaborated on later. However, the inner dialogue created through the process input and output, feedforward and feedback is the human's own projection. To borrow from Searle (1980), the syntax emanating from the tool is not semantic in and of itself; it has no innate intentionality, as would be the case for a human interlocutor.

Drawing a distinction between subject-oriented and object-oriented activity is somewhat complex. In studying object-oriented activity, subject-object relationships are present and need to be understood. This is because social interactions occur in a world of surrounding objects. Similarly, interactions with various objects also occur based on social norms and culture. However, care must be taken in presupposing greater subjective agency with regards to autonomous systems than might actually be the case. The danger is that the social metaphor of human-autonomy interaction is taken for granted, which permits system design where this emphasis is inappropriate to the task or activity, and comes at the expense of pursuing other fruitful metaphors and associated understandings.

The above section has outlined the nature of subject-object relations and the mediational role played by tools. The mediational role played by physical and conceptual tools was claimed to involve both object-oriented and subject-oriented relations with both physical and socio-cultural qualities. These relationships apply to earlier forms of technology as much as autonomous technologies. The theoretical treatment of autonomous technologies as a subject-subject interaction clearly lacks important conceptual precision, and the use of autonomous tools may be conceptualized more appropriately as foremostly "object-oriented." In the coming sections, the dialogic relationship between subject and object, and subject and tool is explored through a description of SSAT components of activity and the important role played by processes of self-regulation in supporting tool-mediated activity.

3.4   HUMAN-AUTONOMY INTERACTION AS A GOAL DIRECTED, SELF-REGULATED SYSTEM

There appears to be something of a conceptual leap emerging concerning the adoption of the metaphor of "teammate" in human-autonomy interaction science. This conceptual leap takes for granted that the greatest metaphorical explanatory power for understanding issues involved in human-autonomy interaction lie with the metaphor of interactions between human and agent, based on human to human intersubjectivity. For example, Chen and colleagues (2014) have proposed the situational awareness-based agent transparency (SAT) model. This model draws on Endsley's (1995) three levels of situational awareness—perception, comprehension, and projection. With the SAT model, Chen and colleagues propose three independent levels that describe the information an agent needs to convey to maintain "transparent" interaction with the human.

Level 1 of the SAT model relates to the goals, action, and plans of the agent. This includes purpose-goal selection; planning, and execution and progress; performance information; and perception of environment and "teammates."

Level 2 of the SAT model relates to the agent's reasoning process, and "motivations" including other environmental constraints and affordances. This use of motivations appears out of place from an SSAT perspective, as motives are considered intrinsic to individuals and social groups.

Level 3 of the SAT model relates to the agent's projections/predictions and associated uncertainty. This includes the projections of future outcome uncertainty and potential limitations, likelihood of success/failure, and history of performance.

In a recent article, Chen and colleagues (2018) updated the SAT model. The "dynamic SAT model" was claimed to better account for the need to describe the "bidirectional flow of information" between the human and the agent "teammate," and better support the operator's role in "shared decision making."

The dynamic SAT model incorporates a description of the information flows that allow for the demarcation of the division of labor for the human and the agent. Chen and colleagues argue that this emphasis on the division of labor helps support the team-based nature of human autonomy interaction. To this end, the dynamic SAT model includes feedforward loops, which indicate that changes in goals create changes in shared reasoning; and changes in goals and reasoning collectively effect changes to shared projections. Conversely, the model also exhibits feedback loops where the effects of changes in projections effect reasoning, and changes in projections and reasoning influence goals and actions.

The updated conceptualization appears to have drawn on cybernetic ideas, however these concepts are not exclusive or unique to an anthropomorphized pair of teammates—whether they both be human or human and agent. The question remains if either a socio-technical human-tool based system, or an extended body/mind-technology system metaphor might be just as applicable in understanding the "bi-directionality" of human-autonomy interaction. These alternatives to the human-agent social interaction metaphor may also provide some new perspectives to the study of human-autonomy interaction.

Gregory Bedny's systemic-structural activity theory has also placed a heavy emphasis on the study of feedback and feedforward mechanisms' role in the processes of self-regulation, applying this knowledge to ergonomics and the psychology of work. This emphasis is a result of drawing on the work of neuroscientist and physiologist Pyotr Anokhin (1935, 1955) and psycho-physiologist Nikolai Bernstein (1935). These researchers were well advanced in understanding mechanisms of feedforward/feedback in human and other biological systems well before Western research in cybernetics, such as that of Wiener (1948). Through the influence of Anokhin and Bernstein, SSAT views activity as having a recursive loop structure organized around feedforward/feedback mechanisms, which enables the evaluation of performance relative to the desired end. This concept of feedforward/feedback mechanisms in SSAT underpins the logically ordered relationship between cognition and behavior (Bedny & Meister, 1997; Bedny, Seglin & Meister, 2000; Bedny, Karwowski & Bedny, 2015; Bedny & Bedny, 2018). Cognitive mechanisms of self-regulation enable the undertaking of actions and tasks that aid in the accomplishment of goals as motivational ends.

Anokhin's (1962) theory of functional systems emphasizes from the physiological perspective that for an organism's successful adaption to an environment, it must be able to forecast different events and predict the consequences of its own reactions and evaluate these reactions based on positive and negative emotional polarity. Anokhin's (1955) discovery of brain mechanisms that reflect current and also possible future events, through which an organism is able to regulate its behavior, led to his concept of "anticipatory reflection." As represented in Figure 3.1, Anokhin's work demonstrated the fundamental connection between neural mechanisms of feedforward and feedback interconnections with a recursive loop structure of organization, whose structure provides for self-regulative functioning.

Bernstein (1966, 1967) examined the processes of feedforward/feedback mechanisms in the self-regulation of the control of movement. Bernstein argued that motor functions make up a group of processes that allow an organism to not only interact within its context, but also act upon it according to its needs. A motor act represents an attempted solution to a problem of action and implies the creation of functional processes or encoded states of a "required future." The undertaking of an action occurs in a context of conflict between an established program of performance and the presence of continual unpredictable changes in the internal and external forces of movement. Feedback and resulting corrections are therefore vital to performance. Bernstein's research into the self-regulative process proposed external and internal

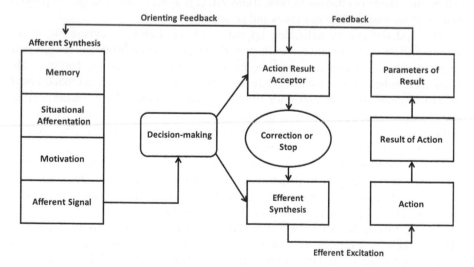

FIGURE 3.1   Representation of the Anokhin's "Theory of Functional Systems"—process of the behavioural act.

paths; the interaction between the paths enables a dynamic self-regulatory system. External paths comprise feedforward and external receptor feedback, which provide meaningful interpretation of events. Internal paths of regulation include feedforward and feedback in systems that are typically unconscious.

Bernstein (1967) also introduced the concept of levels of regulation of movement and actions. Higher conscious levels of regulation function in a governing role to which lower, unconscious levels are subordinated. During adaptive learning and training, the relationship will often be altered, so that actions that have had a high degree of automatization transition from higher conscious control to lower unconscious levels.

Bernstein suggested that as humans learn a movement, they first reduce our degrees of freedom by stiffening the musculature in order to have tight control of movements, then gradually "loosen the muscles up" and explore the available degrees of freedom as the task becomes more comfortable, and from there find an optimal strategy of performance. In terms of optimal control, it has been postulated that the nervous system can learn to find task-specific variables through an optimal control search for the best strategy.

The implications of this work in understanding human interaction with autonomy suggests that problems of human-autonomy interaction lie with the ability of the human to be able to self-regulate such activity using the autonomous tool. Furthermore, many of the mechanisms that humans use for self-regulatory processes lie at the unconscious level. It seems appropriate that research and design into the best forms of regulation of autonomous tool use should take advantage of these mechanisms of self-regulation. Ideally, augmentation of the body and senses should work in congruence with the human's evolved mechanisms of self-regulation. However, at the very least, it appears autonomous systems design might benefit from imposing some predictable constraints on the degrees of freedom involved in autonomous responses to problems thereby acting with some level of program predictability and awareness of the conditions of action, whereby feedback regarding the progress of the action relative to the desired outcome can be more readily assessed and modifications to that progress may be more readily enacted as required.

3.5   HUMAN-AUTONOMY INTERACTION
AS A SYSTEM WITH A STRUCTURE

Recent research into human-autonomy interaction, such as that by Chen and colleagues (2014, 2018), has either tended to focus on the conscious human processes involved in the interaction or the qualities of the autonomous technology's presentation of information to elicit more reliable user performance, or both. However, work on the phenomenon of automation bias and complacency (e.g., Parasuraman & Manzey, 2010) suggests that something of what can lead a user of opaque automation or autonomous technologies into error occurs at the semi-conscious or unconscious level of information processing and attention direction.

Research on the propensity for the human operator to trust or rely on automated or autonomous technology has examined the effect of the features of informational transparency or comprehensibility exhibited by the autonomous technology's interface on that trust or reliance. In this research, unconscious processes have tended to be treated experimentally through something of a "black box" approach where conditions of information presentation are varied without much reference to the user's context of interaction. Much of this context of interaction in terms of the motives of the participants, their tractable prior experience, etc., appears to be largely assumed or at best superficially interrogated. This has led to conflicting results across studies and difficulties in inferring what the relevant mechanisms are that promote safe and effective use of the autonomous technology.

An advantage that SSAT has over a cognitive psychology approach to the study of work is an emphasis on a systemic interpretation of the work context, rather than a process interpretation generally favored by cognitive psychology (Bedny & Karwowski, 2007). Bedny's SSAT proposes that human activity conforms to a generalizable structure that can be analyzed systemically. This systemic-structural interpretation includes a description of the interrelations between the semi-conscious and unconscious components of activity self-regulation, and those that are conscious. In attempting to understand human use of autonomous tools, an understanding of how these self-regulatory components and their content affect human-autonomy interaction may be of value.

An approach to the problems of human-autonomous technology interaction might be to take what is understood of these unconscious mechanisms of self-regulation and their influence on the conscious processes, and apply this to the problem of autonomous tool use. For example, work by Konopkin (1980) highlighted the importance of an operator's mental model, which consists of both conscious and tacit understandings, to the appropriate interpretation of information to support effective task execution. Research that examines not just the quality of the information provided to experiment participants, but also systematically examines participant motives, goals, and formulated strategies of performance may provide insights on the interaction of these components and their effect on information processing in the context of operating autonomous technologies.

Systemic structural activity theory presents the following interdependent components as a general structure of activity:

Activity > Task > Action > Operation > Function Block

The term "activity" refers to the entire system of undertaking something in order to modify an object with reference to a goal that arises out of motive. Tasks are defined as being groups of actions that make up a subunit of activity. Actions involve conscious direction of behavioral and cognitive processes. Most tasks utilize conscious actions and unconscious operations. For example, the task of driving a car requires a combination of conscious actions and subconscious automated operations. The last element in the structure of activity, the function block, refers to the microstructural level of self-regulation. Drawing on the foundational work of Anokhin and Bernstein (1966, 1967) on self-regulation, SSAT emphasizes that all actions and operations are governed by self-regulatory functional mechanisms. These functional mechanisms are key units of analysis to researchers interested in understanding how subjects self-regulate their activity given the work conditions. This description of the structure of activity allows for the description and analysis of human activity and behavior as a system, which is in dynamic interaction with work tools, including autonomous technologies. This creates a system of interdependent elements that are organized and utilized for a specific purpose or goal. Analysis of the systemic-structural nature of activity therefore entails examining the relevant elements of the system and their dynamic interactions and organization. The result of the analysis of function blocks allows the identification of the emergent strategies of activity performance (e.g., Bedny & Karwowski, 2013; Sanda, Johansson, Johansson & Abrahamsson, 2014).

In systemic-structural activity theory, self-regulation is seen as mainly a voluntary conscious process due to activity having a conscious goal, which subjects can modify to help direct their cognition and behavior. In order to analyze the initiation of this goal setting process, SSAT proposes a "model of orienting activity" whereby the subject creates a dynamic mental model of the situation, interprets it, and uses this interpretation to predict future events. Though not equivalent, orienting activity has some apparent similarity to the concept of "situational awareness" (SA; Endsley, 1995) in that it includes perception of the situation, its formation as a comprehensible situation, and the prediction of the near future. Where situational awareness and orienting activity diverge is with regards to the greater emphasis placed on unconscious components of reflection and probabilistic features of situation appraisal that are taken account of by "orienting activity." In terms of orienting activity, SA is one part of the model of the system of self-regulation (Bedny, et al., 2015), which is made up of many other function block components.

Systemic-structural activity theory claims that because mental processes such as sensation, perception, memory, and reasoning are interrelated, to examine these separately in the context of human activity may be overly reductionist. For instance, processes of perception can involve memory and working memory that can be coupled to operative thinking, which in turn can influence other processes. Therefore, the use of functional analysis is suggested as a method of examining activity during task performance, so it can be studied not only in psychological process terms but also in systemic terms, through functional mechanisms or blocks. This provides advantages in studying performance in any work context, such as interaction with autonomous tools—because a function block is context independent (the same function block may be applicable to many contexts).

Each of the function blocks represents a co-ordinate system of sub-functions that has a purpose in activity regulation. Each of the function blocks in the model has feedforward and feedback connections. A researcher might examine the function block called "goal" and direct attention to such aspects of activity as goal interpretation, goal formation, and goal acceptance. The researcher can open another function block called "subjectively relevant task conditions" (dynamic mental model). At this point, the researcher might study aspects of activity such as "operative image" (unconscious reflection of the situation) and "situation awareness" (conscious reflection of the situation) and their relationship. Other function block windows can be opened selectively depending on the task. Function blocks, which are not relevant to the activity or the researcher's questions, can be excluded from the analysis. However, it is also important to analyze the interrelation of the function blocks because these provide insights into the influence the function blocks have on each other, informing how unconscious processes influence conscious processes and vice versa.

The function of orienting activity is to provide the diagnosis about a situation that promotes hypothesis formation about the current and future state of a situation (Bedny & Karwowski, 2004). Orienting activity focuses on identifying the most efficient way of transforming a situation from its current existing condition toward a future desired state. There is also a "general model of self-regulation" (Bedny & Bedny, 2018). The general model of self-regulation includes further functional blocks associated with the executive components of activity involved in the transformation of a situation and the achievement of the goal of the task.

This article is mostly concerned with the issues associated with a subject's initial approach to the use of autonomous technologies in context, and how this influences the development of future strategies of performance. It is also relevant where the situation changes to the extent that a re-orientation to the task is required. Orienting activity may be of particular importance in the context of human-autonomy interaction because there may be limited opportunity for intervention once the technology is enabled. The model of self-regulation of orienting activity is also a useful place to start examining the interrelation of semi- and unconscious function blocks with those operating consciously.

This section gives a broad description of the model of self-regulation of orienting activity (Figure 3.2). The first thing to be noticed in the model is that there are two channels. Channel 1 represents the conscious channel of self-regulation. Channel 2 represents the unconscious channel of self-regulation.

Perhaps the most important functional mechanism in the model is the "image-goal" of the task (block 2). The "image-goal" function block is of particular importance because of the integrative role it plays in the self-regulation of activity. The goal may not always be clearly defined. However, it is usually better defined as task performance progresses. Activity can only exist when a goal of some sort is present for the subject.

The informational model refers to the information available to the subject via the tool interface. The model presents both relevant and irrelevant information to the human. In Figure 3.2, the relevant incoming information is indicated by arrow "a," and the irrelevant information is indicated by arrow "b." The human must actively extract information relevant to the particular task. As can be seen in Figure 3.2, "goal" and "goal-directed set," which is a partially conscious mechanism, as well as several other unconscious mechanisms, will influence what information is seen as relevant.

Incoming information activates the orienting reflex (block 4). The orienting reflex is an automated mechanism of tuning to the external, and affects the general level of activation and motivation of the human (see blocks 4 and 6), as well as afferent synthesis (block 5). The arrow between blocks 4 and 5 indicates the main information that initiates the orienting response, while the other diagonal arrows represent the additional general situational and/or environmental stimulus information that have some influence of the response to the main stimulus (Bedny et al., 2015). Afferent

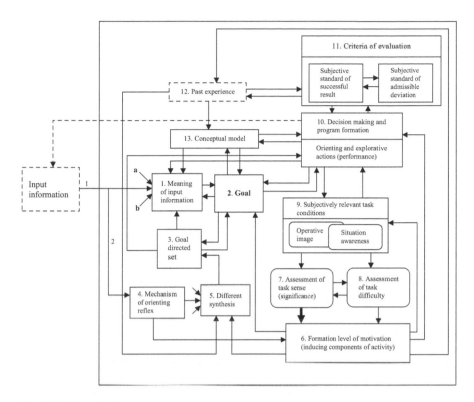

FIGURE 3.2   Model of self-regulation of orienting activity.

synthesis is also influenced by motivational blocks 6 and 12 (past experience). The afferent synthesis function block performs an integrative function, allowing the human to select key stimuli in an environment in relation to their temporal need, past experience, and the specifics of the context. Anokhin (1969) discovered the specific neurons in the brain that conduct these integrative functions.

Afferent synthesis is involved in the formation of the goal-directed set (block 3). The concept of "set" was most coherently developed by Uznadze (1966). Set provides a goal-directed character to an unconscious component of activity. A goal-directed set influences the meaningful interpretation of information (block 1) and goal (block 2), by its interaction with the conscious channel of information processing. However, if processing is largely unconscious, set directly influences block 10 (making decision about situation or strategy of explorative actions). It is important to note that if a set is inadequate, then interpretation of a situation can be inappropriate, making explorative actions associated with block 10 potentially ineffective. For example, the user of a technology may inappropriately invest in unconscious mental or motor explorative control actions due to an inadequately developed goal-directed set (Bedny et al., 2015).

"Goal-directed set" is also directly related to "goal" (block 2). The unconscious set can be transferred to a conscious goal and vice versa. The conscious goal to drive a car to a known destination can be transformed to an unconscious set in order to undertake a conversion with the accompanying passenger. At times when it is required to make a decision about a route, the goal to drive to the destination is temporarily transferred back to a conscious goal. During these unconscious information-processing tasks, goal (block 2) is not yet activated (Bedny, et al., 2015).

Information signals from the environment, which are only possibly integrated into a holistic system or dynamic mental model once the "goal" (block 2) and "subjectively relevant task conditions" (block 9) are activated. Block 9 is responsible for the development of a holistic verbally logical and imaginative mental model of a situation. Function blocks "goal" (block 2), "subjectively relevant task conditions" (block 9), and the "conceptual model" (block 13) include imaginative components and contribute to the development of the mental model. "Subjectively relevant task conditions" (block 9) is responsible for the creation of the dynamic mental model that is suitable to a particular situation. The "conceptual model" (block 13) is relatively stable and reflects possible situations relevant to the task (Bedny et al., 2015).

"Subjectively relevant task conditions" (block 9) provides reflection not only of the current situation, but also of future anticipation and inferences about the past. Block 9 includes two sub-blocks: 1) "operative image," and 2) "situation awareness." These represent the imaginative and conceptual reflections of reality respectively, which partly overlap. The non-overlapping portion of the sub-block 'operative image' provides an unconscious dynamic reflection of a situation, which is unable or only partially able to be verbalized. The reflection contained in the "situational awareness" sub-block is conscious and can be verbalized. Bedny et al. (2015) views "situation awareness," as described by Endsley (2000), as a cognitive psychological construct that in SSAT is applied as a functional component mechanism of activity regulation of the wider regulatory system. The sub-block "situational awareness" functions in tandem with "operative image," enabling self-regulation by constructing a dynamic mental model of the current and anticipated work situation. Taken together the sub-blocks containing function block 9 entail subjectively significant elements of a situation encapsulated in the situation's dynamic reflection. However, these subjectively significant elements are not always objectively important. In the case of a subject's use of an autonomous technology, misattributing subjective significance to unimportant elements can lead to a flawed orientation to a situation.

The importance of "goal" (block 2) as an integrative device is highlighted in SSAT with the observation that although an outwardly identically presented situation can be the same, a change of the goal of the orienting activity results in changes in the dynamic mental model. SSAT contends that goal formation precedes and shapes the development of the dynamic mental model (Bedny et al., 2015). In human-autonomy interaction, the influence of the goal of the subject may not always be accounted for by the designer of a technology with high operational freedom, resulting in a mismatch of intended and actual outcomes.

SSAT emphasizes the emotional-motivational aspects of orienting activity regulation. Other self-regulative models, including Endsley's situational awareness model, do not adequately describe the influence of these aspects. However, these functional mechanisms are considered critically important in SSAT for the development of the reflection of the situation, and particularly for "situational awareness" (Bedny et al., 2015). For example, block 6 "formation of the level of motivation," includes the motivational mechanism that induces activity. Motive provides the energetic stimulus that influences the development of the cognitive goal of activity. Motives influence the conscious and unconscious mechanisms involved in afferent synthesis, set, and the subjectively relevant task conditions model of block 9. Block 6 is firmly functionally linked to the "assessment of task significance" (block 7) and "assessment of task difficulty" (block 8). Assessment of task significance represents the emotionally evaluative features of the situation. Block 7 influences the logical and meaningful interpretation of the situation. The significance of the situation is involved in extracting the appropriate available features of the dynamic model of the situation. In other words, in concert with goal (block 2) and motivation (block 6), function block 7 identifies what is meaningful and relevant in the presented situation (Bedny et al., 2015).

"Assessment of task difficulty" (block 8) represents the subjective evaluation of a task difficulty. As mentioned above, it has emotional-motivational influence on other function blocks influenced by emotional-motivational factors, such as block 7 "assessment of task significance" and block 6 "formation level of motivation." Subjective under- and overestimation of the objective difficulty of the task can occur due to a perceived mismatch between the internal and external resources (perceived complexity, past experience, perceived availability/utility of assistance from others) and the features of the task (Bedny et al., 2015).

"Criteria of evaluation" (block 11) has two sub-blocks: "subjective standard of successful result" and "subjective standard of admissible deviation." These mechanisms are also important in terms of their relationship, and with an influence on, motivation. The subjective criterion for a successful result can deviate significantly from an objective standard. This subjective criterion is modifiable through performance and impacts upon block 10 "decision making and program formation." The relationship between subjective criterion of success and the objective requirements of the task is important because it will affect the satisfaction that arises from the activity. "Subjective standard of admissible deviation" is important as another evaluative mechanism in orienting activity, because it sets the qualitative bounds or range of tolerance of deviations in task performance (Bedny et al., 2015). These emotional-motivational aspects are particularly important to the monitoring and control of autonomous systems.

To provide an example, one might think of an operator monitoring an autonomous robotic system. There might be a requirement for the operator to quickly shift from automatic control to manual input in an emergency. From a self-regulation point of view, incorrect motor actions can stem from the development of an inadequate dynamic mental model of the activity (block 9) derived from the incorrect inputs of "formation of a program of task performance" (block 10).

Inappropriate vigilance in autonomous technology monitoring can worsen goal formation and the formation of dynamic mental models as the activity progresses. This is especially relevant to situations of extended monitoring activity. These factors in themselves can override some of the positive factors of information gathering actions during autonomous operations.

Furthermore, the overestimation of autonomous systems in some situations can lead to decreasing subjective significance of certain signals for an operator's (see block 7) "assessment of sense of task or its significance." An operator's overestimation or underestimation of the significance of autonomous functioning can also be related to the unconscious level of self-regulation "goal-directed set" (block 3). It can also influence meaningful interpretation of information (block 1) and formation of goal (block 2), and based on it (block 9), a "dynamic mental model" as well.

The description above outlines some of the issues of the autonomous technology operations through the model of orienting activity. It particularly emphasizes the systematic exploration of operator's motives, goals, and formulated strategies of performance as well as the way the relationships between function blocks can form an understanding of how information is used to aid human-autonomous technology interaction.

CONCLUSION

This chapter has provided a description of the ontological nature of autonomous technologies from an activity theory perspective in terms of their role as part of a socio-technical system. An outline of systemic-structural activity theory has been also described, with the intent of describing the systemic nature of behavior and cognition in human-autonomy interaction through the perspective of goal-directed, tool-mediated activity. In particular, the article has attempted to provide some critique of the subject-subject interaction metaphor currently favored in much of the research literature on human-autonomy interaction. This article has advanced the argument that a tool use metaphor also provides another useful perspective upon which to undertake research. This chapter has also highlighted the importance of unconscious and partially conscious processes in regulating activity involving human-autonomy interactions.

REFERENCES

Anokhin, P. K. (1935). *The problem of centre and periphery in the psychology of higher nervous activity.* Russia: Gorky Publishers.

Anokhin, P. K. (1955). Features of the afferent apparatus of the conditioned reflex and their importance in psychology. *Problems of psychology, 6,* 16–38.

Anokhin, P. K. (1962). *The theory of functional systems as prerequisite for the construction of physiological cybernetics.* Moscow: Academy of Science of the USSR.

Anokhin, P. K. (1969). Cybernetic and integrative activity of the brain. In M. Cole and I. Maltzman (Eds.), *A Handbook of Contemporary Soviet Psychology.* New York: Basic Books, pp. 830–857.

Bakhtin, M. M. (1982). *The dialogic imagination: Four essays by M. M. Bakhtin.* M. Holquist (Ed.) & C. Emerson, M. Holquist (Trans.). Austin: University of Texas Press.

Bakhtin, M. M., (1986). *Speech Genres and Other Late Essays.* C. Emerson, M. Holquist (Eds.) & V. W. McGee (Trans.). Austin: University of Texas Press.

Bedny, G. Z., (2015). *Application of Systemic-Structural Activity Theory to Design and Training.* Boca Raton, FL: CRC Press.

Bedny. G. Z. & Karwowski, W. (2004). A functional model of the human orienting activity. *Theoretical Issues in Ergonomics Science, 5*(4), 255–274.

Bedny. G. Z., & Karwowski, W. (2007). *A Systemic-Structural Theory of Activity: Applications to Human Performance and Work Design*. Boca Raton, FL: CRC Press.

Bedny, G. Z., & Karwowski, W. (2013). Analysis of strategies employed during upper extremity positioning actions. *Theoretical Issues in Ergonomics Science, 14*(2), 175–194.

Bedny, G. Z., Karwowski, W., & Bedny, I. (2015). *Applying Systemic-Structural Activity Theory to Design of Human-Computer Interaction Systems*. Boca Raton, FL: CRC Press.

Bedny, G. Z., & Bedny, I. (2018). *Work Activity Studies Within the Framework of Ergonomic, Psychology and Economics*. Boca Raton, FL: CRC Press.

Bedny, G. Z., & Meister, D. (1997). *The Russian Theory of Activity: Current Applications to Design and Learning*. Mahwah, NJ: Lawrence Erlbaum Associates.

Bedny, G. Z., Seglin, M. H., & Meister, D. (2000). Activity theory: history, research and application. *Theoretical Issues in Ergonomics Science, 1*(2), 168–206.

Bernstein, N. A. (1935). The problem of relationship between coordination and localization. *Archive of Biological Science, 38*, 1–34.

Bernstein, N. A. (1966). *The psychology of movement and activity*. Moscow: Medical Publishers.

Bernstein, N. A. (1967). *The coordination and regulation of movements* Oxford, UK: Pergamon Press.

Chen, J. Y. C., Procci, K., Boyce, M., Wright, J., Garcia, Andre & Barnes, Michael. (2014). Situation Awareness-Based Agent Transparency. Report No. ARL-TR-6905, Aberdeen Proving Ground, MD, U.S. Army Research Laboratory.

Chen, J. Y. C., & Barnes, M. J. (2015, October). Agent transparency for human-agent teaming effectiveness. In *2015 IEEE International Conference on Systems, Man, and Cybernetics (SMC)*, (pp. 1381–1385). IEEE.

Chen, J. Y. C., Lakhmani, S. G., Stowers, K., Selkowitz, A. R., Wright, J. L. & Barnes, M. (2018). Situational awareness-based transparency and human-autonomy teaming effectiveness. *Theoretical Issues in Ergonomic Science, 19*(3), 259–282.

Endsley, M. R., (1995). Toward a theory of situational awareness in dynamic systems. *Human Factors, 37*, 32–64.

Endsley, M. R. (2000). Theoretical underpinnings of situational awareness: A critical review. In M. R. Endsley and D. J. M. Garland (Eds.), *Situational Awareness Analysis and Measurement*. Mahwah, NJ: Lawrence Erlbaum Associates, pp. 3–32.

Engeström, Y. (1987). *Learning by expanding: An activity-theoretical approach to developmental research*. Helsinki: Orienta-Konsultit.

Hollan, J., Hutchins, E., & Kirsh, D. (2000). Distributed cognition: toward a new foundation for human-computer interaction research. *ACM Transactions on Computer-Human Interaction (TOCHI), 7*(2), 174–196.

Konopkin, O. A. (1980). *Psychological mechanisms of the regulation of activity*. Moscow, Russia: Science Publishers.

Leont'ev, A. N. (1978). *Activity, consciousness and personality*. Englewood Cliffs, NJ: Prentice Hall.

Nosulenko, V. N., Barabanshikov, V. A., Brushlinsky, A. V., & Rabardel, P. (2005). Man-technology interaction: Some of the Russian approaches. *Theoretical Issues in Ergonomic Science, 6*(5), 359–383.

Parasuraman, R.,& Manzey, D. H. (2010). Complacency and bias in human use of automation: An attentional integration. *Human Factors, 52*(3), 381–410.

Rubinstein, S. L. (1957). *Existence and consciousness*. Moscow: Academy of Pedagogical Sciences.

Rubinstein, S. L. (1935). *Problems of General Psychology*. Moscow: Academy of Pedagogical Sciences.

Sanda, M-A., Johansson, J., Johansson, B., & Abrahamsson, L. (2014). Using systemic struc-
    tural activity approach in identifying strategies enhancing human performance in
    mining production drilling activity. *Theoretical Issues in Ergonomic Science, 15*(3),
    262–282.
Schaeffer, K. E., Chen, J. Y., Szalma, J. L., & Hancock, P. A. (2016). A meta-analysis of fac-
    tors influencing the development of trust in automation: Implications for understanding
    autonomy in future systems. *Human Factors, 58*(3), 377–400.
Searle, J. (1980). Minds, brains, and programs. *Behavioral and Brain Sciences, 3*(3), 417–424.
Stetsenko, A. (2007). Being-through-doing: Bakhtin and Vygotsky in dialogue. *Cultural
    Studies in Science Education, 2,* 746–783.
Uznadze, D. N. (1966). *The psychology of set.* New York: Springer Science + Business Media.
Vygotsky, L. S. (1978). *Mind in society: The development of higher order psychological pro-
    cesses.* Cambridge, MA: Harvard University Press.
Wiener, N. (1948). Cybernetics: *Or control and communication in the animal and the
    machine.* Cambridge, MA: MIT Press.

4 Modeling and Mobile Device Support of Goal Directed Decision-Making under Risk and Uncertainty

Alexander M. Yemelyanov
Department of Computer Science,
Georgia Southwestern State University, Georgia, US

CONTENTS

4.1   INTRODUCTION

Decision-making in situations of uncertainty is a complex problem, with the risk of potential losses of money, health, reputation, etc., always present. It is conventionally assumed that decision-making under uncertainty considers situations in which several outcomes are possible for each course of action, and the decision-maker is unable to estimate the probability of occurrence of the possible outcomes. However, when the decision-maker is able to calculate these probabilities of occurrence, decision-making is considered to be performed under risk (Knight, 1964). The probabilistic approach of defining uncertainty relates directly to a lack of information on probabilities of outcomes; despite this, uncertainty is reduced when the probabilities of outcomes are specified—in other words, when a problem shifts from the category of uncertainty into the category of risk. In the current work, the uncertainty of an outcome is viewed from a broader perspective and is associated both with a lack of information regarding the value of the outcome, and with a lack of information regarding the possibility of its occurrence. It is worth noting that an outcome's uncertainty is related to the uncertainty of the goal, while the uncertainty of its possibility is related to the uncertainty of the conditions of the problem. The necessity of considering possibility instead of probability is related to the fact that possibility is more broadly understood than probability, and apart from informational (statistical) characteristics reflected at the decision-maker's conscious level, possesses emotional (energy) characteristics, reflected at their unconscious level. Finally, emotional characteristics reflect the subjective complexity (or difficulty) of obtaining the considered outcome, which is an important characteristic of its uncertainty, directly influencing the decision that is made. It is important to consider that when evaluating uncertainty and risk, the analysis of subjective values and subjective possibilities of outcomes should not be quantitative, as is typically observed to be the case in the probabilistic approach, where these factors have relative monetary or statistical value. Instead, they should reflect a vagueness inherent in the decision-maker's perception of uncertainty and risk. As an alternative for representing uncertainty, implementation of possibility and fuzzy set theories have been suggested, which both relate to vague linguistic variables, such as "high" or "often" (Aven et al, 2014). The presence of uncertainty in the occurrence of outcomes contributes a principle correction to the construction of the model of the decision-making problem. As will be demonstrated in the present paper, such a model can be constructed by an individual only in the process of decision-making.

Provided below is an analysis of the existing viewpoint on decision-making under risk and uncertainty with the focus on the motivational approach from the position of the Systemic-Structural Activity Theory. It also delves deeper to illustrate how the major provisions of this theory, related to the concepts of goals, self-regulation, internal feedback, etc., help to recognize the mechanisms of decision-making, as well as to create effective systems for decision support. This chapter also covers measurement of the level of motivation; rules of motivation and self-regulation; and the self-regulation model of decision-making. Based on this model, the Performance Evaluation Process (PEP) is proposed, in which the goal, selection criteria, and mental model are formed during decision-making. Express Decision (ED)—a mobile application for quick everyday decision-making—is developed. ED uses PEP for solving problems, which are typically solved intuitively, and in which individual biases and emotional factors may be an important influence.

4.2 DECISION THEORIES

4.2.1 NORMATIVE AND DESCRIPTIVE THEORIES

Decision theories present two relatively prominent theoretical approaches to the analysis of decision-making under conditions of risk and uncertainty. Decision theories assume that people's preferences depend on two factors: the value people attribute to the outcomes of different courses of action and the probability that each of the outcomes will occur. With this in mind, the *normative* approach places emphasis on how to make the best decisions by deriving an algebraic representation of preference from abstracted behavioral axioms, whereas the *descriptive* approach uses this algebraic representation with the incorporation of people's preferences for safety or risk. Decision theories have usually limited their consideration to cognitive phenomena such as choice and risk. The two main theories of choice are expected utility theory and prospect theory. Although both of these theories have different assumptions about subjective representations of objective outcomes and probabilities, they present decision rule as an algebraic function of the probabilities and values of the outcomes of each alternative.

Expected utility theory (EUT) (von Neumann and Morgenstern, 1944) is a major normative theory of decision-making under risk, in which the probability distribution of the outcomes is known. EUT is constructed based on a set of axioms, for example, transitivity of preferences, which provide criteria for the rationality of choices. According to this approach, decisions can be abstracted and represented as the selection of a single course of action X described by the value of the possible outcomes $\{x_1, x_2, ..., x_n\}$ and the associated probability $\{p_1, p_2, ..., p_n\}$ that each outcome would occur, if the action was selected. The option X that has the highest expected utility $EU(X) = p_1 \cdot x_1 + \cdots + p_n \cdot x_n$ will be selected. For example, consider a prospect with two options: (A) a certain outcome valued at \$100, and (B) a risky option with an 80% chance of \$100, a 15% chance of \$200, and a 5% chance of receiving nothing. The expected utility calculation recommends that one should take the second option, because $EU(B) = \$110 > \$100 = EU(A)$. Although EUT allows for individual differences in attitudes regarding risk by using subjective values of outcomes and their subjective probabilities, it does not describe any sort of psychological mechanism that underlies risk attitudes, such as satisfaction with success and fear of failure, as well as individual differences in motivation, such as a need for security and avoidance of failure (Larrick, 1993).

Descriptive theories of decision-making accept the algebraic representation of decision rule, but incorporate known limitations of human behavior. Kahneman and Tversky (1979, 1981) noted that people exhibit patterns of preference, which appear incompatible with expected utility theory. They produced a descriptive model, called prospect theory, which modifies expected utility theory so as to accommodate these observations. Prospect theory is a behavioral economic theory that describes the way people choose between probabilistic alternatives that involve risk, where the probabilities of outcomes are known. Prospect theory introduced the notion of reference dependence, in which outcomes are not evaluated absolutely but relative to some benchmark or reference point. In this context, relative to the reference point, outcomes were evaluated differentially based on whether they were seen as gains (utility function $UG(x)$) or losses (utility function $UL(x)$). The concept of loss aversion was also proposed, in that the marginal utility of a constant change is greater for losses than for gains (a \$10 loss is more aversive than a \$10 gain is pleasant). Once the decision-maker evaluates all prospects, he or she then chooses the prospect that offers the highest overall value V, which is calculated in a manner analogous to the expected value $EU(X)$: $V(x,p) = w(p_1)v(x_1) + \cdots + w(p_n)v(x_n)$. However, decision weight $w(p_i)$ replaces probability p_i, and subjective value functions $v(x_i)$ replace value x_i. It is particularly worth noting that functionvs $v(x)$ и $w(p)$ are individual for each decision-maker, and serve as adjustment factors of that individual's perception of value and probability of outcomes. In view of this, as the authors of the prospect theory point out, decision weights are not probabilities, because "they do not obey the probability axioms and they should not be interpreted as measures of degree or belief" (Kahneman and Tversky, 1979, p. 280). With the incidence of uncertainty in the occurrence of outcomes, as it can be assumed from further analysis, the role of these weights escalates, since they present a subjective difficulty of the alternative and the individual's emotional reaction to this difficulty. Then again, decision weights in prospect theory remain psychologically undetermined (Johnson and Busemeyer, 2010); therefore, for the occurrence of situations with uncertainties (unknown event probabilities), prospect theory as a theory for decisions under risk (known event probabilities) loses its descriptive properties, becoming misleading and logically contradictive.

As an illustration of the aforementioned theory, let us analyze the following decision-making problem in a hypothetical life and death situation presented by Tversky and Kahneman (1981, 1984), and widely discussed in several different papers on risk psychology (Bless et al, 1998), etc. Participants were asked to choose between two alternative programs to combat an unusual Asian disease that is expected to kill 600 people. This decision-making problem was presented to participants with positive framing (survival format), i.e., how many people would live, and with negative framing (mortality format), i.e., how many people would die. Results are presented below in the form of prospects (the percentage who choose each option is indicated in parentheses).

Positive framing:
If Program A is adopted, 200 people will be saved. (72%)
If Program B is adopted, there is a 1/3 probability that 600 people will be saved and 2/3 probability that no people will be saved. (28%).

Negative framing:
If Program C is adopted 400 people will die. (22%)
If Program D is adopted, there is a 1/3 probability that nobody will die and 2/3 probability that 600 will die. (78%)

Tversky and Kahneman claim that positive and negative framing result in different descriptions of the same problem, where programs C and D are undistinguishable in real terms from programs A and B. At the same time, experimental results show that most participants chose programs A and D, despite the fact that in terms of consequences, these choices are contradictory. This is because programs A, C and B, D are equivalent. This is a violation of the logical principle of extensionality in decision-making, which states that making a decision in a problem should not be affected by how the problem is described (Bourgeois-Gironde and Giraud, 2009). Tversky and Kahneman, admitting the fallacy of such a choice, explained it by employing the general properties of peoples' attitude toward a risk: people are expected to show a risk-seeking preference when faced with negatively framed problems, and risk aversion when presented with positively framed ones. It is clear that such an explanation is superficial and does not provide the answers to many questions, such as why participants were being illogical with their answers and why programs B and D—with the same statistical outcomes (200 living and 400 deceased)—were differently evaluated by participants. In reality, in the analysis of the Asian disease, participants made decisions in two different problems with two different goals: Problem 1 (positive framing) has one specific goal—"save all 600 lives" and Problem 2 (negative framing) has another specific goal—"do not allow any living patient out of 600 to die." Considering this, program D motivated to "save lives," which is primarily associated with the medicinal effects of treatment to combat a disease, while program D motivated to "prevent loss," which is often associated with prophylactic measures to prevent illness. This conclusion also comes about due to the fact that the given information regarding survivors is typically present in programs regarding treatment, while information regarding the deceased is generally found in programs related to preventive measures. And since with identical expected outcomes (200 survivors and 400 deceased) present in both programs, prevention is generally associated with fewer efforts (difficulties) than any sort of treatment undertakings, it can therefore be suggested that program D is more preferable than program B. Such a conclusion is also consistent with the answers obtained from the participants of the experiment: 78% > 28%.

The absence of a goal leads to a situation in which the same decision is capable of electing different choices depending on how the problem is framed (edited). At the same time, it should be noted that although the goal is not explicitly indicated, it still serves as a reference point in the current example (Yemelyanov, 2017). Indeed, when the goal stresses to "save all 600 lives" (positive framing), the reference point is that "600 died," which is "a state of affairs in which the disease is allowed to take its toll of 600 lives"; at the same time, when the goal stresses to "not allow any living patient out of 600 to die" (negative framing), the reference point is therefore "0 died"—in other words, "a state of affairs in which no one dies of the disease." However, in this specific scenario, it is essential to highlight that the concept of a "goal" is not actually mentioned anywhere throughout the example, but is instead only evident through the reference point; in other words, the idea that the goal "stresses" something is used to indicate the fact that the goal itself is not stated explicitly, yet its effects are nonetheless quite evident within the context of the example.

Thus, in the example analyzed above, the goal was considered only for its evaluative component, which was demonstrated in the separation of outcomes into positive and negative categories. Another significant motivational component of the goal, reflecting the difficulties of its attainment (Bedny, 2015), was not reflected in the given example, which did not allow the conclusion that program D is more preferable than program B. This fact provides a clarification regarding why the prospect theory decision weights are not probabilities. It happened because these weights reflect difficulties, related to the treatment of patients and to the prevention of disease.

In 1992, Tversky and Kahneman developed a new version of prospect theory, called cumulative prospect theory (CPT), which employs cumulative rather than separable decision weights and extends the theory in several respects. One of these fundamental differences is demonstrated by the fact that this theory makes room for different weighting functions for gains and for losses. Two principles, "diminishing sensitivity" and "loss aversion," are invoked to explain the characteristic curvature of the value function and the weighting functions. A review of the experimental evidence, as well as the results of a new experiment, confirm a distinctive four-fold pattern of risk attitudes: risk aversion for gains and risk seeking for losses of high probability; risk seeking for gains and risk aversion for losses of low probability (Tversky and Kahneman, 1992). According to the authors' claim, unlike PT, CPT can be applied to both risks and uncertainties with any number of outcomes. Köbberling (2002) created a CPT calculator, which is a computer program for calculating the cumulative prospect theory value of prospects with a maximum of four outcomes. In order for a decision-maker to use this calculator for the purpose of decision-making, they must first specify the value and weighting functions by identifying the following individual parameters: power for gains (0.88), power for losses (0.88), loss aversion (2.25), probability weighting parameter for gains (0.61), and probability weighting parameter for losses (0.69). The brackets indicate the medians of the corresponding parameters, obtained from the experiment, which was carried out by Tversky and Kahneman to obtain detailed information about the value and weighting functions. It should be noted that it is practically impossible to determine the indicated parameters (and in particular, the level of risk aversion) before the resolution of a problem that is uncertain both in terms of the goal and of the difficulties associated with its achievement. Therefore, it is difficult to agree with Tversky and Kahneman, who state that CPT can be applied to both risky and uncertain prospects.

In conclusion, it should be recognized that the formation of the goal is not given sufficient attention in the decision theories of goal formation. The only criterion of choice is the maximization of the goal function, which is represented by a linear algebraic formula. The solution of the problem is preceded by the construction of its model, which is only possible for well-structured problems with risks of a probabilistic nature and quantitative representation for the subjective value. As soon as risks of a different nature arise, such as those associated with uncertainty in a specified goal, or with difficulties in attaining it, neither the theory of utility nor its further development, with the consideration of the human factor (prospect theory), are sufficiently productive. Computer functions now become beneficial only for the purpose of conducting mathematical calculations on pre-constructed algebraic formulas, in which the individual characteristics of a person are taken into account only in the form of separate numerical parameters, which tend to be selected without proper psychological justification.

4.2.2 MULTI-CRITERIA DECISION THEORIES

Multi-criteria decision theories provide a general and systematic framework supporting complex decision-making situations with multiple and often conflicting objectives (i.e., criteria, attributes). One of the main theoretical approaches within these theories is multi-attribute utility theory (MAUT; Keeney and Raiffa, 1976) that built a multi-attribute utility function, which captures the relative weights of the attributes, their interactions, and the decision-maker's risk attitude toward uncertainty in each of the individual attributes. MAUT is designed to handle the tradeoffs among multiple objectives (for example, such as the price, customer reviews, size, etc., when purchasing a smartphone). The goal of MAUT is to construct a utility function of the form $u(v_1, v_2, \ldots, v_k) = f(v_1, v_2, \ldots, v_k)$, where v_k is the k-th attribute of concern.

Analytic hierarchy process (AHP) is a multi-criteria decision theory that provides a general and systematic framework supporting complex decision-making situations with multiple and often conflicting objectives (i.e., criteria, attributes). The AHP method was designed by Thomas Saati (1980). It includes four steps: problem modelling, weights valuation, weights aggregation, and sensitivity analysis (Ishizaka and Labib, 2011). When using AHP, the problem is decomposed into a hierarchy of more easily comprehended sub-problems, each of which can be analyzed independently, according to a hierarchy in which the upper level is the goal of the decision. The second level of the hierarchy represents the criteria and sub criteria, and the lowest level represents the alternatives. As soon as the hierarchy is built, the decision makers systematically assess its various elements by comparing them to each other two at a time (pairwise comparison), with respect to their impact on an element above them in the hierarchy. In making the comparisons, the decision makers use their judgments regarding the outcomes' relative importance. For example, with the goal of *selecting the vehicle that best meets one's objectives*, the following factors could be considered as relevant criteria: initial cost, maintenance cost, prestige, and quality. The quality criterion could be divided into the sub-criteria of safety, frequency of breakdown, performance, and design. And, in turn, the design sub-criterion could be divided into the sub-sub-criteria of exterior design and interior design. Once the criteria, sub-criteria, and sub-sub-criteria have been determined, pairwise comparison of the factors will need to be made with respect to the goal. For example, the user should compare the relative importance of prestige and maintenance cost with respect to the goal. In order to make the pairwise comparisons, a 9-point linear scale was used. The scale allows determining whether one criterion compared to another is equally important (1), slightly more important (3),…, or absolutely more important (9). Ultimately, each of the five criteria above, along with their sub- and sub-sub-criteria, will need to be compared to each other. When pairwise comparisons of all criteria with respect to the goal will be finished, the pairwise comparisons of all alternatives with respect to each criterion should be considered. All pairwise comparison numerical values are gathered in the comparison matrix. Linear algebra is used with the purpose of calculating the relative weight of each criterion or sub-criterion. This allows determining the vector of weights as the normalized eigenvector of the matrix, associated with the largest (principle) eigenvalue that measures the degree of inconsistency. In the next step, AHP aggregates the local priorities across all criteria in order to determine the relative global priority of each alternative. AHP applies an additive aggregation with a normalization of the sum of the local priorities to unity. The AHP method is supported by Expert Choice software, which helps for the selected model to provide weights valuation (collecting the consistent pairwise comparisons), weights aggregation (determining the global priority), and sensitivity analysis (validation of the suggested model). A short overview of AHP is provided below, which is important to its subsequent consideration in the current work.

In this way, in AHP, the model of the problem must have independent criteria and cannot be changed within the process of making a decision. AHP solves the given problem by using quantitative methods of linear algebra, based on a previously created multi-criteria evaluation model. After building an evaluation model, this model remains unchanged during all subsequent steps of the AHP method: weights valuation, weights aggregation, and sensitivity analysis. In AHP, a rank reversal can occur due to a lack of sound psychological evidence for the data, which is collected from the user and then further extrapolated with the use of linear algebra models. In this case, the use of linear scales to measure human responses, as well as the use of linear algebra models for their subsequent transformation into the final decision regarding the relative priorities of the alternatives, are not based upon sound psychological evidence, which can thus result in rank reversal. AHP turns out to be more successful when applied to well-structured (or quantitatively described) problems, as opposed to ill-structured problems, which have unclear goals and incomplete information. AHP only determines the relative preferences of existing alternatives, without linking to their subsequent performance. This means that even the highest-ranked alternative, from the perspective of AHP, might not be selected for performance, perhaps because of a lack of necessary motivation for its execution on behalf of the decision-maker. It should be noted here that the AHP is limited to a problem in which the goal and all criteria for success are determined in advance.

4.2.3 MOTIVATIONAL THEORIES

The motivational approach assumes that when people make decisions, not only should the subjective values of outcomes be considered, but the affective components of decisions and the motivational factors that enhance them in achieving a goal should be considered, as well. One of the most basic aspects of motivation is that people do not see outcomes as neutral, but rather, they categorize them as either a success or failure and then experience positive or negative emotion based on their categorization. The reference point of the value function plays a key role in this categorization, because it divides the space of outcomes into a positive and negative region (Heath, Larrick, and Wu, 1999).

The motivational theories propose that the choices people make are often the result of a compromise between two competing desires. On one hand, people wish to choose the option that maximizes their outcomes (in terms of the value of the outcome); on the other hand, they want to avoid making poor decisions that may cause feelings of failure or disappointment. The tension between these two motives pushes decision-makers toward a specific level of risk-seeking or risk-averse behavior. In Lewin's (1939) field theory, the motivational force (S) is considered a function of two basic indicators of the theory—the valence (V) of the event (the strength of its attractiveness or repulsion) and the expected probability (P) of occurrence of an event with this valence: $S = f(V, P)$. This field theory position became widespread throughout various models that formed the basis for developing new branches of field theory, such as cognitive valence theory, expectancy theory of motivation, achievement motivation theory, etc., all of which found various practical applications. For example, according to Vroom's (1964) expectancy theory of motivation, when deciding among behavioral options, individuals select the option with the greatest amount of motivational force $MF = Expectancy \times Instrumentality \times Valence$. Atkinson (1957), in his analysis of achievement-oriented activity, proposed that preferences for different probabilities of achieving success or avoiding failure are related to individual differences in motivation. The several variables are combined multiplicatively to obtain a value termed "resultant motivation strength," which indicates tendencies to either approach success or avoid failure: $(M_s \times P_s \times I_s) + (M_f \times P_f \times I_f)$. This model incorporates six variables: the motive to achieve success (M_s) and the motive to avoid failure (M_f); subjective probabilities of success (P_s) and failure (P_f); and incentive values of success (I_s) and failure (I_f). In addition, there are two criteria: valence (success and failure); and two sub-criteria: (incentive value of success and incentive value of failure).

It must be noted however that neither Lewin nor his successors could clarify specific functional dependences that existed among the indicators. They considered motivational strength to be a simple algebraic combination of positive and negative motives, values of success and failure and their probabilities, without any proof of how these dependences are psychologically determined (Kotik, 1994).

4.3 ACTIVITY THEORIES

Based on the activity and applied activity theories, G. Z. Bedny and his associates developed the systemic-structural activity theory (SSAT), which provides a systematic view of decision-making under risk and uncertainty (Bedny and Meister, 1997; Bedny and Karwowski, 2006; Bedny, Karwowski, and Bedny, 2015; etc.).

4.3.1 SELF-REGULATION MODEL OF THE THINKING PROCESS

SSAT considers activity as a system, which integrates not only cognitive and behavioral, but also emotional-motivational components. As it is presented to us, the description of the thinking process proposed by SSAT is by far the most productive approach in the study of problem-solving processes. Below, we will illustrate the advantages of this approach, as well as describe its practical application to decision-making problems in situations of uncertainty. SSAT contributes key interpretations into the understanding of fundamental psychological concepts that describe human thinking, such as goals, motives, mental states, mental feedback, etc., and it considers thinking from a functional analysis perspective, when it is presented as a self-regulating system with external and internal mental feedback.

In SSAT, the *goal* is considered to be a basic concept of problem-solving activity. At the start of making a decision, the goal may have a very general form. Only during the process of problem solving does the goal gradually become clearer and more specific, and it may even be corrected if necessary during the course of activity. The problem-solving process begins when the *initial mental model* of the problem is created. However, the initial mental model of the problem often is unable to facilitate the attainment of a desired result on its own. Therefore, after understanding the problem at hand, the subject divides the problem into *sub-problems*. The subject initiates the formation of sub-problems by formulating various *hypotheses*. Each hypothesis has its own potential goal. Based on the comparison and evaluation of such hypotheses, a subject selects one and formulates the first sub-goal associated with the selected hypothesis. Problem solving includes the continuous *reformulation (disaggregation)* of a problem and the development of its corresponding mental model. If the received data is evaluated *positively (positive feedback)*, the thinking process cycle is complete. However, if the result is evaluated *negatively (negative feedback)*, internal or external information should be added to continue this process on a lower level (or to search for other alternatives of solving the problem). In other words, in SSAT, unlike cognitive psychology, the decision-maker is able to regulate their behavior not only externally, but also internally by using the inner mental plane. The self-regulation model of solving problem under uncertainty (Figure 4.1) shows how the *initial mental model* of the problem was transformed into the *final mental model*. For this purpose, a problem should be disaggregated to such terminal levels of sub-problems, which can be *positively evaluated*.

The levels of motivation for solving problems are formed based upon the *imaginative, verbally-logical*, and *emotionally-evaluative analyses*, along with a

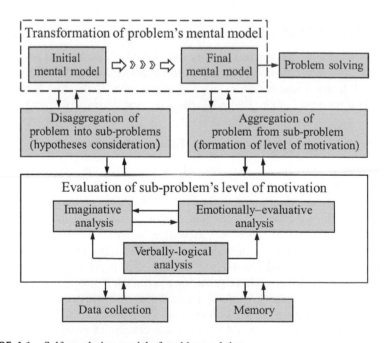

FIGURE 4.1   Self-regulation model of problem-solving.

collection of appropriate external and internal information. When the evaluation process for each positively evaluated sub-problem is finished, the aggregation process begins. Within this process, all determined levels of alternatives' motivations in sub-problems will be aggregated backward according to the previously created structural hierarchy (problem, sub-problems, sub-sub-problems, etc.) within the disaggregation process. The aggregation process results in the formation of the final mental model of the problem, with the levels of motivation available for all alternatives. This allows the individual to make a decision concerning whether the alternative with the highest level of motivation will be executed, or if its level of motivation is not high enough, which would then require new alternatives to be added for consideration.

4.3.2 Motivation Evaluation Process

In the process of problem solving, the subject disaggregates a problem and assesses the obtained results according to the level of *significance* (subjective importance) of the outcomes and the level of its *difficulty* (subjective complexity). The relationship between difficulty and significance determines the level of activity motivation. In the decision-making process, the alternative with the highest level of motivation will finally be chosen for execution.

There are complex relationships between the assessment of difficulty and the significance of the task (Bedny & Karwowski, 2006). If an individual evaluates the problem as a personally significant and difficult one, the level of activity motivation that is required for problem performance would increase. At the same time, if an individual evaluates the problem as a difficult, but not personally significant one, the level of motivation can drop, or the individual can reject the problem altogether. In Chapter 6, Gregory Bedny and Inna Bedny specified this relationship in Figure 6.6 in the following rules.

a. If the difficulty of the task is not adequate for the individual, then motivation is negative.
b. If the difficulty of the task is adequate for the individual and there is only positive significance, then motivation is positive.
c. If the difficulty is adequate for the individual and interaction between positive and negative significances is positive, then motivation is positive.
d. If the difficulty of the task is adequate for the individual and interaction between positive and negative significances is negative, then motivation is negative.

The motivation evaluation process (MEP) for solving problems that we suggest in a current paper implements the motivation rules mentioned above and SD-Frame (Figure 4.2) to evaluate the motivational levels of sub-problems in the process of solving a problem.

The goal splits the outcomes of the alternative into positive and negative categories and thus determines the levels of their significances, respectively, as either positive or negative. Positive significance and a positive component of difficulty create a

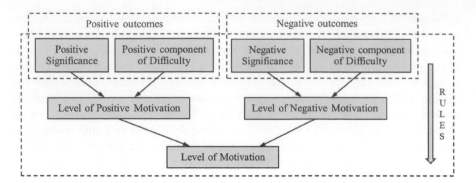

FIGURE 4.2 SD-Frame of mental evaluation of motivational level.

level of positive motivation, and negative significance and a negative component of difficulty create a level of negative motivation. It should be noted that SSAT does not consider the different components of difficulty. Here, a positive component of difficulty characterizes the difficulty associated with gaining positive outcomes, and the negative component of difficulty characterizes the difficulty associated with avoiding negative outcomes. We implemented this separation for difficulty in the event when more detailed rules are needed to describe the relationship between level of motivation and its indicators. Finally, the level of positive motivation and the level of negative motivation determine the overall level of motivation.

SD-Frame is an abstract, schematic structure, which is more general than the actual model of motivation. It presents a mental frame or template that consists of a core (fixed information about the structure of formation of the level of motivation) and slots (containing variable information about the levels of positive and negative significances and difficulties). The possession of slots for significance and difficulty makes the SD-Frame adaptable and flexible to accommodate a formation of a motivation level at any intermediate mental state of a sub-problem. When the intermediate sub-problem, with its sub-goal, is adopted (positively evaluated), the corresponding levels of significances and difficulties can be assigned to the slots, according to the existing motivation rules, which allow the resultant levels of an alternative's positive, negative, and integrated motivation to be determined. If the intermediate problem is so complex that the levels of significances and difficulties cannot be assigned, and the decision-maker is still motivated to continue the decision-making process, then the sub-problem, according to the motivation rules, should be further simplified into sub-sub-problems by using hypotheses to such an extent that the SD-Frame can be adopted. During the process of aggregation, the levels of motivation in the sub-problems, which are determined in their SD-Frames, are then combined into the final level of motivation of the problem.

In the analytic hierarchy process theory analyzed above, the model of the decision-making problem was created before the individual solved the problem, so consequently it did not consider such important factors as the potential correction of the mental model of the problem during the process of solving it. The inclusion of changes is regulated by internal mechanisms of self-regulation when the requested data are extracted from memory; these data are able to introduce fundamental changes into the results of solving a problem. Unfortunately, the cognitive approach, which recommends the creation of models before an individual has solved a problem, orients itself based only upon external feedback, which is essentially a link with the preliminary (and therefore, a largely inadequate) constructed model, but not with the individual (and their own mental model). Thus, the AHP—as all other multi-criteria decision theories in which the criteria are not included in the model during the process of decision-making, but instead, before it—is limited only to the application of the problems with previously determined criteria of selection. Another important concern of the implemented models is the fact that within them, the factor of uncertainty in the occurring outcomes is largely described with the help of probabilities. Even though this is convenient for subsequent mathematical computations of the model, it is absolutely unfeasible when the guidelines of the problem possess an internal or emotional characteristic. In the mind of the decision-maker, the factor of uncertainty in the attainment of a goal, as well as the achievement of desired outcomes and the avoidance of undesired outcomes, do not possess a probabilistic character, but instead a problematic (related to difficulties) one.

4.4   MEASURING THE LEVEL OF MOTIVATION

4.4.1   IL-FRAME

Decision-making is an essential part of the problem solution. There is no solution to any problem without making a decision. In the decision-making process, alternatives are compared in terms of expected outcomes. The best alternative is the alternative with a level of motivation not only comparatively better, but also high enough for the realization of the chosen alternative. Thus, measuring the level of motivation is an important part of the decision-making process.

According to research by Bedny and Karwowski (2006) on thinking activity, forming the level of motivation in the decision-making process passes through three stages: (1) preconscious, (2) goal related, and (3) task evaluative. At the third stage, the mental model of the decision-making problem is formed, and a conscious motivated choice of the most preferable alternative occurs through an evaluation of difficulty and significance of expected outcomes. The level of motivation is formed after the formation of the goal. It should be noted that the goal is a fundamental element in decision-making. The goal determines the criteria for success and divides the outcomes into positive and negative categories—in other words, it determines their valence. A change in the goal leads to a change in the valence of outcomes, which in turn leads to a change in the problem being solved and an awareness of the best solution to the problem. At the initial stage of decision-making, when the goal is not clearly defined, the valence of outcomes is a comprehensive criterion of success in the choice made by a person. When measuring the motivational level of an alternative, the level of positive motivation (motivation to get positive outcomes) and the level of negative motivation (motivation to avoid negative outcomes) are measured. The level of positive motivation is determined by the level of significance of positive outcomes and the level of difficulty in obtaining them, and the level of negative motivation is determined by the level of significance of negative outcomes and the level of difficulty to avoid them. With this in mind, the evaluation of significance of positive (negative) outcomes reflects the level (intensity) of their positive (negative) importance for the individual. Evaluation of difficulty depends on the valence of outcomes, and for positive outcomes reflects the level of subjective possibility to attain these outcomes and for negative outcomes, the level of subjective possibility to avoid them. Since the decision takes place in the conditions of uncertainty of the outcomes, we assume that subjective possibility can be reflected by subjective perception/feeling of their likelihood.

In relation to this, the level of *positive motivation* is determined by *intensity (I+)* of positive outcomes *(positive intensity)* and *likelihood (L+)* of attaining them *(positive component of likelihood)*, while the level of *negative motivation* is determined by *intensity (I-)* of negative outcomes *(negative intensity)* and *likelihood (L-)* of avoiding them *(negative component of likelihood)*.

These assumptions allow the use of studies by Mikhail Kotik (Chapter 7; Kotik and Yemelyanov, 1983; Kotik, 1984), in which he measured the attractiveness (preference) of emotion-inducing events based on the intensity and likelihood of their outcomes. Essentially, this measured positive (negative) preference reflects the level of positive (negative) motivation to attain positive (and avoid negative) outcomes, which will be demonstrated below. In these studies, likelihood (*L*) and intensity (*I*) were measured on the verbal fuzzy scales of "weak-strong" and "seldom-often," respectively, and each of these scales contains nine levels. Bocklisch et al (2012) noted that using words becomes especially useful in most everyday situations, when a subjective belief or uncertainty cannot be precisely verbalized in quantitative terms. Measured in this way, intensity and likelihood are conveyed through an emotional-evaluative perspective.

It should be noted here that emotions play an important role in decision-making (Pfister and Böhm, 2008; Lerner et al, 2015; etc.). Moreover, emotional processes guide decision-making. The *somatic marker hypothesis*, formulated by Antonio Damasio (1994), proposes that a decision is made largely from the *images* of future positive and negative outcomes. These images are "marked" by positive or negative feelings linked directly or indirectly to somatic states. He hypothesized that somatic markers improve the decision-making process and the ability to make adequate decisions. Slovic et al (2007) consider *affect heuristic*, a mental shortcut that allows people to make decisions and solve problems efficiently based on positive or negative feelings. "Affect" means the specific quality of "goodness" or "badness," which is experienced as a feeling state (with or without consciousness), and demarcates a positive or negative quality of a stimulus. Affective responses occur rapidly and automatically.

In the future, it will be convenient for us to present the results obtained by Mikhail Kotik in the following mathematically stricter form (Yemelyanov, 2018). *Positive preference (P+)* is considered a composition of *positive intensity (I+*; subjective value of the positive outcomes in the context of attaining the goal) and *positive component of likelihood* or shortly—*positive likelihood (L+*; subjective expectations of attaining *I+* in the context of existing conditions): $P+ = S+ (I+, L+)$. *Negative preference (P-)* is considered a composition of *negative intensity (I-*; subjective value of the negative

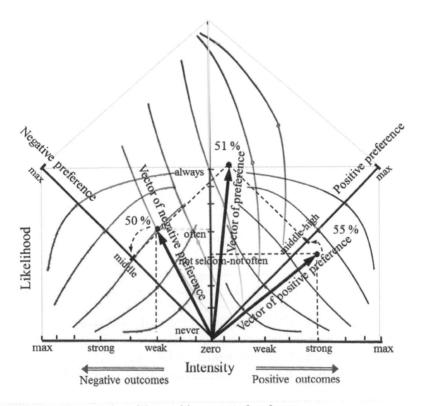

FIGURE 4.3 Construction of the resulting vector of preference.

outcomes in the context of attaining the goal) and *negative component of likelihood* or shortly—*negative likelihood* (L-; subjective expectations of avoiding I- in the context of existing conditions): P- = S-(I-, L-). *Preference (P)* is considered to be a function of marginal utility functions P+ and P-: P = S (P+, P-). Figure 4.3 demonstrates how the integral level of preference of 51% is formed based upon positive (55%) and negative (50%) levels of preference, determined by their positive ("strong") and negative ("weak") subjective intensities, as well as by their positive ("not seldom-not often") and negative ("often") subjective likelihoods, respectively.

The distinct features of these functions include the fact that they are not, firstly, numerical additive functions, which are traditionally used in MAUT, and secondly, that they reflect the level of motivation necessary to choose the most preferable alternative. At the same time, the positive and negative intensities represent positive and negative subjective importance (significance) of outcomes, respectively; while positive and negative likelihoods represent both the possibility of attaining positive outcomes (positive component of difficulty) and the possibility of avoiding negative outcomes (negative component of difficulty), respectively. The validity of these functions when making decisions under uncertainty with integrated positive and negative outcomes (pros and cons) is discussed in Kotik and Yemelyanov, 1992: in 77% of cases, the preference function correctly determines the best alternative,

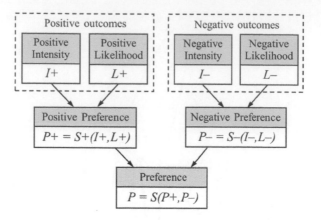

FIGURE 4.4   IL-Frame of measuring the preference level.

while in 66% of cases, it correctly identifies the priority ranking. This allows us to assume that the level of preference that is determined in this way reflects, to a certain extent, the motivational level in decision-making. In this way, the likelihood-based approach for describing difficulty makes it possible to use the IL-Frame (Figure 4.4), instead of the SD-Frame in decision-making problems, until more precise methods of measuring the level of preference may appear, which in turn reflect the corresponding level of motivation.

It should be noted that the preference function defined by the IL-Frame has three main properties of motivation indicated in Bedny, 2015. The preference function provides: (1) *connection with emotions* by using soft verbal characteristics), (2) *directness to the goal* by considering "intensity" as a subjective importance of the positive or negative outcomes in the context of attaining the goal, and (3) *energy in reaching the goal* by considering "likelihood" as a power of possibility to attain positive or avoid negative outcomes in the existing external and internal conditions. It is important to note that this motivational preference is different from the standard evaluative preference traditionally used in MAUT. The following demonstrates the use of the IL-Frame in the performance evaluation process and in Express Decision software.

4.4.2   RULES OF MOTIVATION

In the process of building the mental model of the problem under analysis, the level of motivation is formed dynamically with a complex relationship between the factors of difficulty and significance. The factor of significance is decisive in the formation of the mental model, while the factor of difficulty is decisive in the formation of the motivational level. Since our purpose is to use the factors of significance and difficulty as the primary criteria in the decision-making process of selecting the alternative with the highest motivational level, the rules described above in the

TABLE 4.1
Numerical levels of significance, difficulty, and motivation.

Level	Significance S+/S−	Difficulty D+	Difficulty D−	Motivation M/M+/M−
1	min	always	min	min
2	extremely weak	extremely often	extremely seldom	very low
3	very weak	very often	very seldom	low
4	weak	often	seldom	middle
5	not weak-not strong	not seldom-not often	not seldom-not often	high
6	strong	seldom	often	very high
7	very strong	very seldom	very often	max
8	extremely strong	extremely seldom	extremely often	−
9	max	min	always	−

motivation evaluation process are not adequate enough. Therefore, we resolve to use functional relationships between levels of motivation, intensity, and likelihood in order to obtain additional relationships among levels of motivation, significance, and difficulty. In other words, we decided to present the results of Kotik's experimental work reflected in the IL-Frame through variables of *significance* and *difficulty*. This takes into account the existence of an inversely proportional relationship between the positive component of difficulty and the likelihood/possibility of obtaining positive outcomes, and a directly proportional relationship between the negative component of difficulty and the likelihood/possibility of avoiding negative outcomes. The following demonstrates how such an approach allows definition of functional relationships among levels of motivation, significance, and difficulty, while formulating new rules of motivation for decision-making. Table 4.1 shows the numerical levels that are used for positive (S+) and negative (S-) significance, for positive (D+) and negative (D-) components of difficulty, and for positive (M+), negative (M-), and integral motivation (M).

The interpolated relationships for positive and negative motivation are shown in Figure 4.5 and Figure 4.6 respectively.

Figure 4.5 presents the relationship among positive component of difficulty (D+), positive significance (S+), and positive motivation (M+):

$$D^+ = c_1^+ \cdot S^{+-c_2^+ \cdot M^+} \tag{4.1}$$

where $c_1^+, c_2^+ > 0$ are constants for the given level of positive motivation.

Formula (4.1) indicates that the level of positive motivation is a *power* to (a) *decrease* the level of difficulty of attaining positive outcomes when their level of significance *increases,* and (b) *increase* the level of difficulty of attaining positive outcomes when their level of significance *decreases.*

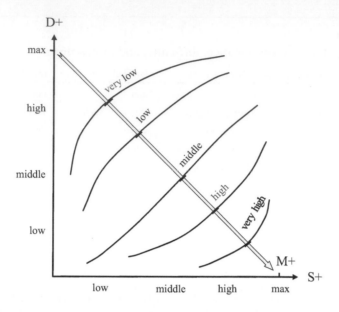

FIGURE 4.5   The relationship among positive component of difficulty (*D*+), positive significance (*S*+), and positive motivation (*M*+).

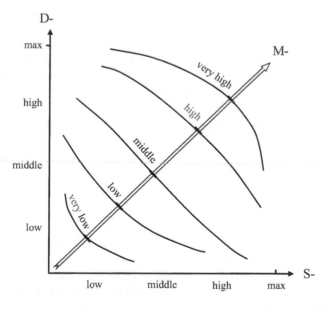

FIGURE 4.6   The relationship among negative component of difficulty (*D*-), negative significance (*S*-), and positive motivation (*M*-).

Figure 4.6 presents the relationship among the negative component of difficulty (D-), negative significance (S-), and negative motivation (M-):

$$D^- = c_1^- \cdot S^{-c_2^- \cdot M^-} \tag{4.2}$$

where $c_1^-, c_2^- > 0$ are constants for the given level of negative motivation.

Formula (4.2) indicates that the level of negative motivation is a *power* to (a) *decrease* the level of difficulty of avoiding negative outcomes when their level of significance *decreases*, and (b) *increase* the level of difficulty of avoiding negative outcomes when their level of significance *increases*. The obtained curves illustrate the notion that motivation provides the *energy* that drives the individual in overcoming difficulties to accomplish a goal (attain positive and avoid negative outcomes), which reflect one of the key properties of motivation. Thus, the preference function defined by the IL-Frame provides energy in choosing the best alternative.

According to the curves, shown in Figure 4.5 and Figure 4.6, the rules of motivation in the self-regulative decision-making process can be formulated. These rules determine the ratio of the levels of motivation, difficulty, and significance, depending on the valence of the expected outcomes. Presented below are these primary rules.

Rules of motivation

Rule 1. If $(M^+ \geq high)$ and $(D^+ \geq high)$, then $w(S^+) > w(D^+)$.
If the levels of positive motivation and the positive component of difficulty are high, then the factor of significance has a greater weight than the factor of difficulty: an increase (decrease) in significance level by one rank is equivalent to a decrease (increase) in difficulty level by more than one rank.

Rule 2. If $(M^- \geq high)$ and $(D^- \geq high)$, then $w(D^-) > w(S^-)$.
If the levels of negative motivation and the negative component of difficulty are high, then the factor of difficulty has a greater weight than the factor of significance: a decrease (increase) in difficulty level by one rank is equivalent to a decrease (increase) in significance level by more than one rank.

Rule 3. If $(M^+ \geq high)$ and $(D^+ \leq low)$, then $w(D^+) > w(S^+)$.
If the level of positive motivation is high and the level of the positive component of difficulty is low, then the factor of difficulty has a greater weight than the factor of significance: a decrease (increase) in the level of difficulty by one rank is equivalent to an increase (decrease) in the level of significance by more than one rank.

Rule 4. If $[(M^+ = middle)$ and $(D^+ = middle)]$, then $w(D^+) = w(S^+)$.
If the levels of positive motivation and difficulty are intermediate, then factors of difficulty and significance have the same weights: an increase (decrease) in the level of significance by one rank is equivalent to a decrease (increase) in the level of difficulty by one rank.

Rule 5. If $(M^- \geq high)$ and $(D^- \leq low)$ then $w(S^-) > w(D^-)$.
If the level of negative motivation is high and the level of the negative component of difficulty is low, then the factor of significance has the greater

weight that the factor of difficulty: a decrease in significance level by one rank is equivalent to a decrease in difficulty level by more than one rank.

Rule 6. If $[(M^- = middle)$ and $(D^- = middle)]$ then $w\,(D^-) = w\,(S^-)$.

If the levels of negative motivation and difficulty are intermediate, then the factors of difficulty and significance have the same weights: an increase (decrease) in the level of significance by one rank is equivalent to an increase (decrease) in the level of difficulty by one rank.

Because the current work is focused on the development of decision support tools, of particular interest to us are Rule 1 and Rule 2. Rule 1 regulates the maintenance of a high level of positive motivation to obtain difficult-to-reach positive outcomes: a high level of positive motivation is maintained to a greater degree by increasing the level of feeling of positive significance (subjective value) of these outcomes, rather than by reducing the feeling of their difficulty. Rule 2 regulates the reduction of a high level of negative motivation to evade hard-to-avoid negative outcomes: reducing a high level of negative motivation is less supported by reducing the degree of negative outcome than by reducing the feeling of their difficulty.

4.5　SELF-REGULATION MODEL OF DECISION-MAKING

Decision-making is considered a self-regulative thinking process driven by a motivation to attain a goal (Bedny, Karwowski, and Bedny, 2015). The foundation for decision-making is a continuous reformulation of a problem, and the development of such a mental model that can determine the level of motivation for selecting the best of the available alternatives is based on the expected outcomes. The initial mental model of a problem often cannot facilitate the achievement of desired outcomes. In this situation, after understanding the problem at hand, the decision-maker divides a problem into sub-problems. The decision-maker then begins creating sub-problems by formulating various hypotheses. Each hypothesis has its own potential goal. Based on the comparison and evaluation of such hypotheses, the decision-maker selects one and formulates the first sub-goal associated with the selected hypothesis. Comparing a new sub-goal with an existing mental model of the problem allows transformation of the original mental model into a new one that is adequate for the new sub-problem. If the problem is evaluated positively (positive feedback), the thinking process cycle is complete. However, if the result is evaluated negatively (negative feedback), internal or external information should be added to continue this process on a lower level (or to search for other alternatives of solving the problem). In SSAT, unlike cognitive psychology, the decision-maker is able to regulate his behavior not only externally, but also internally by using the inner mental plane. This process is not straightforward, and includes a continuous cycle of updating the mental model with feedback and feedforward controls. It is necessary to note that in traditional dynamic decision-making (Brehmer, 1992) only external feedback from the environment is considered. The description of the suggested self-regulation model of decision-making is presented below. In this model, a key role is assigned to the factors of significance and difficulty.

The self-regulation model (SRM) of decision-making includes two sub-models: formation of mental model (FMM) and formation of the level of motivation (FLM).

Execution of SRM is driven by the motivation to attain a goal and includes the execution of FMM and FLM, as well as the regulation of their interaction by using feedback and feedforward controls.

The design strategy for FMM implements a divide-and-conquer algorithm (D&C). The divide-and-conquer technique (Levitin, 2011) uses a recursive breakdown approach in decision-making: decompose the problem into smaller sub-problems, solve them, and then recombine their results to solve the bigger problem. This division of the problem into sub-problems may span several levels deep until a basic (ad hoc) level of certainty will be reached, at which point the problem can be positively evaluated based on IL-Frame. In other words, the problem will contain only those outcomes for which the decision-maker will be able to determine their respective positive (or negative) intensity and likelihood, which in turn will allow determining the positive (or negative) motivational level (preference). It should be noted that the efficiency of the divide-and-conquer algorithm increases when people apply hypotheses and split the problem into two mutually exclusive hypotheses. In FMM, feedback control is used to verify whether the current state of the individual's mental model is capable of either evaluating the problem based on IL-Frame or choosing the best alternative. The feedback is *positive* (+fb_FMM) when the individual can perform the verification, and *negative* (–fb_FMM) when the individual cannot perform it. Feedforward control leads to an upgrade of the existing mental model. With this purpose, by considering various hypothetical situations and alternative solutions, the problem is divided into sub-problems with corresponding sub-goals.

The design strategy for FLM implements a dynamic programing algorithm (DP). The dynamic programing technique (Levitin, 2011) is used to solve a problem by breaking it down into smaller and simpler sub-problems. This method is applied to solve problems that have the properties of overlapping sub-problems and that fulfill the principle of optimality. The first requirement is satisfied because the divide-and-conquer method splits problems into sub-problems. The principle of optimality is also satisfied, in the case that the formation of the level of motivation indicates that this level can accomplish the individual's goal: if the level of motivation for solving a problem accomplishes the individual's goal, then the level of motivation for solving sub-problems will also accomplish the individual's corresponding sub-goals. In FLM, feedback control verifies whether the level of motivation for choosing an alternative is created. Feedback is *positive* (+fb_FLM) when the level of motivation is created, but *negative* (–fb_FLM) otherwise. Feedforward control allows to predict the level of the alternative's motivation after changing verbal characteristics (manipulated control inputs) in IL-Frame.

Feedback control provides a connection between FMM and FLM models. It is regulated by the factor of difficulty, which determines the individual's self-efficacy (Bandura, 1997) in attaining the goal. Feedback control is corrective, connected to the individual's past experience, and provides robustness and error elimination in formation of the mental model and the level of motivation (Yemelyanov, A.M. and Yemelyanov, A.A., 2017).

Feedforward control produces upgrades in FMM and FLM. It is regulated by the factor of significance, which determines the directness to the goal (Bedny, 2015). Feedforward control is predictive and leads to an upgrade in the existing mental model or level of motivation in order to enhance its ability to solve the problem and obtain the desired outcomes.

Within SRM, both FMM and FLM, there are two concurrently-running processes that are self-regulated by feedback and feedforward controls and driven by motivation to attain the goal. FMM implements the divide-and-conquer algorithm for mental model formation, while FLM implements the dynamic programing algorithm for motivation level formation. The level of motivation for selecting an alternative forms dynamically, according to the mental model that forms recursively. It is worth noting that using the dynamic programming algorithm to determine the level of motivation demonstrates just how powerful motivation really is. Many combinatorial decision problems that typically require exponential time and space to be solved by standard algorithms can be solved in polynomial time and space using dynamic programming. Therefore, self-regulation in decision-making optimizes the use of time and memory resources. It is worth noting that in the dynamic programming of the level of motivation, the *power of motivation* itself is demonstrated.

Kotik (1974) illustrated that self-regulation increases the reliability of the goal-directed activity. He described two types of self-regulation: self-regulation in the area of information processes (*information-based self-regulation*) and self-regulation in the area of energy (emotional) processes (*energy-based self-regulation*). The complexity in decision-making is compensated by the intensification of information processes, as well as by the energy reaction of the brain. All of this generates energy mobilization of the organism, aimed at bringing it into readiness for intensive spending forces and overcoming the difficulties. Zarakovsky and Pavlov (1987) showed that emotions have an inducing function that provides energy for switching, reinforcing, compensating, and organizing functions in self-regulation.

According to behavioral and neuroscientific research, cognition and rational decision-making are not entirely the product of rational information processing and symbol manipulation, but instead require the involvement of emotion. Domasio's somatic marker hypothesis, discussed above, suggests a mechanism by which emotional processes can self-regulate decision-making. When making decisions, these somatic markers and their induced emotions are consciously or unconsciously associated with prior positive and negative outcomes that allow us to quickly evaluate these outcomes based on past experiences. This is a demonstration of energy-based self-regulation in which the formation of the level of motivation (FLM) is activated based on internal positive feedback from the formation of the mental model (FMM). This means that entirely emotionally-driven decisions can be advanced in a purely rational way by comparing positive and negative outcomes of each of the hypothetical decision options. The latter suggests that using IL-Frame for assessing positive and negative outcomes will be equally productive for both information-based and energy-based self-regulation.

In conclusion, we formulate the following rules of self-regulation, which will be used in the decision support system described below.

Rules of self-regulation
SR1. +fb_FMM ⇒ FLM.
Positive feedback on formation of the mental model activates formation of the
level of motivation (D&C algorithm in FMM calls DP algorithm in FLM).
This means that, if the difficulty of evaluation of outcomes in the IL-Frame
is adequate, then the level of motivation is formed.

SR2. –fb_FMM ⇒ ff_FLM.

Negative feedback on formation of the mental model activates feedforward control of formation of the level of motivation (D&C algorithm in FMM upgrades DP algorithm in FLM). This means that if the difficulty of evaluation of outcomes in the IL-Frame is not adequate, then the significance of the problem determines whether this problem should be split into sub-problems (hypotheses) for their further evaluation in the Il-Frame, or if the decision-making process should be terminated.

SR3. + fb_FLM ⇒ FMM.

Positive feedback on formation of the level of motivation activates formation of the mental model (DP algorithm in FLM calls D&C algorithm in FMM). This means that if the difficulty of evaluation of the level of motivation is adequate, then the mental model is formed and a decision can be made.

SR4. –fb_FLM ⇒ ff_FMM.

Negative feedback on formation of the level of motivation activates feedforward control of formation of the mental model (DP algorithm in FLM upgrades D&C algorithm in FMM). This means that if the difficulty of evaluation of the level of motivation is not adequate, then significance determines whether the goal and/or alternatives of the problem should be modified and the decision-making process repeated or completely terminated.

4.6   PERFORMANCE EVALUATION PROCESS

Performance Evaluation Process (PEP) implements the self regulation model of decision-making (Yemelyanov, 2018). The performance evaluation process, unlike other decision-making methods, not only allows a decision about which of the available alternatives is most preferable, but also making a decision regarding which alternative can be chosen for execution. Another distinctive feature of PEP is that this method does not require the decision-maker to form a mental model of the problem *before* solving it, but instead calls for the decision-maker to form this mental model in the *process* of arriving at the solution itself. In other words, an individual is not required to present a clear concept of the goal, nor to formulate any selection criteria for the construction of a model of the problem that would then assist in solving it. PEP allows an individual to build a mental model of a problem and to solve it using the *self-regulation model*. This model, based on the *self-regulation rules*, dynamically forms the levels of motivation for selecting alternatives, according to the mental model that is recursively formed. Consequently, this allows the decision-maker to be efficiently led to the right choice through the construction of a problem model during the process of making a decision.

The performance evaluation process includes the following three stages.

1. **Problem decomposition and mental model formation.** The *decomposition* process starts from the evaluation of the decision-making problem by applying the IL-Frame. If for each alternative the intensities and likelihoods of the positive and negative outcomes can be determined, this means

that the problem can be positively evaluated, does not need further alternatives decomposition, and so the next stage, aggregation, will be applied. At the *aggregation* stage, the level of preference for each alternative will be measured with the IL-Frame, which allows a decision regarding which alternative is more preferable in the final *decision-making* stage. If the problem is unable to be positively evaluated—for example, when the intensities of positive outcomes cannot be determined—the problem should then be divided into sub-problems with corresponding sub-goals. For a decision-making problem, this means that the decomposition process is required at the alternatives' levels, with further evaluation and measurement of each alternative by using the IL-Frame. As a result of the decomposition, the problem should be divided into sub-problems, until the point that such a level of detail is reached that the positive and negative outcomes will be assessed by their levels of intensity and likelihood—so that in the aggregation stage, this will allow determining (measuring) the problem's preference level. For this purpose, hypotheses are used. Each hypothesis creates such a sub-problem that the alternatives are evaluated from the position of the outcomes (positive or negative), which are more specific (less uncertain) than the previous group of outcomes. If for such outcomes, their intensity and likelihood can be evaluated on the given scale, this signifies that further specification of the sub-problem and its outcomes is not required. Otherwise, a new hypothesis should be considered with even more specified outcomes. The hypothesis allows us to assess the intensity(s) and likelihood(s) of the specific outcome(s), which in turn allows us to determine the level of a problem's preference, according to the IL-Frame. The choice of hypotheses is determined by the decision-maker and helps them to construct a mental model of the problem being solved. Hypotheses are typically considered for mutually exclusive situations (i.e., the occurrence of some outcomes and their non-occurrence) in order to simplify the assessment of outcomes' degrees of influence on the higher preference levels of sub-problems or alternatives. As a result of decomposition, a *decision tree* (DT) is established for each alternative, and this reflects the functional structure of all the sub-problems examined. In this case, nonterminal vertices determine the sub-problems examined, and terminal vertices contain estimates of intensity and likelihood for positively evaluated sub-problems.

2. **Problem aggregation and formation of the level of motivation.** In the process of aggregation, a preference level will be determined (measured) for each alternative, which reflects the level of motivation for choosing a particular alternative. Aggregation includes *three stages*. In the *first stage*, by using the IL-Frame, the preference levels of all positively evaluated sub-problems should be determined. In the *second stage*, it is necessary to separately determine the positive and negative preference levels for each alternative. In order to do this, we use special rules for aggregation that present hierarchical and timing dependencies of outcomes. The timing dependence of outcomes is determined by hypotheses. In the *third stage*, by using the IL-Frame, the positive and negative preference levels of each

alternative are combined into its cumulative level of preference. Thus, the result of problem aggregation is the construction of a *goal-motivational decision tree* (GMDT) for each alternative (Yemelyanov, Baev, Yemelyanov, 2018). This tree differs from DT in that all sub-problems are assigned their respective measured levels of preference. The level of preference of the entire alternative aggregates the levels of preference of the examined sub-problems.

3. **Making a Decision.** In this process, the preference levels of each of the alternatives are compared, and a decision is made regarding selecting an alternative for execution. An alternative with a sufficient level of preference for execution is selected. A low level of preference can occur when the problem is too complicated or not significant enough. The specification of the goal and the addition of new alternatives make it possible to arrive at an acceptable solution. If both alternatives have a sufficiently high level of preference, but their preference levels are equal or very close to each other, then positive and negative preference levels should both be evaluated. If the levels of motivation for choosing all alternatives are low, then a new alternative should be considered.

Therefore, the performance evaluation process allows to determine the preference of alternatives, and then to compare them and make a decision based on which alternative is more preferable.

4.7   DECISION SUPPORT

Decision-making is no longer viewed simply as a rational process in which logical thinking determines the best means to achieve a goal. Researchers from different fields of cognitive science have shown that human decisions and actions are influenced to a far greater degree by intuition and emotional reactions than was previously thought. Neurological research demonstrates that not only are information and emotional processes inherently connected, but also that without emotion, logical reasoning is impossible and decisions cannot be made. According to Damasio's somatic marker hypothesis, decision-making is guided largely by *images* of future positive and negative outcomes. According to Slovic's affect heuristic approach, a decision with affective response is determined by positive ("goodness") and negative ("badness") feelings. Such valence (positive or negative) in imaging and feeling in decision-making is consistent with Benjamin Franklin's well-known procedure for making difficult decisions by finding a balance between weighted pros and cons (Franklin, 1975). The aforementioned establishes the existence of emotional rationalities in decision-making based on this valence. In this way, even in emotionally-driven decision-making, there exists a certain rationalism that influences the division of outcomes into positive and negative. The decision support system (Express Decision) we developed in this work is based on this rationalism.

Furthermore, it should be noted that when the mathematical model is built for the decision-making problem (as, for example, in AHP), then we make a decision according to this model. In other words, this model replaces the human decision-maker in making decisions, which can lead to inadequate results in the case of ill-structured problems. Our goal is to develop such a decision support system that will become integral to the decision-maker, who will not feel the need to separate their decisions from the decisions obtained with the help of this system. The support system's aim is to help the decision-maker build such a mental model that would lead them to make the best decision for them self. The model is created within the decision-maker's mind, and the role of the support system is to assist the decision-maker in building such a model, which can then help to arrive at a solution most relevant to the decision-maker's needs. The foundation of support should be based upon the mechanisms of emotional and logical rationality. The following decision support system Express Decision, implements PEP, and meets the requirements of system support formulated above.

4.7.1   EXPRESS DECISION

Express Decision (ED), a mobile web application, is guided by PEP to provide an individual with decision-making support in making quick decisions in complex problems. ED supports the decision-maker by transforming their mental model of the decision-making problem from the initial to the final state. By implementing the IL-Frame, ED helps to evaluate the problem, measure its positive, negative, and resulting levels of preference, and create DT and GMDT trees.

ED supports the following processes.

a. Formation of Mental Model (construction of the decision tree).
   ED uses IL-Frame as a template for assessing on the verbal scale the intensity of expected positive ($I+$) outcomes and their likelihood ($L+$), and the intensity of expected negative ($I-$) outcomes and their likelihood ($L-$). These four verbal characteristics are manipulated variables in the self-regulation process of decision-making.
b. Formation of the Level of Motivation.
   ED uses experimentally determined relations $P = S(P+, P-) = S(S+ (I+, L+), S- (I-, L-))$ to combine levels of positive and negative preferences of outcomes into cumulative level of preference (motivation) for alternative.
c. Interaction between FMM and FLM (feedback and feedforward controls).
   The decision-maker is guided by ED using the self-regulation rules SR1–SR4. ED checks the completeness of collected data and the consistency of the constructed mental model, and conducts sensitivity and what-if analyses.

Express Decision is developed as a progressive web application. The core functionality of the app utilizes the PEP algorithm augmented with an intuitive user interface that allows to run the app on both *mobile* and *desktop* platforms. ED has a verity of applications (Yemelyanov, 2017; Yemelyanov, Baev, and Yemelyanov, 2018) and using it is no more difficult than using a regular calculator.

4.7.2 APPLYING EXPRESS DECISION TO MAKE A DECISION ABOUT A FACEBOOK FRIEND REQUEST

Here in a real life example we demonstrate how Express Decision assists in making a sound decision regarding accepting, rejecting, or considering a Facebook friend request. Below we will present how the formation of the problem's mental model and the level of motivation to choose the most preferable solution is carried out with the help of the rules of self-regulation.

Gena has a decision-making problem; however, instead of rushing to solve it on the fly, she has decided to turn to the assistance of ED to help her make the decision that is most appropriate to her particular situation. A couple months ago, Gena was accepted for a position as an analyst in a mid-size marketing company. Gena was able to establish professional friendly relations with her boss, Robert. As a side gig, Gena is a photographer at music shows and posts her photos to her Facebook page, which is only visible to her friends. Her boss in the marketing company, Robert, recently came across some of the photos that Gena has taken that had been shared on others' Facebook pages; he is an ardent music lover and becomes interested in seeing more of her photography. Robert sends Gena a friend request with the hopes that she'll accept it so that he'll have access to all her photos and related work. However, Gena is concerned that if she accepts his request, he will also gain access to all her private information, along with some graphic content that he might find inappropriate or offensive, and this could negatively affect her career prospects. Gena tries to weigh the pros and cons of either accepting or rejecting Roberts's friend request, and wonders what she should do.

The uncertainty of such decision-making is largely influenced by the fact that Gena cannot predict how a new friend's access to her account could impact her career prospects if their friendly business relationship might come to an end in the future. On one hand, Gena does not want to delete the request from her boss for fear of jeopardizing her current relationship with him, since friendly relations with him would only contribute to her career growth. On the other hand, she does not want any personal information he could retrieve from her account to be used against her, in case the relationship with her boss somehow deteriorates or does not work out over time. Evidently, the correct decision between "accept" and "reject" is very important (significant) for Gena because it is directly tied to her professional growth and the future of her career. At the same time, each option has its own share of pros and cons, and finding a reasonable compromise between them proves to be a difficult task for her. This determined the fact that she was highly motivated to carefully evaluate each of the alternatives for the subsequent selection of the best one among them. In the given situation, Gena worried about making a mistake in her selection, and decided to rely on more than her intuition alone. This is why she turned to Express Decision. She knows that Express Decision is a mobile device application no more difficult to use than a standard calculator, and that it can quickly assist anyone with their mixed feelings regarding evaluating and selecting the best solution for them. The only requirement is a general understanding or awareness (in broad terms, at the very least) of what it is that you ultimately want to accomplish with your selection—you need to have a general awareness of the *goal* you wish to realize. For Gena, the ultimate goal is continued *career growth*, which is why she chose to use Express Decision with this specific goal in mind, since it will help her solve the problem of whether or not she should *accept* her boss's friend request or *reject* it. We will denote this for brevity as **problem (*decision making: accept friend request vs. reject friend request*)**. According to the performance evaluation process used in ED, Gena must first name the alternatives (or options) available to her. Then, each alternative should be evaluated on the IL-Frame by four verbal characteristics: *positive intensity* and *its likelihood*, and *negative intensity* and *its likelihood*. In terms of this framework, subjective intensity is measured on a fuzzy verbal scale from *extremely weak* to *extremely strong*, while subjective likelihood is measured on the scale from *extremely seldom* to *extremely often*. Assume that these characteristics were determined by the following reasoning for each alternative:

Problem 1: *Evaluation of Alternative 1 (accept friend request)*

Pros (+): Gena knew that accepting the request could *strongly* (*I5*) contribute to her career growth, since her boss also uses Facebook for business purposes, but she rated her chances as *seldom* (*L3*), since she uses Facebook relatively infrequently and often doesn't log into it for days.

Cons (–): On the other hand, it is not entirely unlikely (*L4—not seldom-not often*) that her photography and private information could be used against her—if not now, then sometime down the line—and this could *very strongly* (*I6*) negatively impact her career prospects.

Because Gena does not experience any difficulties in selecting adequate verbal characteristics for the IL-Frame, *Alternative 1* is considered to be positively evaluated. This means that the *initial* mental model state allows Gena to evaluate *Alternative 1* (**+fb_FMM₁**), which in turn allows ED to support Gena in the formation of the level of motivation for choosing this alternative (**SR1: +fb_FMM$_1$ ⇒ FLM$_1$**).

- Level of positive motivation (preference) is 52% ("middle"): $S+ (I5, L3) = 0.52$.
- Level of negative motivation (preference) is 71% ("high"): $S-(I6, L4)) = 0.71$
- Level of motivation is 37% ("low"): $S (S+ (I5, L3), S-(I6, L4)) = 0.37$.

This indicates that for Gena, accepting Robert's friend request has a few select pros, but there are many more significant cons involved, which ultimately allows for ED to determine that Gena's level of motivation for selecting alternative 1 is 37%, which is "low." Gena accepts the level of motivation for choosing *Alternative 1* (**+fb_FLM$_1$**)—since she has always firmly believed that mixing work with pleasure is likely to severely jeopardize professional relations—which in turn upgrades her mental model state (**SR3: +fb_FLM$_1$ ⇒ FMM$_1$**).

This is why she decides against correcting the IL-Frame and instead turns to the evaluation of *Alternative 2*.

Problem 2: *Evaluation of Alternative 2 (reject friend request).*

Pros (+): Gena assumes that if she rejects the request, thus denying Robert the right to gain access to her private account information, this could *often* ($L5$) yet *very weakly* ($I2$) contribute to her career growth.

Cons (–): Gena only associates the cons to her career with the potential consequences of a rift in friendly relations with her boss. However, these consequences are *not apparent* to Gena, since her current mental model state does not allow her to evaluate alternative 2 (**–fb_FMM₂**). Because of this, Gena has difficulty measuring the negative consequences from Robert's perspective and therefore, *Alternative 2* is considered to be negatively evaluated. At the same time, because Gena is highly motivated to evaluate *Alternative 2*, the new challenges that have arisen in light of this situation do not deter her.

Gena simplifies problem 2 by splitting it into two sub-problems, 2.1 and 2.2, which relate to two possible and mutually-exclusive *hypothetical situations* (Robert could either *maintain* or *refuse* his friendly relations with Gena) with the purpose of determining their levels of motivation (**SR2: –fb_ FMM₁ ⇒ ff_FLM₂₁_FLM₂₂**). Problem 2.1 evaluates *Alternative 2* assuming the first hypothetical situation, in which Robert continues to *maintain friendly relations* with Gena after having his friend request rejected by her, while problem 2.2 evaluates *Alternative 2* assuming the second hypothetical situation, in which Robert *refuses to maintain friendly relations* with Gena after having his friend request rejected by her.

Problem 2.1: *Evaluation of Alternative 2.1 (reject friend request and maintain friendly relations).*

Pros (+): Gena realizes that there is a chance to preserve friendly relations with Robert even after rejecting his friend request, since they still both have a strong mutual interest in music. This provides a relatively good ($L4$) chance for her to be able to maintain strong ($I5$) positive prospects beneficial to her career growth.

Cons (–): Gena understands that negative outcomes could occur not only from a falling out with her boss, but also from maintaining friendly relations with him. She recalls a scenario with one of her friends, in which the friend's boss received a rejection to his invitation for the friend to attend a concert with him. Although outwardly the boss seemed to maintain good relations with this friend, he was still inwardly upset with her rejection, which ultimately ended up affecting the friend's annual evaluation, albeit weakly. Due to this, Gena decided to analyze as *seldom* ($L3$) the likelihood of a *weak* ($I3$) negative impact on her career growth if she were to reject the friend request from Robert, although he would still appear to maintain friendly relations with her. Here we will note that initially, Gena only associated the cons to her career with the consequences of a potential rift in friendly relations with her boss. Using ED enabled Gena to also see the potential disadvantages to her career in the case that she is able to maintain friendly relations with her boss.

Because Gena does not experience any difficulties in selecting adequate verbal characteristics for the IL-Frame, *Alternative 2.1* is considered to be positively evaluated. The current mental model state allows Gena to evaluate *Alternative 2.1* (**+fb_FMM$_{21}$**), which in turn allows ED to support Gena in formation of the level of motivation for choosing this alternative (**SR1: +fb_FMM$_{21}$ ⇒ FLM$_{21}$**).

- Level of positive motivation (preference) is 57% ("middle"): $S+(I4,L5)=0.57$.
- Level of negative motivation (preference) is 38% ("low"): $S-(I3,L3))=0.38$.
- Level of motivation (preference) is 58% ("middle"): $S\ (S+\ (I4,\ L5),\ S-(I3,\ L3)) = 0.58$.

Gena agreed with this evaluation of her level of motivation (**+fb_FLM$_{21}$**), since she's keen on staying professional and maintaining some distance with her boss, which in turn upgrades her mental model (**SR3: +fb_FLM$_{21}$ ⇒ FMM$_{21}$**). This is why she decides against correcting the IL-Frame and instead turns to the evaluation of *Alternative 2.2*.

Problem 2.2: *Evaluation of Alternative 2.2 (reject friend request and refuse to maintain friendly relations).*

Pros (+): Gena understands that abruptly ending friendly work relations with Robert would likely cause him to evaluate her professional abilities from a more objective, and perhaps even slightly unfavorable, position. Moreover, Robert would have far less of a justification to try to advance to a more personal level in their relations, beyond what would be comfortable for Gena and acceptable professionally. With all this, Gena associated *very realistic (L5)*, although *relatively insignificant (I3)*, prospective career growth.

Cons (–): At the same time, Gena has not discounted the *likely (L4)* possibility of a *very strong (I6)* negative impact on her career growth if she were to reject the friend request from Robert, and he consequently ends up refusing to maintain friendly relations with her.

Because Gena has not experience any difficulties in selecting adequate verbal characteristics for *Alternative 2.2* (**+fb_FMM$_{22}$**), this alternative is positively evaluated, which in turn allows ED (**SR1: +fb_FMM$_{22}$ ⇒ FLM$_{22}$_FLM$_{2}$**) to

1. form the level of motivation for choosing Alternative 2.2 (**FLM$_{22}$**):
   - Level of positive motivation (preference) is 52% ("middle"): $S+\ (I3,\ L5) = 0.52$.
   - Level of negative motivation (preference) is 71% ("high"): $S-(I6,\ L4)) = 0.71$.
   - Level of motivation (preference) is 37% ("low"): $S\ (S+\ (I3,\ L5),\ S-(I6,\ L4)) = 0.37$.

2. form the level of motivation for choosing *Alternative 2* (**FLM₂**) by
aggregating the preference scores of *Alternatives 2.1* and *2.2*:
- Level of positive motivation (preference) is 52% ("middle").
- Level of negative motivation (preference) is 52% ("middle").
- Level of motivation (preference) is 47% ("middle").

In this way, the conducted analysis indicates that Gena was more motivated
to reject her boss's friend request (47%, "middle" level) than to accept it
(37%, "low" level). Although Gena had initially leaned toward *Alternative
2 ("reject")*, as it seemed more preferable than *Alternative 1 ("accept")*,
she was surprised to learn that ED established a relatively low (47%) level
of motivation for *Alternative 2*, which she did not consider sufficient for
its selection (**–fb_FLM₂**). In relation to this, Gena decides to re-evaluate
Alternative 2.2 by changing the potential for cons from *"very strong"*
(I6) to *"strong"* *(I5)* (**SR4: –fb_FLM₂ ⇒ ff_FMM₂**). The current mental model state allows Gena to re-evaluate *Alternative 2* (**+fb_FMM₂**),
which in turn allows ED to form a new level of motivation for choosing
Alternative 2 (**SR1: +fb_FMM₂ ⇒ FLM₂**).

Although ED reflected this change by increasing the level of motivation for *Alternative 2* from 47% to 53% ("middle"), for Gena this
increase was not sufficient for making a motivated selection in favor
of *"reject"* (**–fb_FLM₂**).

In the resulting situation, when Gena is unable to make a motivated
decision in favor of either alternative, *"accept"* or *"reject,"* Gena
decides not to terminate the decision-making process (because the
right choice was significant for her), and instead to turn to evaluate
her third and final option, *Alternative 3* (wait and consider the friend
request) (**SR4: –fb_FLM₂ ⇒ ff_FMM₃**).

Problem 3: *Evaluation of Alternative 3 (wait and consider friend request)*

Pros (+): Gena believes that waiting and considering the request *strongly* *(I5)*
and *very often* *(L6)* contributes to her career growth, since the cost of a
mistake in the case of an incorrect choice between "accept" and "reject" is
relatively great. If anything, Robert could resend his friend request if he so
desires—if he does not forget about it altogether.

Cons (–): On the other hand, this alternative, although *seldom* *(L3)*, is nonetheless able to cause *tangible* *(I4)* negative effects, since Gena's apparent laxity
in her personal correspondence could contribute to Robert perceiving her as
being similarly irresponsible from a professional standpoint.

Because Gena does not experience any difficulties in selecting adequate verbal
characteristics for IL-Frame (**+fb_FMM₃**), *Alternative 3* is positively evaluated, which in turn allows ED to support Gena in the formation of the level
of motivation for choosing this alternative (**SR1: +fb_FMM₃ ⇒ FLM₃**).
- Level of positive motivation (preference) is 76% ("very high").
- Level of negative motivation (preference) is 48% ("middle").
- Level of motivation (preference) is 63% ("high").

Gena accepts the level of motivation (63%) for choosing *Alternative 3* (**+fb_FLM₃**), which in turn upgrades her mental model state (**SR3: +fb_FLM₃ ⇒ FMM₃**). With the help of ED (**SR1: +fb_FMM₃ ⇒ FLM**), this *final* mental model state allows Gena (**+fb_FMM₃**) to compare all alternatives and make a decision in favor of *Alternative 3*.

Indeed, there is a significant positive to the uncertainty present in *Alternative 3* (waiting to either "accept" or "reject" her boss's friend request), which puts Gena in a more neutral position, allowing her to buy some time and better consider both sides of the situation—i.e., the different possible pros and cons of accepting versus rejecting her boss's friend request. Therefore, Gena has decided for herself that she is content with her selection of *Alternative 3*, which involves waiting and considering Robert's friend request. Figure 4.7 presents a screenshot of the resulting screen.

In this way, Express Decision has enabled Gena not to rush into making a final decision relying on her intuition alone. Instead, ED has let her make a well thought-out decision closely aligned with her own values and preferences within the span of just a few minutes. Express Decision helped Gina *build a mental model* to solve a decision-making problem by providing her with an option that was the best fit for her circumstances.

More specifically, using ED has allowed Gena to *clarify her goal* and *the criteria* of her decision problem. For example, as a result of splitting problem 2 into two sub-problems—problem 2.1 and problem 2.2—Gena was able to find additional pros in the situation, when her boss decides to refuse to maintain friendly relations, as well as pinpoint additional cons in the situation when her boss decides to maintain friendly relations; both of these insights helped Gena to better understand the

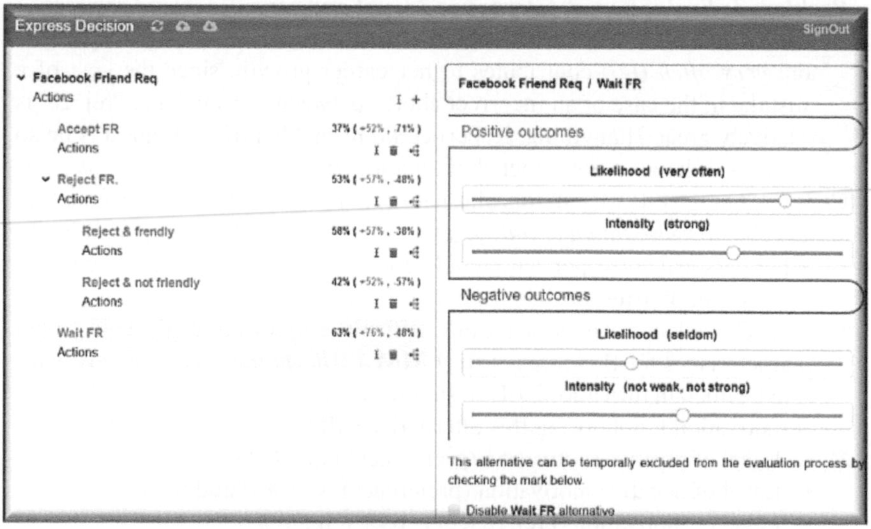

FIGURE 4.7   Screenshot of ED resulting screen (Facebook friend request).

ultimate goal, i.e., her continued career growth. ED also enabled Gena to correct the criteria of success in her decision-making. At the beginning of solving her problem, Gena had a specific criteria of success, such as the fact that she associated her career growth with maintaining friendly relations with her boss. In the process of decision-making with the help of ED, Gena was forced to change this criterion, since she came to understand the complexity of the association between friendly work relations and career growth.

ACKNOWLEDGMENTS

This work would not have been possible without the generous help of Gregory Bedny. He inspired me to get involved with such an undertaking, openly shared his thoughts, ideas, and suggestions, and helped me considerably throughout the process of brainstorming, writing, and idea conceptualization. Many thanks go out to him and his wife, Inna Bedny.

REFERENCES

Atkinson, J. W. (1957). Motivational Determinants of Risk-Taking Behaviors. *Psychological Review*, 64, 359–372.

Aven, T., Zio, E, Baraldi, P., and Flage, R. (2014). *Uncertainty in Risk Assessment: The Representation and Treatment of Uncertainties by Probabilistic and Non-Probabilistic Methods*. Wiley Publishing.

Bandura, A. (1997). *Self-Efficacy. The Exercise of Control*. New York: W. H. Freeman.

Bedny, G., Meister, D. (1997). *The Russian Theory of Activity: Current Application to Design and Learning*. Mahwah, NJ: Lawrence Erlbaum Associates.

Bedny, G. and Karwowski, W. (2006). *Systemic-Structural Theory of Activity: Applications to Human Performance and Work Design*. Boca Raton, CRC Press/Taylor & Francis Group.

Bedny, G. (2015). *Application of Systemic-Structural Activity Theory to Design and Training*. Boca Raton, CRC Press/Taylor & Francis Group.

Bedny, G., Karwowski, W., and Bedny, I. (2015). *Applying Systemic-Structural Activity Theory to Design of Human-Computer Interaction Systems*, Boca Raton, CRC Press/ Taylor & Francis Group.

Bless, H., Betsch, T., Franzen, A. (1998). Framing the Framing Effect: The Impact of Context Cues on Solutions to the "Asian disease" problem. *European Journal of Social Psychology*, 28 (2), 287–291.

Bocklisch, F., Bocklisch, S.F., and Krems, J.F. (2012). Sometimes, often, and always: Exploring the vague meanings of frequency expressions. *Behavior Research Methods*, 44 (1), 144–157.

Bourgeois-Gironde, S. and Giraud, R. (2009). Framing Effects as Violations of Extensionality. *Theory and Decision*, 67 (4), 385–404.

Brehmer, B. (1992). Dynamic decision making: Human control of complex systems. *Acta Psychologica*, 81(3), 211–241.

Damasio, A.R. (1994). *Descartes' Error: Emotion, Reason, and the Human Brain*. New York: Grosset/Putnam.

Franklin, B. (1975). *To Joseph Priestley*. In Willcox, W.B., *The papers of Benjamin Franklin: January 1 through December 31, 1772*, 19. New Haven: Yale University Press, 299–300.

Heath, C., Larrick, R. P., and Wu, G. (1999). Goals as Reference Points. *Cognitive Psychology*, 38, 79–109.

Ishizaka A. and Labib A. (2011). Review of the Main Developments in the Analytic Hierarchy Process, *Expert Systems with Applications*, 38 (11).

Johnson, J. G. and Busemeyer, J. R. (2010), Decision Making Under Risk and Uncertainty. *Wiley Interdisciplinary Reviews: Cognitive Science*, 1 (5), 736–749.

Kahneman, D. and Tversky, A. (1979). Prospect Theory: An Analysis of Decision Under Risk. *Econometrica*, 47 (2), 263–291.

Kahneman, D. and Tversky, A. (1984): Choices, Values, and Frames. *American Psychologist*, 39 (4), 341–350.

Keeney, R. and Raiffa, H. (1976). *Decisions with Multiple Objectives: Preferences and Value Trade-offs*. New York: John Wiley & Sons, Inc.

Knight, F. H. (1964). *Risk, Uncertainty, and Profit*. New York: Sentry Press.

Kotik, M. A. (1974). *Self-regulation and reliability of operator*. Tallinn, Estonia: Valgus.

Kotik, M. A. (1984). A Method for Evaluating the Significance-as-Value of Information. *Studies in Artificial Intelligence. Scientific Notes of Tartu University*, 688, 86–102. Tartu, Estonia: Tartu University Press.

Kotik, M. A. (1994). Developing Applications of "Field Theory" in Mass Studies. *Journal of Russian East European Psychology*, 2 (4), July – August, 38–52.

Kotik, M. A. and Yemelyanov, A. M. (1983). A Method for Evaluating the Significance-as-Anxiety of Information. *Studies in Artificial Intelligence. Scientific Notes of Tartu University*, 654, 111–129. Tartu, Estonia: Tartu University Press.

Kotik, M. A. and Yemelyanov, A. M. (1992). Emotions as an Indicator of Subjective Preferences in Decision Making. *Psychological Journal*, 13 (1), 118–125.

Köbberling, V. (2002). Program for calculating the cumulative-prospect-theory value of prospects with at most four outcomes (http://psych.fullerton.edu/mbirnbaum/calculators/cpt_calculator.htm).

Larrick, R. P. (1993). Motivational Factors in Decision Theories: The Role of Self-Protection. *Psychological Bulletin*, 113 (3). 440–450.

Lerner J. S., Li, Y., Valdesolo, P., and Kassam, K.S. (2015). Emotion and Decision Making, *Annual Review of Psychology*, 66, 799–823, 2015.

Levitin, A. (2011). *Introduction to the Design & Analysis of Algorithms*. Pearson.

Lewin, K. (1939). Field Theory and Experiment in Social Psychology. *American Journal of Sociology*, 44 (6), 868–896.

Pfister, H. R. and Böhm, G. (2008). The multiplicity of emotions: A framework of emotional functions in decision making. *Judgment and decision making*, 3 (1), 5–17.

Saati, T. (1980). *The Analytic Hierarchy Process*. New York: McGraw-Hill.

Slovic, P., Finucane, M., Peters, E., MacGregor, D. (2007). The affect heuristic. *European Journal of Operational Research*, 177, 1333–1352.

Tversky, A., Kahneman, D. (1981). The Framing of Decisions and the Psychology of Choice. *Science*, 211 (4481), 453–458.

Tversky, A., Kahneman, D. (1992). Advances in prospect theory: Cumulative representation of uncertainty." *Journal of Risk and Uncertainty*. 5 (4), 297–323.

von Neumann, J., Morgenstern, O. (1944). *The Theory of Games and Economic Behavior*. Princeton, NJ: University Press.

Vroom V. H. (1964). *Work and Motivation*. New York: Wiley.

Yemelyanov, A. M. (2017). The Model of the Factor of Significance of Goal-Directed Decision Making. In C. Baldwin (Ed.), *Advances in Neuroergonomics and Cognitive Engineering, Advances in Intelligent Systems and Computing*, 586, Springer International Publishing, 319–330.

Yemelyanov, A. M. (2018). Decision Support of Mental Model Formation in the Self-Regulation Process of Goal-Directed Decision-Making under Risk and Uncertainty. In H. Ayaz and L. Mazur (Eds.), *Advances in Neuroergonomics and Cognitive Engineering, Advances in Intelligent Systems and Computing*, 775, Springer International Publishing, 225–236.

Yemelyanov, A. M., Baev, S, and Yemelyanov A. A. (2018). Express Decision–Mobile Application for Quick Decisions: an Overview of Implementations. In H. Ayaz, L. Mazur (Eds.), *Advances in Neuroergonomics and Cognitive Engineering, Advances in Intelligent Systems and Computing*, 775, Springer International Publishing, 255–264.

Yemelyanov, A. M. and Yemelyanov, A.A. (2017). Applying SSAT in Computer-Based Analysis and Investigation of Operator Errors. In C. Baldwin (Ed.), *Advances in Neuroergonomics and Cognitive Engineering, Advances in Intelligent Systems and Computing*, 586, Springer International Publishing, 331–339.

Zarakovsky, G.M. and Pavlov, V.V. (1987). *Laws of Functioning Man-Machine Systems.* Moscow, Russia: Soviet Radio.

5 Self-Regulation Models of Perceptual Process and Memory

Gregory Z. Bedny
Essex County College, New Jersey, US

Inna Bedny
United Parcel Service, Atlanta, US

CONTENTS

5.1 INTRODUCTION

The study of the interaction between a human and a machine, a computer, or an external world is impossible without studying cognitive processes. Sensation, perception, memory, and thinking are stages of cognition that intervene between the external and internal impacts and human responses to its influence. Based on cognition, a human can reflect external influences and creates a "mental model of reality." A human develops the goal, program of performance, and cognitive and behavioral actions that are involved in task execution based on this model. In their applied studies, cognitive psychologists concentrate their efforts on various mental processes that are relatively isolated, while in applied and systemic-structural activity theories, mental processes are investigated as interdependent mental processes. Cognition also interacts with emotionally-motivational processes, and such interaction is the critical factor in functioning of cognition. Such interdependence is understandable because cognitive processes are a system of interdependent cognitive actions (perceptual actions, mnemonic actions, thinking actions, etc.). Internal actions transform information and images while motor actions transform real objects. Human cognitive and behavioral actions are organized into tasks. Therefore, from the SSAT perspective, task is a logically organized system of mental and behavioral actions that are utilized for achieving the goal of task.

SSAT considers cognition as a process (cognitive approach), and as a structure or system of cognitive actions (activity approach). Cognition also can be considered from the functional perspective as a self-regulative system, where various stages of self-regulations are presented as functional mechanisms or function blocks (Bedny, 2015; Bedny, Karwowski, I. Bedny, 2015).

Cognitive actions are involved in the purposeful mental transformation of the external situation or its mental representation into the adequate mental model of the situation according to the task goal. Mental operations, which are components of cognitive actions, can also be viewed as independent cognitive mechanisms that are involved in the transformation of an external situation into its mental representation. These components of mental activity are usually unconscious. Such transformation also depends on emotionally-evaluative (significance) and motivational factors. Of course, motor actions play an important role in this process. They can perform not only transforming but also cognitive functions.

The feedforward and feedback connections that facilitate the gradual transformation of the external situation into an internal mental model that is adequate to the existing goal of activity are sustained by the mechanisms of self-regulation. These processes are important not only at the level of perception but also at the level of more complex cognitive transformations of reality.

Coding, recoding, and decoding of information take place from the point of view of the tasks performed by the subject in these processes. Thus, a subject develops the mental picture of situation, and utilizes the adequate responses based on the analysis of the incoming information.

For example, when an operator receives visual information, distinct essential characteristics that are important for solving a particular problem can be extracted not only using external fixation but also by utilizing internal mental scanning extracted from the memory of the identical situations. These strategies are not always conscious or verbalized. According to Pushkin (1978), such mechanisms provide extraction of "nonverbalized operational meaning" or a "situational concept of thinking." The same external situation can constantly change in the subject's mind. This continual change of the situation in the mind of a subject in light of its external constancy is called the "gnostic dynamic."

In this chapter, we describe the sensory-perceptual process and memory from the perspective of SSAT where not only activity in general, but also separate cognitive processes are considered as goal directed self-regulative systems. Activity self-regulation process includes conscious and unconscious levels. However, the conscious level of self-regulation plays a leading role in development of the adequate strategies of activity during task performance. In the process of self-regulation, a subject has the ability to formulate goals, to change the ways of achieving goals, and so on. SSAT offers various models of self-regulation of activity in general, and models of cognitive processes in particular. For example, we developed the model of orienting activity. The main purpose of orienting activity is to reflect the situation, and to create the mental model of the situation. It precedes the execution stage. The general model of self-regulation also includes execution. We also developed self-regulation models of such cognitive processes as attention and thinking (G. Bedny, Karwowski, I. Bedny, 2015; G. Bedny. 2015). In this chapter, we present self-regulation models of perceptual process and memory. Below we consider the self-regulation model of the perceptual process.

5.2   SELF-REGULATION MODEL OF PERCEPTUAL PROCESS

We've developed a self-regulative model of perceptual process based on our analysis of sensation and perception. Analysis of sensation and perception in cognitive psychology and applied activity theory (AAT) demonstrates that sensation and perception are tightly interconnected, and can be divided into a sequence of sub-processes or stages. They can be presented as functional mechanisms or function blocks. Each function block interacts with other blocks through feedforward and feedback connections. Each function block represents a coordinated system of sub-functions that serve a specific purpose in the regulation of activity. In AAT, perceptual process is divided into four stages: detection, discrimination, identification, and recognition. Demarcation between these stages is not always clearly defined. The stages can be presented in the following way (Zaporozhets, et al., 1967; Zinchenko, Velichkovsky, Vuchetich, 1980):

Signal → Detection → Discrimination → Identification → Recognition

According to applied and systemic-structural activity theories, all stages of perceptual process are performed by conscious perceptual action that can be verbalized or unconscious perceptual operations. However, in this area of research there is no clearly defined terminology. Scientists do not distinguish the stages of perceptual process from perceptual actions and operations. In some studies, detection, discrimination, identification, and decoding are considered stages. In other ones, these stages are interpreted as types of perceptual actions. However, in any particular situation each stage can include various actions and operations, which can be classified into groups according to the names of these stages. In the works of the above-mentioned authors the term "action" is used in a very general manner. Discussions usually concentrate on the relationship of hand movements and eye movements, the relationship between vision and sense by touch, the role of object-oriented practical actions in mental development, etc. In these studies, such terms as mental and practical actions are not clearly specified. For example, if we are talking about perceptual or mnemonic actions we have to know the beginning and end of actions, their clear specification, relationship, and logical organization. The aforementioned authors utilize such terms as "perception as an action," "memory as an action," etc. Perception, memory, and other cognitive processes usually include a number of various cognitive actions and operations. Therefore, instead of such terms as "perception as an action" or "memory as an action," we use such terms as "perception as activity" or "memory as activity" and so on. Really, perception, memory, and other mental processes usually include not one but a number of mental actions during performance of a specific task.

SSAT allows us to clearly distinguish the differences between various actions and operations in the perceptual process. For example, decoding is a system of actions and operations that are involved in the creation of a perceptual image and its interpretation. Identification is a system of actions and operations that provides a comparison stimulus to the templates already stored in memory. Thus, perception is often not an action of creating perceptual image as has been stated by Zinchenko and his colleagues (1980). The perception of a complex stimulus can involve a system of various perceptual actions and operations that are necessary for the creation of a perceptual image. Moreover, perceptual process is closely related to other mental processes, and its isolated consideration may be only conditional. For instance, interesting data was obtained by Clifford P. J., Tong C. F. (2007). They conducted experiments in which they utilized the binocular rivalry as a tool to reveal the perceptual and mnemonic contents of mental imagery. Observers were either shown or instructed to imagine one of two oriented patterns, several seconds prior to the presentation of an ambiguous rivalry display consisting of both competing patterns. The presentation of low luminance patterns strongly biased the perception of subsequent rivalry displays, in favor of the previously seen pattern. Amazingly, mental imagery of a specific pattern led to the equally potent bias effects. In addition, the longer periods of mental imagery led to progressively stronger bias effects, mimicking the effects of prolonged viewing of a physical pattern. This study confirmed the interaction of mental processes.

Perception is closely linked to the goal of activity, its motives, and significance of perceived information. Rubakhin (1974) conducted the experiment that involved the process of deciphering army reconnaissance codes. Two groups of average-skilled operators had to decipher the codes. The first group was told that they were determining the position of a "strong" enemy, whereas the second group was informed that they were to locate a "weak" enemy. Deciphering information about the weak enemy resulted in a decrease in correct identification, but deciphering the strong enemy resulted in a sharp increase of false alarms.

According to SSAT, the model of perceptual process has to include the goal of perceptual process, significance of information (emotionally-evaluative component), and motivation. This is especially true for complex vigilance tasks when perception is the main component of human activity. Thus, according to SSAT the model of perceptual process should be presented as a goal-directed self-regulated system. This system includes considered stages with feedforward and feedback interconnections and some additional functional blocks as presented in Figure 5.1.

Let us consider the first four stages:

Detection is the initial stage of perceptual process, and is presented as such in this model. At this stage a subject can only answer the question of whether there is a stimulus in her or his visual field. The detection stage can be simple and can be performed in a very short time. However, it is well known in psychophysics that detection of a signal in noisy conditions can sometimes be a complex task. Under normal conditions, this initial stage is quickly transformed into the next stage of perceptual process.

Determining sensory threshold is one of the main aspects of activity analysis at this stage of studying the perceptual process. At this stage specialists try to determine the absolute or difference threshold by using methods that have been developed in psychophysics. However, in practical situations, this laboratory method might be inapplicable. Sensory threshold data is related to performance in extreme conditions that require exerting efforts for detecting a signal. The notion of the "operative threshold" was introduced in AAT. According to Dmitrieva (1964), the operative threshold is 10 to 15 times greater than the absolute and difference thresholds.

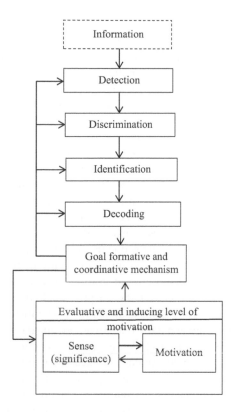

FIGURE 5.1 Self-regulation model of perceptual process.

The operative absolute and difference thresholds reflect the optimal efforts for detection or discrimination of various stimuli during task performance. In task analysis, the operative threshold is determined by multiplying the value of absolute or difference thresholds by 10 to 15. Signal detection is determined not only by sensitivity of the sense organ, but also by the criteria used by a person. This is why in psychophysics there is the notion of the "threshold region," meaning that threshold is not a constant value. Bardin et al. (1982) demonstrated that the observer, while performing the task of signal detection in noisy conditions, can use additional criteria corresponding to difference axes of sensory space. For example, for signal detection in a noisy environment, when two signals are very difficult to discriminate on the basis of loudness, the subjects began to use additional criteria. They began to perceive previously unnoticed qualities in the acoustic stimuli, and used these as discrimination criteria. For example, they reported that the sound seemed to be dimmed, resonant, dull, and so on. According to SSAT this means that a subject changes his or her strategy of detection of a signal based on mechanisms of self-regulation. Even this brief overview of detection stage demonstrates a number of questions that can arise at this stage of perceptual analysis. The next stage is signal discrimination.

Discrimination is the selection of individual features in the object in accordance with the goal of a task at hand and the formation of the perceptual image. At this stage, sensory features of a stimulus are selected in accordance with the specifics of a stimulus or situation, and requirements of a perceptual task. In the subsequent step selected features are integrated into a structurally organized system. The end result of this stage is formation of an image of a stimulus or an object. This stage is the combination of active cognitive processes and innate mechanisms of the perceptual system. The considered stage of the perceptual processes is particularly important when a subject is involved in perceiving an unfamiliar object. A subject develops a counter of an object and constructs its perceptual image.

Another type of perceptual process is identification and recognition. For example, the ability of subjects to relate the perceptual object to previously presented objects can be considered the recognition stage. A subject matches current stimulus with the one that is stored in memory. During recognition, the perceptual process is integrated with the functions of memory. Recognition should be distinguished from identification.

Identification always suggests dividing all presented stimuli into two classes: those that are identical according to all features to templates stored in memory (positive identification) and those that are not identical to templates according to at least one feature (negative identification). Sometimes a template can be presented externally. Identification is not accompanied by the discovering of the inner contents of the stimulus. Recognition is required for this purpose.

Decoding or *recognition* can be considered as the ability of a subject to relate the newly presented object to previously encountered objects. In recognition, perceptual processes are integrated more intensively with the function of long-term memory and sometimes involve thinking operations. Conscious processes become important at this stage of perceptual process. According to Shekhter (1967), recognition is distinguished from identification because it also involves categorization (classification) of an object and its verbal designation. For instance, recognition includes selection of verbal templates that can correspond to their visual equivalents in the visual perceptual process. Success of perception depends to some extent on the number of verbal templates, compatibility of verbal and visual templates, capacity of working memory, past experience, and training. This fact is important for developing a training method for the operators. Therefore, the last stage of the perception process includes decoding that involves discovering the relationship between the sign and the object, which is denoted by this sign. Decoding is a process of transforming an image of a signal into an image of an object. An abstract code makes this process more difficult. The speed, precision, and reliability of such a process depends on the relationship between "the alphabet of signals" and "the alphabet of objects." For abstract codes, associating a signal with its meaning takes longer and requires greater mental efforts. Requirements for the presentation of information can change depending on which stage of information reception is the most important.

Sometimes it is difficult to draw clear boundaries between such stages as detection, discrimination, and identification for the perception of familiar objects. In these situations, the genetic method suggested by Vygotsky helps to clearly define the stages described above.

According to SSAT, consideration of cognitive processes or activity as the goal-directed self-regulative system is regarded as a functional analysis of activity (Bedny, Karwowski, I. Bedny, 2015; Bedny, 2015). Informational and energetic components of activity are interdependent in the functional analysis of activity. Therefore, the presented model of perceptual process includes cognitive and emotionally-motivational mechanisms. This model also includes goal formative and coordinative blocks. Decoding stage, and specifically the goal formative and coordinative stages, are tightly connected with the thinking process. The last one is important not only in creation of the goal of perceptual process or sometimes independent perceptual task, but also in the coordination of functioning of all other blocks.

The evaluative and inducing level of the motivation block consists of two sub-blocks that influence each other. One sub-block "sense" (significance) is responsible for evaluation of personal significance of the received information. The sub-block "motivation" refers to the inducing components of perceptual process. The regulative integrator coordinates functioning of the above-mentioned information processing blocks. Hence, the model of self-regulation of perceptual process demonstrates that this process is goal-directed, and integrates informational and energetic components of activity.

5.3 STUDY OF MEMORY IN THE FRAMEWORK OF GENERAL ACTIVITY THEORY

The cognitive process that facilitates storing and keeping information in our mind for retrieval when necessary is known as memory. Sternberg (1999) defines memory as the means by which we draw on our experiences in order to use this information in the present. We can remember episodes, general facts, and skills or procedures. Memory is first an active goal-directed process. Many failures of memory are associated with its first stage of remembering, and are caused by failure to acquire information. A subject has been exposed to some information but did not pay attention to it. We need to consciously pay attention to the presented information in order to obtain it. Therefore, memorization is not just copying an event or facts. Instead, acquisition of information involves other cognitive processes associated with the goal of the mnemonic task. The next aspect of memory is storage of information. To be remembered, our experience must be recorded in our mind. This record, known as memory trace, should be held in enduring form for later use. This information can either gradually fade away or stay in memory forever. The final aspect of memory is retrieval. This process is extraction of required information from memory for using it.

Activity theory differentiates voluntary and involuntary memory. In cognitive psychology there are such terms as deliberate or intentional memorization and incidental memorization that often occurs without awareness of the memorization process.

Cognitive psychologists analyze characteristics of various mental processes as relatively isolated ones, while in activity theory mental processes are considered interdependent. Cognitive processes include memory interactions with other cognitive processes, and this interdependence influences memory. Memory also interacts with emotionally-motivational processes, and such interaction is a critical factor in memory functioning. This interdependence is understandable because cognitive processes are a system of interdependent cognitive actions (perceptual actions, mnemonic actions, thinking actions, and so on). Mnemonic actions are basic elements of memory that transform units of information held in our memory. Internal actions transform information and images while motor actions transform real objects. Memorization is dependent not so much on the nature of material but rather on how it is presented. Organization of material may, in fact, assist in its utilization and enhance memorization. For example, an experiment by P. Zinchenko (1961) suggested that classification depends on the goal of activity. This study utilized pictures and numbers on cards. Subjects were instructed to organize the cards either by pictures or by the numbers they saw on these cards. Those instructed to organize the cards by the pictures were unable to recall the numbers. In fact, some insisted that there were no numbers on the cards, while those instructed to organize the cards by the numbers could not recall the pictures.

This experiment demonstrated that memorization depends not only on the particular feature of the stimulus, but also on how the material is presented. Motor components of activity also help to organize external material. In other words, memorization of material is stipulated by motives, goals, and the method of activity performance. Effective organization of various cognitive and motor actions and, specifically, the way mnemonic actions are organized influence memorization.

There are mnemonic tasks the goal of which is to memorize some information. There are various types of tasks that involve memory functions. These can be tasks in our everyday life, tasks that are involved in our professional activity, and so on. We pay special attention to tasks that are involved in human performance. Every task that involves memory functions includes mnemonic actions. Mnemonic actions include manipulation of information in working memory, extracting information from long-term memory, maintaining information in short-term memory, and so on.

In activity theory memorization, storage and retrieval of information depends on what a subject is doing with the corresponding material. In other words, memory functioning depends on the activity the subject in involved in. The goal and motives of activity are of the particular importance. Memory is closely related to other mental processes. Memory is used in perception, thinking, language, decision-making, etc. This creates some difficulties in the study of memory as an independent cognitive process. For example, recognition can be considered as the stage of the perceptual process. At the same time, recognition can be seen as the function of memory.

Thus, in general activity theory specialists recognize various types of memory that are dependent on the characteristics of the activity.

There are three basic criteria for the classification of memory:

1. according to the nature of mental activity, memory is divided into motor, emotional, imaginative, and verbal-logical
2. according to the goal of activity, memory is divided into voluntary and involuntary
3. memory is divided into sensory, short-term, long-term and operational or working memory.

General activity theory began to distinguish memory according to the duration of storing the required information after accumulating the ideas of cognitive psychology. Detailed studies of these types of memory are primarily conducted within the framework of AAT (Repkina, 1967; Zinchenko, et al., 1980, and so on). For instance, motor memory is involved in keeping and reproducing information about our movements, and emotional memory is memory about our feelings. Imaginative memory is associated with our images such as pictures of nature, sounds, smells, tastes, and verbally logical memory includes our thoughts. Some similarity with such classification can be found in cognitive psychology. Special attention should be paid to voluntary and involuntary memory because the concept of goal is especially important in activity theory. These types of memory depend on the goal of activity. Memorization and reproduction of information, when there is no special goal to remembering or recalling something, is called involuntary memory. Voluntary memory takes place in cases when there is a goal to remember or recall something. Involuntary and voluntary memories are two stages in the development of memory. It should be noted that not only voluntary, but also involuntary memory is important in human life. Thanks to involuntary memory, people accumulate a large amount of information without much mental effort.

In another experiment, first grade school children and university students were given five simple arithmetic tasks. Unexpectedly for both groups of subjects, they were asked to recall numbers that were presented in each task. It was discovered that first graders remembered almost three times more numbers presented in these tasks than the university students. This is because for the first grade schoolchildren the ability to add and subtract numbers is not yet an acquired skill. They do arithmetic tasks by using conscious purposeful actions. Manipulation with numbers during performance of arithmetic tasks for school children has been connected with the goal of activity. They also were more motivated to perform these tasks. University students on the other hand. used automated mental operations during performance of such simple calculations and therefore were not motivated to remember the numbers. These two examples clearly demonstrate that memory functioning depends on the goal of task, motivation, and the method of task performance.

Memory is an active process. It is well known that memory has reconstructive features. When reconstructing memory, subjects utilize their past experience depending of the goal of the task and the subjective significance of information and motivation. Event recollection and incorporation of new information in the memory reconstruction process depends not only on experience, but also on the subjective significance of information, the goal of recollection, and on human activity in general. All kinds of memory are interdependent. As can be seen, the study of memory in the framework of general activity theory has its own specifics. We have briefly described some studies of memory in general activity theory. Initially, the term cognitive psychology was not utilized in general activity theory, and the term mental processes has been used instead. A new stage of memory studies has emerged due to the influence of cognitive psychology. This stage of memory studies is critically important for AAT. The interdependence of memory studies in applied activity theory and cognitive psychology is considered in the following chapter.

5.4   STUDY OF MEMORY IN APPLIED ACTIVITY THEORY

Here we consider the self-regulative model of memory developed in the framework of system-structural activity theory. The self-regulation model of memory can be used in task analysis. We will briefly consider how memory is studied in the AAT. There are different theories of memory. In AAT scientists distinguish two kinds of activity (Zinchenko, Munipov, 1979). The first kind of activity involves short-term memory associated with the presentation of a small amount of information per unit of time to an operator immediately followed by decision-making and performance. This kind of activity is called "informational search with immediate performance."

The second kind of activity is characterized by the fact that informational search does not end with the immediate performance of actions. If for instance, an item of information is presented to an operator, he or she evaluates the importance of the elements of this information, then puts the elements in the specific order, and finally keeps it in the working memory during task performance. This kind of activity is called "informational search with delayed performance." Putting elements in a specific order requires keeping the defined order in memory, and then altering this order according to the changing situation. This kind of task is more complex from the perspective of memory involvement.

As we have mentioned, such concepts as sensory memory, short-term, long-term, and working memory are of particular importance for the AAT.

Thus, let us consider some basic ideas in studying these types of memory. Models of memory that contain these types of memory systems are linked to the information processing models in cognitive psychology. According to the model in Figure 5.2, we encode information, or convert information in a form that our brain can use. At the next stage, we store the information or retain it over time, and finally retrieve the information for using it.

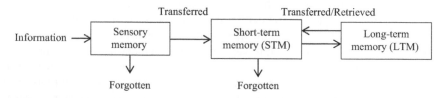

FIGURE 5.2   Three-box model of memory system.

This is the simplified three-box model of memory, which is well known in cognitive psychology. Sensory memory has a relatively high capacity and a short retention period (up to ½ second for visual and 2 seconds for auditory information). This memory contains sensory data. Short-term memory has limited capacity and holds information for a brief period, perhaps up to 30 seconds. Subjects can use conscious effort to keep information longer. Long-term memory has unlimited capacity. It is the storage of our thought or ideas that can be permanent. Information can be retrieved into short-term memory for future use and analysis of incoming sensory information, or performance of mental operations. Information is organized and indexed in the long-term memory (Sperling, 1960; Atkinson, Shiffrin, 1977; Lindsay, Norman, 1992; and so on). We will utilize the model in Figure 5.2 in future discussions. This three-component model and some other information-processing models have been long dominant in the study of memory. However, our brain does not work like a computer. The three-box memory model emphasizes sequential processing of information, yet a human brain can perform many operations simultaneously. Even contemporary computers can perform multiple operations in parallel utilizing multiple processors. This is the reason some scientists utilize a parallel distributed processing or connectionist models. Such models represent the content of memory as a connection of the tremendous number of interacting units that can be presented as a network with operations performed not only in sequence but also in parallel.

Similar to cognitive psychology, scientists in applied activity theory developed models of memory that deal with sensory, short-term, and long-term memory. In AAT, operative or working memory is distinguished from short-term memory. V. Zinchenko and Munipov (1979) suggested separating two kinds of work activity. The first (short-term memory) is associated with presentation of a small amount of information per unit of time with immediate decision-making and performance. Such activity is called "informational search with immediate performance" where working memory is not strongly involved. For the second kind of activity called "informational search with delayed performance," an operator has to recall information, and keep it in memory during task performance. Working memory plays a critical role in this kind of task performance.

Working memory is the temporary, attention-demanding kind of memory. Operative or working memory involves intense consciousness. We use operative or working memory for examining, comparing, and transforming information according to the task goal. Operative or working memory uses new information that can be provided by the external input or information recalled from long-term memory. According to SSAT, working memory is tightly connected with our thinking. Thinking is also tightly connected with other types of memory. Thinking is responsible for operations that are called central executive mechanisms of working memory in cognitive psychology. Information from operative or working memory can be transferred to the long-term memory. It can also transfer some information into short-term memory for more detailed analysis of input information. This memory uses information in verbally logical and imaginative form. Thus, operative or working memory, together with thinking, is responsible for serving actually performed cognitive and motor actions, and operations that are involved in task performance.

Perceptual actions are involved in performance of the tasks with visual information. The type of memory considered above is critical when an operator has to keep intermediate information in memory and perform various mnemonic actions in order to do that. A rehearsal mechanism (covert speech) is an example of such actions. Operative units of memory known in cognitive psychology as chunks are used for performance of mnemonic actions. Mnemonic actions always have the conscious goal of actions. In contrast, mnemonic operations are performed automatically without awareness. Therefore, information processing in memory is considered a system of conscious mnemonic actions or unconscious automatized mnemonic operations. We also want to emphasize the fact that the transfer of information from one type of memory to another is mainly carried out in sequence. However, the processing of information in each of the types of memory can be executed not only in sequence, but also in parallel.

Thus the memory as a system includes the processes of control and monitoring, which is involved in the organization, transformation, and structuring of information in separate blocks of memory, and the distribution of the flow of information between different memory blocks. These transformations are connected with the function of thinking, decision-making, and mechanisms of attention. Various actions, including mnemonic ones, are elements of activity that are the main units of analysis of not only long-term memory, but also short-term memory. Cognitive actions as well as automatized mental operation are used in memory functioning. They are specifically important for functioning of the short-term memory. It is important to mention here that our recalling has reconstructive features. This is furrther evidence that memory functioning should be considered a special type of human activity. Mnemonic tasks include verbal, imaginative, and symbolic actions.

Zinchenko (1980) and his colleagues have shown that there is certain difference between symbolic and imaginative memory. Storing images is based on perceptual actions. Imaginative memory can operate independently of short-term memory. The latter is predominantly long-term memory considering the methods of storing and retrieving information. Verbally logical actions play a leading role in symbolic memory. Symbolic memory includes mechanisms of both short-term and long-term memory. At the same time, imaginative and symbolic memory are interconnected and function as a system. For example, it was revealed that verbal and semantic coding plays an important role in memorization of visual information (Baddeley, 1976). During performance of various tasks, subjects develop a mental model of a task that includes imaginative and verbally logical components. Thus, operative or working memory plays a critical role in the creation of such a model and it should be distinguished from short-term memory. The last plays a leading role in purely mnemonic tasks.

The above-mentioned study demonstrated that the motivational factor plays an important role in functioning of not only long-term memory, but also in functioning of sensory-memory and short-term memory. The emotional significance of stimulus influences perception (Bruner, 1958). It has also been revealed that emotions influence memorization (Reykovski, 1974). The selectivity of memorized material based on its significance indicates the presence of emotionally evaluative mechanisms in the functioning of memory (Bedny, Karwowski, 2007). Thus, various studies clearly demonstrated that performance of mnemonic tasks, or tasks that involve functioning of memory, should be considered a special type of activity that includes such components as goals, motives, and actions or automated operations. Such an interpretation of memory processes is also applicable to sensory-memory and short-term memory. In the latter case, automated data processing methods are of particular importance. At the same time, there is no clear understanding of such notions as task, goal, actions, and operations, as can be seen in the studies of perceptual process in AAT. These basic concepts should be more rigorously defined (Bedny, I. Bedny, 2018).

5.5   SELF-REGULATIVE MODEL OF MEMORY IN SYSTEMIC-STRUCTURAL ACTIVITY THEORY

Considering that activity is regulated by a subject, an individual can use various strategies to perform tasks wherein the memory is important. Thus, the construction of the memory model should take into account the possibility of self-regulation of the mnemonic activity. The material presented above allows proceeding to the consideration of a proposed model of memory that can be applied in analyzing tasks that involve memory functions. The model presented below reflects only some basic aspects of memory functioning during task performance (see Figure 5.3). At the same time, it helps to select adequate theoretical data, which can be useful when studying the functioning of memory during performance of a specific task.

The presented model can be interpreted primarily as a multi-store model of memory. There are other concepts of memory. For example, incoming information is processed by a central memory processor according to the levels-of-processing effect (Craik, Lochart, 1972). This processor has a number of processing levels. The deeper the level of processing the better the information is stored in memory. Our model is adequate even for such a concept of memory. We can consider our memory blocks as relevant to levels of processing information in memory.

Of course, these blocks are not strictly identical to the levels of processing described by Craik and Lochart. However, it is clear that memory storage or levels of processing in the model can be described similarly.

The material presented above clearly demonstrates that memory is involved in functioning of all cognitive processes. Therefore, consideration of memory separately from other cognitive processes is to some degree artificial. Thus, in our model there are blocks that depict the close connection of memory with such cognitive processes as perception, thinking (intellectual block), and emotionally-motivational processes (evaluative and inducing levels of motivation). We start analysis of the presented model with the consideration of memory interaction with perception and thinking.

The relationship of sensory processes with perception is well established. This is demonstrated in our model. Such stages as detection and discrimination are considered here as sensory processing stages, and in our model, they are associated with the functioning of sensory memory (SM). However, identification, recognition, and decoding are related to perceptual process. This process also involves short-term memory functioning. Perception always imposes some structural organization on incoming information. Moreover, such structural organization of incoming information often cannot be prevented and is performed automatically. This means that perception includes automatic processing mechanisms. Perception also includes decoding and/or interpretation of information. During the interpretation of information a subject gives a certain

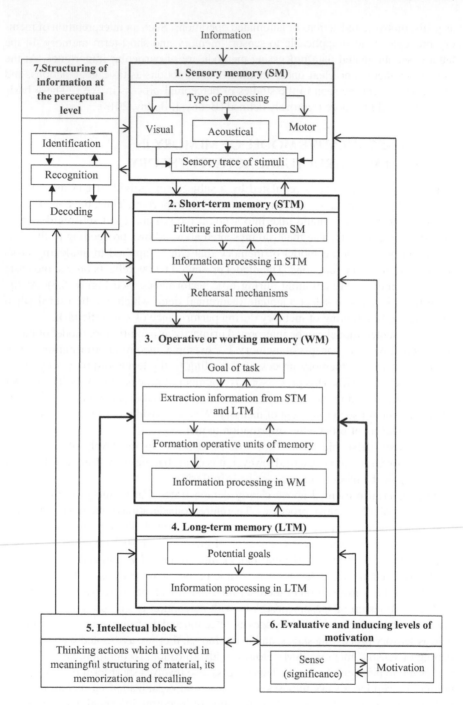

FIGURE 5.3 Self-regulation model of memory.

meaning to a signal. For example, when a spot appears on the screen of the ship's radar an operator not only identifies it as a spot, but interprets it as a friendly or enemy ship. Interpretation is always carried out in accordance with a certain system of rules. Therefore, the special instructions that are given to an operator are very important. The context plays the key role in the interpretation of the incoming information.

Two streams of information interact during perception. One comes from the external environment and the other one from the depth of memory. Constant revision of expectations, application of some logical rules during perception, decisions about received signals and assignment of a certain meaning to these signals presupposes the existence of special mechanisms that are responsible for these functions. According to cognitive psychology, such mechanisms include processes for monitoring and control that govern the distribution of the flow of information between different memory stores. Various authors assign different names to these mechanisms. They can function at conscious and unconscious levels of processing. The functioning of these mechanisms indicates that they belong to the thinking process. Thinking is an important machinery, which works with information in our memory. All this indicates that the memory model should depict the relationship of memory with perception and thinking. Thus, in our model we have two function blocks that reflect it: block 7 (structuring of information at the perceptual level) and block 5 (intellectual block). Block 7 is tightly connected with block 5, which is involved in the structuring of information. These blocks are also interconnected with sensory memory (block 1) and short-term memory (block 2). Due to the interaction of block 1 and block 2 with intellectual and perceptual blocks, the functioning of sensory and short-term memory has a goal-directed character. This influences the specificity of the selection and storage of information at the early stages of memory functioning. The intellectual block can affect automated operations that predominate in the sensory memory to a lesser extent. The interaction of intellectual block 5, perceptual block 7, and emotionally-motivational block 6 with sensory memory block 1 and short-term memory block 2 demonstrates that these two types of memory cannot be considered as rigid, totally involuntary, and programmed processes because consciousness, motivation, thinking, and so on can affect these types of memory. This also was proved by Zinchenko and his colleagues who studied visual short-term memory. They wrote (Zinchenko, et al., 1980, p. 68):

> *These processes are not so hard-programmed and invariant as it was assumed in the model of visual memory suggested by Sperling. [The] Amount of information that can be reproduced depends not only on the physical conditions (exposure time of a stimuli), but also on the individual characteristics of subjects, as well as on the level of their motivation.*

This group of scientists (Zinchenko, et al., 1968) also showed that not only verbal repetition, but also the duration of attention concentration affects the retention of the visual stimulus information in visual memory. Intellectual block 5 also interacts with block 3 (operative or working memory) and block 4 (long-term memory). Intellectual block 5 has an important role in the creation of semantic memory (Lindsay, Norman, 1977). Semantic memory plays the leading role in the functioning of the long-term memory.

Collected in cognitive psychology data also confirms that different cognitive processes are involved in the functioning of other types of memory. There is data that proves that it is possible to store information in short-term memory not only acoustically but also semantically. Thinking is especially important for the functioning of short-term memory and working memory. The most active processing of information takes place in working memory. The thinking process gives conscious awareness of information in this type of memory. The possibility of storing information semantically or in terms of its meaning clearly demonstrates the role of thinking in the functioning of memory. The level of human memory processing also emphasizes the importance of thinking in the functioning of memory. The deep processing means the significant involvement of thinking in memory functioning. At the same time, other factors also play a specific role in memorization. For example, the significance (emotionally-motivational factors) of information plays a key role in memorization. Various factors influence each other during memorization.

Analysis of how block 5 interacts with other blocks of memory demonstrates that the intellectual block includes various mechanisms that monitor described function blocks of memory and govern the distribution of the flow of information between different memory stores. This block reflects the functioning of operative or working memory and long-term memory, which is of particular importance.

Information in long-term memory is organized as a complex structure that interconnects units of information and creates our past experience. Recalling and extracting information from long-term memory involves thinking and utilizing mental strategies that are appropriate for task performance. Hence, the interaction of intellectual block (block 5) and long-term memory (block 4) is performed through working memory block (block 3). The last block is specifically important for not purely mnemonic tasks, when a subject is retrieving information for practical purposes, performing problem-solving tasks, and making decisions. Thus, the study of memory functions in task performance is the study of mental strategies and the mechanisms of thinking that guide working memory and long-term memory. At the same time, it does not directly affect sensory memory. This block influences block 1 indirectly through the function of block 7. Block 6 (evaluative and inducing level of motivation) has two sub-blocks. One reflects the evaluation of significance of the situation or its separate characteristics (sense or subjective significance). The other is the inducing or motivational sub-block. This block affects all function blocks of memory. Thus, not only cognitive but also emotionally-evaluative and motivational mechanisms are involved in the selection, storing, and extraction of information from our memory. Feedforward and feedback connections between various blocks demonstrate that our memory functions as a goal-directed self-regulative system. Thanks to self-regulation, subjects can utilize various strategies of selection of information, storing it in memory, and recalling it. Strategies of memory depend on subjects' past experience, individual differences, and specifics of the task. In each block of this model information can be processed sequentially or in parallel.

The model presented here emphasizes the special role of operative or working memory in task analysis. SSAT distinguishes operative or working memory from short-term memory. Operative memory signals mnemonic processes serving activity that are directly involved in the specific task performance. Goal is very important for this type of memory, as well as consciousness or awareness that is a key aspect of this type of memory function. Long-term memory has potential goals that can be extracted from it and transferred into operative or working memory. Connections between blocks 5 and 6 and blocks 3 and 5 demonstrate the importance of intellectual and emotionally-motivational mechanisms in the functioning of memory and specifically of working memory during task performance. Block 3 and 5 are major mechanisms in the development of a mental model of the task or situation. In order to analyze the mnemonic mental model in greater detail, we need to concentrate our attention on operative thinking. Intellectual block 5 is involved not only in the identification and structuring of information on the perceptual level. It is also important in the interaction with operative or working memory, and, in particular, it is involved in classifying data and organizing it in operative or working memory. This block is responsible for developing a strategy of memorizing and reproducing information, the construction of the program of execution of mnemonic actions depicted in block 3. Verbally-logical components of activity are especially important at this stage of memory functioning. Block 5 also plays a role in decision-making related to the implementation of verbal or motor actions during the task performance. We do not consider this aspect of task performance because in developing the memory model we focused on internalization aspects of information processing.

In the considered model intellectual block 5 has various feedforward and feedback connections with block 3 (WM) and block 4 (LTM). According to Norman (Lindsay, Norman, 1977) there are three main components in memory: the database, the interpretive system, and the monitor. The monitor guides the interpretive system and examines the database. This sets up the strategies that are used to assess information. The interpretive system performs examination of the database structure. According to our model, the interpretive mechanism and the monitor cannot be reduced to performing purely mnemonic functions. These mechanisms are tightly connected with intellectual block 5. Using basic terminology of general psychology, this is the thinking processes. Thus, memory cannot be considered an independent system. It includes other mechanisms associated with various cognitive processes that are integrated together into the self-regulative system.

Each functional block presents a basic mechanism that include various psychological operations. The main units of analysis of each function block are cognitive actions and operations, and specifically the mnemonic actions. Such concept as operative units of memory have profound significance in studying memory. Operative units of memory are the semantically holistic entities that are used by mnemonic actions during task performance. Working memory is an important mechanism in the formation of operative units of mnemonic actions. In cognitive psychology, as we have mentioned above, instead of operative units of memory psychologists utilize the term chunks. These two terms have similar meanings. However, there are some differences in the understanding of this term. Operative units of memory are tightly connected with concepts of action and the goal of action. In applied and SSAT there are also such terms as operative units of thinking and operative units of perception. These units have also been called operative units of information (OUI). Operative units of information are formed during acquisition of the specific skills. Such units can be understood as contextually defined entities or items (image, concept, statement, and so on) formed through experience that enables a subject to mentally manipulate them using various cognitive actions. While external motor actions manipulate material objects, cognitive actions manipulate mental items (operative units of information) in accordance to the goal of actions. It is well known that the capacity of short-term memory is between five and nine items when subjects concentrate their attention, and this limited capacity has a short time span. However, information can be stored longer in working memory, which is a combination of short-term memory and long-term memory. Temporary storage of operational information that is required for performance of a specific task is facilitated through continuous interaction of short and long-term memory, and transformation of data from one type of memory to another. Working memory provides chunking material, rehearses it, relates new information that is already stored in memory, structures information, and so on. All this mnemonic activity is performed by using conscious mnemonic or other cognitive actions, and unconscious mental operations. Operational memorization (keeping required information until the task is completed) can be represented as the interaction of working and long-term memory during task performance. Not only voluntary, but also involuntary reproduction of new information is achieved by operative or working memory functions. The reproduction of information depends on the content of the activity performed by the subject. The long-term storage of information is also possible through the functioning of operative or working memory. Voluntary and involuntary memorization and reproduction at the level of operative or working memory depends largely on the goal of the activity undertaken, the significance of information for the subject, strategies of activity, and individual characteristics of the subject.

The significance of each function block depends on the specifics of the task being analyzed. The interaction of blocks 5, 7, 1, and 2 plays the main role when analyzing the relatively simple tasks that require observation of input data and automated operational responses. And when considering the more complex tasks, the interaction of blocks 5, 3, ands 4 plays the leading role. In some cases, the information processed by the subjects during the task performance can have different subjective significance for different individuals. Then, block 6 (evaluative and inducing levels of motivation) is of particular importance in the analysis of memory functioning. The proposed model also shows that memory should be considered as a self-regulative system. Hence, special attention should be paid to the study of the strategies of memory during performance of the specific task. The offered memory model allows presentation of the most important information about how memory works during task performance in a structural way. Through this model, data obtained in psychology about the functioning of the memory can be utilized more efficiently when analyzing the tasks performed by the operators.

CONCLUSION

The activity of operators in the human-machine system occurs in four consecutive stages: receiving information, evaluating and updating information, decision making, and implementing the accepted decision. In this work, we concentrated on how information is received, processed, and stored in memory. So in this chapter we presented two self-regulative models of cognitive processes. One model describes perceptual process and the other one describes functioning of memory. Such models are useful in task analysis. They help to predict preferable strategies of task performance where receiving information and executing mnemonic function are specifically important. In SSAT the basic mechanism of self-regulation models is a functional block. Each block interacts with other blocks through feedforward and feedback connections. The function blocks can be considered as specific stage of information processing. Each block presents a coordinated system of functions that are important at the particular processing stage. Each function block determines the scope of questions that should be studied at this stage of analysis of the considered cognitive process. The main units of analysis of each function block are cognitive actions and operations. For example, sensory actions and perceptual actions are important for the study of perceptual processes while mnemonic actions are most important when studying memory functions. The obtained information can be compared with data received by analyzing other function blocks.

Not all function blocks should always be used in the analysis of cognitive processes. For instance, the task might involve the perceptual process that requires detection of a weak stimulus in the noisy background. This stage may interfere with the receiving of the meaningful stimuli. Here, the analysis of the first function block (detection) is required. Moreover, the detection stage of the perceptual process can arise as an independent stage of analysis. This stage can immerge as an independent task, which is called detection of signal in the presence of noise. However, if the presented signal can be easily detected by an operator the function "detection" can be omitted from the analysis.

Summarizing the study of perceptual process and memory in the framework of the SSAT, we should consider them as specific stages of activity or in some cases as independent types of activity. For example, in mnemonic tasks memory is the main factor in the task performance.

The same as in perception, the mnemonic tasks include goals, emotionally evaluative components, motives, means of memorization, mnemonic, and other cognitive actions. The offered for memorization material is not a stimulus but an object with which the subject interacts in the process of memorizing. Each block of the model of memory performs some function in the task performance. Due to the feedforward and feedback connections this model presents a goal-directed self-regulative system. Each memory block in this model should be primarily analyzed from this theoretical perspective.

The study of cognitive processes and of memory in particular clearly shows the interconnection between cognitive psychology and systemic-structural activity theory.

The models presented here are goal-directed and have emotionally-motivational blocks, feedforward and feedback connections between blocks, and therefore are self-regulative models. The models are adequate to the analysis of receiving stages of information and mnemonic functions during performance of various tasks. Psychological processes, including the stage of receiving information, and processing it in memory, depend on goals, motives, and the specificity of the task. These models help to discover strategies of task performance. Cognitive psychology failed to describe the enormously important role of the subject's activity during the functioning of various cognitive processes and the particular role of a goal, self-regulation, strategies, and interdependence of cognitive processes. In our model, we address these aspects of perception and memory functioning. At the same time, the models presented here, as well as other models of self-regulation developed in SSAT, can be instrumental in task analysis.

REFERENCES

Atkinson, R., Shiffrin, R. M. (1977). Human memory. A proposed system and its control processes. In G. H. Bower, (Ed.) *Human memory. Basic processes.* New York: Academic Press.

Baddeley, A. D. (1976). *The Psychology of Memory.* New York: Basic Books.

Bedny, G.Z., Karwowski, W. 2007. *A systemic-structural theory of activity. Application to human performance and work design.* Boca Raton: CRC, Taylor and Francis.

Bedny, G. Z., Bedny, I. (2018). *Work Activity Studies Within the Framework of Ergonomics, Psychology and Economics.* Boca Raton: CRC, Taylor and Francis.

Bedny, G. Z., Karwowski, W., Bedny, I. (2015). *Applying systemic-structural activity theory to design of human-computer interaction systems.* Boca Raton: CRC Press, Taylor and Francis Group.

Bardin, K. V. (1982). The observer's performance in a threshold area. *Psychological Journal,* 1, 52–59.

Clifford P. J., Tong C. F. (2007). Perceptual and mnemonic contents of mental imagery revealed by binocular rivalry. *Journal of Vision,* 7(9).

Craik, F.I., Lockhart, R. S. (1972). Levels of processing: A framework for memory research. *Journal of Verbal Learning and Verbal Behavior.* 11, 671–684.

Dmitrieva, M. A. (1964). Speed and accuracy of information processing and their dependence on signal discrimination. In B. F. Lomov (Ed.) *Problems of engineering psychology.* (pp. 121–126). Leningrad Association of Psychology Publishers.

Leont'ev, A. N. (1978). *Activity, consciousness and personality.* Englewood Cliffs: Prentice Hall.

Lindsay, P. H., Norman, D. A. (1977). *Human information processing. An introduction to psychology.* Second edition. Harcourt Brace Jovanovich, Publishers. San Diego, New York, London.

Pushkin, V.V. (1978). Construction of situational concepts in activity structure. In Smirnov, A.A. (Ed.). *Problems of General and Educational Psychology* (pp. 106–120). Moscow, Russia: Pedagogy.

Shekhter, M. S. (1967). *Psychological problems of recognition.* Moscow: Pedagogical Publishers.

Sperling, G. (1960). The information available in brief visual presentations. *Psychological Monograph.*74 (Whole No 11).

Sternberg, R. J. (1999). *Cognitive psychology* (2nd ed.). Fort Worth, TX: Harcourt Brace College Publishers.

Strelkov, U. K. (2007). Operationally-meaningful structures of the professional experience. In V. A. Bodrov (Ed.). *Psychology of professional activity.* Moscow: Logos Publisher, 261–268.

Vygotsky, L.S. (1978). *Mind in society. The Development of Higher Psychological Processes.* Cambridge, MA: Harvard University Press.

Zinchenko, P. I. (1961) *Involuntary memorization.* Moscow: Pedagogy.

Zinchenko, V. P., Munipov, V. M. (1979). *Fundamentals of ergonomics.* Moscow: Moscow University Publishers.

Zinchenko, V. P., Vergiles, N. U., Vuchetich, B. M. (1980). *Functional structure of visual memory.* Moskow: Moskow State University Publishers.

6 The Study of Work Motivation and the Concept of Activity Self-Regulation

Gregory Z. Bedny
Essex County College, New Jersey, US

Inna Bedny
United Parcel Service, Atlanta, US

CONTENTS

6.1 INTRODUCTION

In this paper we will consider the concept of work motivation from the systemic-structural activity theory (SSAT) perspective. SSAT is the original high level generality theory. The study of human performance, and work motivation in particular, is central to ergonomics. However, the concept of work motivation is not well developed in this area.

Motivation should be considered the source of energy that drives activity toward reaching its formulated goal. Ergonomists and work psychologists should distinguish individual motives from motivation in general. Motivation has a broad meaning and includes a hierarchy of individual motives that can be conscious, semi-conscious, or even unconscious.

The functions of motives is fulfilled by needs, sets, desires, or the like that are transformed into motives by providing direction toward reaching the goal of activity.

Motivation is always connected with emotions. In cognitive psychology there is no clear-cut distinction between these two notions. However, in activity theory (AT) these two notions are defined differently. In AT, motivation is emotions plus the directness of activity toward a specific goal. In general, motivation should be considered as an energetic component of activity and cognition as its informational component. Usually positive emotions are linked to success and negative emotions are associated with the performance degradation. Nevertheless, in some situations positive emotions can have a negative effect on performance. This can occur when feedback about success is conveyed with very strong positive emotions that influence the activity regulation. The same can be observed when success requires a great deal of effort after which relaxation is necessary.

In some cases, negative emotions can improve performance. This can be observed when failure follows a sequence of successful performances. However, such improvement has a short duration.

This brief analysis demonstrates that emotions cannot be considered just as disorganizational components of activity. In general, emotions perform an important adaptive function in human behavior.

Motives are associated with various needs that can be categorized hierarchically. The basic needs are associated with the *self*: self-preservation, food, shelter, love, etc. At a somewhat less personal level is the need for self-recognition and feelings of worth, and at a more abstract level, a concern for the society in which one functions.

Needs, however, do not directly generate activity; they merely create dispositions that are directed toward their satisfaction. Activity only derives from those cases in which the person consciously imagines specific objects that can satisfy their needs. Such an image then becomes the goal.

In applied and systemic-structural activity theory we distinguish inducing and regulatory components of emotions. Inducing components have just one function: to direct a person to achieve a specific goal. The regulatory components of emotions have four functions: switching, reinforcing, comprehending, and organizing. If an individual has several concurrent motives, the switching function enables them to concentrate on the behavior that is most closely related to the goal of the activity and has the highest subjective value for the individual. The reinforcing function provides rewards, and thus reinforces the desired behavior. For example, rewards increase the response while punishment decreases it. The adequate compensation enables emotions to be transformed into the increased level of motivation. For instance, time constraints increase the emotional intensity of the operator and motivates him or her to mobilize efforts to achieve the desired goal. It results in an increased speed of performance. The organizing function of emotions promotes recognition of any conscious discrepancy between existing and required methods of achieving a goal, and thus leads to an adequate organization of activity. This function is connected with the selection of correct strategies of activity depending on the emotional state of the person and the requirements of the task.

Therefore, activity is the function of the interaction of an individual and a situation in which this individual is involved. The significant cause of activity is the interaction of cognitive and motivational factors. This paper is dedicated to these aspects of activity.

6.2 GOAL AND MOTIVES IN COGNITIVE PSYCHOLOGY AND IN ACTIVITY THEORY

According to AT, a goal is connected with motives and creates the vector motive $\rightarrow$ goal that lends activity its goal-directedness.

The more significant the goal is, the more value it has for the person, and the more the person is motivated to reach this goal. If a situation has negative

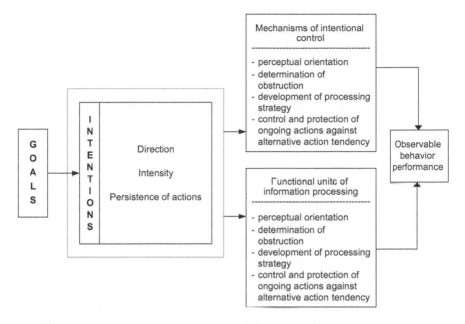

FIGURE 6.1   Concept of goal in cognitive psychology.

meaning for a person, they will be motivated to avoid such a situation. Cognitive psychology and AT have somewhat different conceptions of the attributes a goal possesses. Locke and Lathman (1984) stated that a goal has both cognitive and affective features, Pervin specified that a goal can be weak or intense, and Locke and Lathman (1990) also considered a goal to be a motivational component of behavior. Presumably, the more intense the goal is, the more one strives to reach it. Hence, a goal pulls activity.

The difference in the understanding of the role a goal plays in the structure of activity is clear in the work of Kleinback and Schmidt (1990). They described the volitional process, and considered a goal to be a source of inducing behavior (Figure 6.1). The direction of the volitional process goes from a goal to behavior. These authors do not separate goal from motives.

In AT the goal is only an informational component that does not have intensity. It can be precise and detailed or unclear and general, but not intense. From the SSAT point of view, the more intense the motive is, the more efforts will be expended to reach the goal (Figure 6.2).

FIGURE 6.2   Concept of goal in activity theory.

Energetic and informational components of activity are interconnected, but they are not similar. In AT the difference between informational and energetic components of activity has both theoretical and practical meaning. Motivation is viewed as a dynamic component of activity.

6.3   BRIEF DESCRIPTION OF THE MOTIVES OF WORK ACTIVITY

There are various classification systems for the motives of work activity. We will describe one such useful system that has been created by Tomashevsky (1971). He identified five basic types of motives:

1. **The motive to obtain profit.** This has the purpose of being rewarded for the results of one's efforts. It can be material profiting such as salary, wages, awards, etc., and social profiting such as self-esteem, social appreciation, or popularity. The clear connection between the result of the efforts and the gained profit raises the level of motivation. It is important to inform the employee about the results of their work, and for this information to be properly distributed in time.
2. **The motive of being safe.** The employee is motivated to avoid danger while working. The danger can be physical (harm to one's health), the danger of losing wages or rewards, and the social danger of losing respect.
3. **The motive of convenience.** This is the motive to choose the easiest way of task performance, i.e., spending the minimum of physical and mental effort. Often, such methods are not the easiest ones but they are the most adequate formed by an employee's skills. The individual means of safety are often not evaluated by their efficiency but rather by how convenient it is to use them.
4. **The motive of satisfaction.** This motive is to gain satisfaction from the process and results of work. It depends on an individual values, interests, etc. Such a motive is especially strong when the profession matches the interests of the individual and is evaluated as a prestigious one. In such cases, some components of activity can give better satisfaction than the others.
5. **The motive of leveling with others.** This is striving to perform on the same level as the other members of the group. The opinion of the group is more important than reward or punishment. The employee wants to be on the same level as the other members of their group in terms of performance.

For different individuals some motives are more important than others. The role of each motive depends on its importance for the individual. If an employee does not clearly understand the danger of their work, does not appreciate the role and importance of the safety tools and measures, the usefulness of safety measures is lowered.

Skills influence the strength of motives because advanced skills make it easier to fulfill motives. If the motivation is strong enough, the undesirable events seem less probable than they really are. Strong motivation can distort the adequacy of reflection of the real situation (Mcclelland, 1953).

6.4   THE CONCEPTS OF MOTIVATION
IN ACTIVITY THEORY

An image or a verbal representation of a future desired result does not necessarily constitute a goal but rather emerges as a goal only when it is joined with motivation and when a subject is involved in an activity for achieving this result. For example, a student wants to get a good grade. Such a desire does not create a goal and is not a motive. Only when a student starts to perform a number of tasks will they eventually be able to reach a final goal. Therefore, only when a future desired result is joined with motivation will this imagined result be transformed into a goal, and desires, wishes, etc., become motives. It is clear that the goal does not exist apart from the motivation. There are no unmotivated goals. Goals and motives are interdependent, but not the same. The vector motive → goal that they create is the critically important mechanism of goal-directed activity.

We view vector motive → goal as a source of purposeful human activity. The goal cannot exist outside of this vector. Each component has its own specificity in human activity. The vector can be viewed as a two-component sub-system of activity that defines not only the course of activity, but also the energetic characteristic of its orientation. The ability to maintain performance over time, and the resistance confounding factor are determined by this vector.

The concept of "spatial vector" or simply "vector" is mainly used in physics and engineering to represent directed quantities. Multiple physical quantities and directions can be represented by the length and direction of an arrow. Their magnitude and direction can be added to other vectors according to the rules of vector algebra. Of course, it is important to distinguish understanding of the vector in psychology from its meaning in physics or engineering. Similarly to a vector in physics, a psychological vector has intensity and direction. Moreover, motivation can be metaphorically represented as a resultant vector. So if we assume that human activity is determined by the interaction of two motives that have opposite directions, then there is a "conflict of motives." Directness toward a goal or avoidance of one depends on the intensity of a corresponding motive. Naturally, unlike in psychology, in physics a vector has physical units of measure. Motive and motivation in psychology are hypothetical constructs. Hence the vector motive → goal is a qualitative psychological concept. It has only some similarity with a vector in physics. It reflects the intensity of activity directness to the achievement of a certain goal and an ability to sustain this directness. In the absence of such a vector, goal directness of activity disappears. Therefore, the vector motive → goal is one of the basic concepts of AT. It clearly manifests the interaction of energetic and informational components of activity. We can outline the following critical points of this important concept: motive and goal are not the same; motives are energetic and goal is a cognitive mechanism; the more intensive the motive, the stronger the subject's desire to reach the goal; the vector motive → goal gives activity its directed character, and sustains this directness in time, etc.

In AT, goals and motives are considered interrelated mechanisms. However, even in AT, where these two concepts have fundamental meaning, some leading scientists still misinterpret their analysis. For example, Leont'ev (1978) wrote about the possibility of a shift of the motives to the goal of activity. According to Leont'ev, in such cases the motive and goal coincide. He stated (1981) that when a person realizes a conscious motive, the goal might overlap with the motive. Therefore, in this situation the vector orientation of activity is lost. Lomov (Lomov, 1981; Lomov Ed., 1986, p.173) also claimed that under certain conditions, motive can be shifted to the goal. However, this contradicts the notion of a vector motive → goal. Shifting of the motive to the goal leads to this vector shrinking and becoming a point, and the activity goal directness is lost. Information and energy are interdependent but not the same. Why a person performs the action and what they want to achieve during performance of the action is not the same. If a person wants to drink water and there is different tableware on the table, they move their hand to the glass, not to the fork, because the person is keenly aware of the desire to drink water. The thirst has been transformed into a conscious motive. In this situation the goal of the motor action (grasp the glass) and conscious motive (I want water) are not the same.

Freud (1916/1917) was the first to introduce a concept of energy into psychological studies. According to Freud, people are complex energetic systems. Energy is necessary for the functioning of mental processes. The idea that energy is an important component of mental functioning has been borrowed from biology and physics. However, interaction between energetic and informational aspects of human functioning has not been extensively covered in Freud's works. Today such concepts as energy and information are clearly defined in psychology. Information processing is a cognitive function. The concept of energy that has its roots in the neural system electro-physiological function has been utilized in the studies of emotionally-motivational aspects of behavior or activity. Although these two concepts are distinguishable phenomenon in AT, they are considered to be tightly connected. There are several types of informational-energetic interconnections in human activity (Vekker, Paley, 1971). The first type has been shown in psychophysical studies where it has been discovered that an increase in the intensity of external stimuli causes the rise of sensory qualities. Another type of connection is related to the reticular activating system of the brain (RAS), which is responsible for regulating wakefulness and sleep-wake transitions. For instance, Kahneman's (1973) view on attention includes the concept of energy. The success of cognitive processes depends not only on physical characteristics of the information presented to the subject, but also on the level of activation of their neural system. A third group of interconnection is linked to the emotionally-motivational components of activity and to the specificity of information processing. These groups are codependent. Currently, cognitive approach dominates in ergonomics, where a person is considered purely an information-processing system that picks up information from external situation and memory, because the relationship between cognition and the energetic aspects of activity is not sufficiently studied. However interpretation of information also depends on the emotionally-motivational state of a person. Moreover, information can't be transmitted without energy.

Motivation is an intentional or inducing component of activity that is tightly connected with human needs. A need is an internal state of individuals that is less than satisfactory and produces a feeling of a desire for something (Carver, Scheier, 1998). Some needs are biological in nature. Other needs such as achievement, power, etc.,

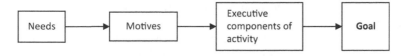

FIGURE 6.3   Relationship between needs, motives, and goal in the activity structure.

are secondary or psychological needs. Human needs are not just a result of biological evolution, but also of an experience acquired through human culture.

If needs are connected to the goal, they are transferred into motives of activity. Motives are an inducing force that catalyzes the person's desire to reach the activity goal. So needs operate through the construct called motives. Motives derive from needs but are closer to our activity or behavior. For example, a need for food as a physiological state can be transferred into a motivational state called hunger. This state is experienced cognitively and affectively. The external stimuli creates a desire to obtain or avoid something. If someone has received recognition, this can trigger their own motives to gain further recognition. Hence, internal needs can activate motives to engage in a particular kind of activity to achieve a conscious goal. The relationship between internal needs, motives, and a goal can be presented as follows (Figure 6.3).

Motivation is another important construct that includes diverse motives that have a hierarchical organization. Some motives can be more important than others. Some of them are conscious, other semiconscious or even unconscious. The relationship between motives is typically dynamic and is modified during activity. The goal is often induced not by one, but by several motives. The activity result can coincide with the goal or deviate from it, and a subject has to correct their actions or activity strategy.

Deviation of the result from the goal can be considered an error if this deviation is outside of the subjective criteria of success, which is not always the same as the objective one. Sometimes, activity result that deviates from its goal can be useful for a subject when this result is a desired but unexpected one.

The principle of the vector motive → goal functioning can be summarized as follows: When a person formulates a goal, the motivational tension is set up. Once the subject reaches the goal, the ongoing motivational tension is relieved. Motivational tension is analogous to the forces that direct our activity to reach the goal. Activity can be habitual and does not require significant motivational forces to reach a goal. A person might be unaware of these motional forces. Motivational forces are capable of automatically activating human activity. A person can be aware of activity's goal without consciously realizing what the motivational factors are. The conscious goal-directed activity is a voluntary process. A latent or potential goal can be activated by a situation and arise immediately in our consciousness. This is an involuntary goal formation process. In human activity, the goal formation process is mostly the voluntarily conscious one. In some cases, conscious formulation of a goal can arise as an independent task. In contrast, automatic human behavior can be triggered outside of goal awareness. Motivational tendencies can be directly activated by the environment. As a result, an automatic behavioral response is triggered by environmental situation without consciousness.

There is another aspect of motivation that is associated with the role of emotions in the motivational process. According to some authors (Zarakovsky, Pavlov, 1987), emotions reflect the relationship between our needs, and real or possible success in satisfying them.

Motivation is distinguished from needs, wishes, desires, intentions, etc., by its connection with the goal of activity. Not just human needs, but also wishes, desires, intentions, etc., are transformed into motives when they are joined with the goal.

A person's motives can be divided into two groups: sense formative and situational (Leont'ev, 1978). Sense formative motives are relatively stable and determine a person's general motivational direction. Situational motives are associated with immediate ongoing activity and performance of a specific task. So situational motives are more flexible. The sense formative motives are tightly connected with the features of personality and might be important in the professional selection of people for various jobs. The situational motives are involved in task performance. Therefore, they are central in task analysis.

Motivation is always connected with emotions. Some scientists write that there is no clear-cut distinction between emotions and motivation, and it is an unresolved issue in psychology. However, according to the activity approach, motivation is emotions plus the directedness of activity to achieve a specific goal. Emotions can be positive or negative. However, even negative emotions cannot be considered as only disorganizational factors. In some situations negative emotions can improve performance. Emotion according to Simonov (1982) is a specific "currency" of brain or a universal measure of utility. Our feelings (happiness, anger, outrage, etc.) present quality and level of our needs in relation to a possibility of satisfying them. These characteristics of motivation can be measured.

Zarakovsky (Zarakovsky, Pavlov, 1987) described functions of emotions in activity regulation. Inducing components of emotion has only one function: to direct a person to achieve a conscious goal. The regulatory aspects of emotions have four functions: switching, reinforcing, compensation, and organization. The switching function enables a person to concentrate on activity that is most closely related to a goal that has the highest subjective value for a particular person. The reinforcing function provides rewards, and thus reinforces the desired behavior. The compensative functions facilitate changes in emotional tension and therefore allows transfer into a higher level of activity regulation. However, this can sometimes result in activity disorganization. The organizational function of emotions contributes to the creation of an orienting reaction and reflection of mismatch between available and required ways of activity organization.

The presented analysis of relationship between the goal and motives demonstrates that actions during task performance can be successful and unsuccessful. Each action can be evaluated in relation to the goal of action and in relation to the goal of the task. Evaluation of actions or activity always includes emotional components or subjective significance of the obtained result. From an AT perspective, the task always includes a motivational component and there is no such a thing as an unmotivated task performance.

Some authors use such terms as "anti-goal" or "work avoidance goal" (Carver, Scheier, 2005; Pintrich, 2003, etc.). There are no goals of this kind in AT. A subject may or may not accept a formulated by instructions goal. Moreover, in response to a presented goal a subject can formulate their own goal, which contradicts the objectively presented goal. "Anti-goal," "work avoidance goal," etc. are just new goals, and thus a subject formulates a plan for the new type of activity or task, and a new vector motives → goal that determines the direction of activity is formed. One goal can contradict another, but this is not an "anti-goal," because such a goal is created based on the same mechanisms.

Let us consider a hypothetical example. There is a family with children and the father of these children is a gambler. Sometimes after receiving his paycheck, the father spends all his money in the casino. At the same time, he feels responsibility for the family and tries not to spend the money and instead keeps it to take care of his family. Here, we can distinguish between two types of motivation and two goals. The positive motivation and the first goal create the vector motive(s) → goal 1 (keeping money for the family). The negative motivation creates another vector motive(s) → goal 2 that directs the father's activity to spending money in the casino. This situation is represented by two vectors (see Figure 6.4).

One can see that the above two vectors have opposite directions but are mutually exclusive. The second goal has its own vector and cannot be considered an anti-goal. The mechanisms of goal formation of this vector are the same. The difference lies in the fact that the significance of goals for a subject, and motivation to achieve them, is different.

The process of goal selection according to Zarakovsky (Pavlo, Zarakovsky, 1987) with our modification can be presented in the following manner. The long-term memory contains units of memory (engrams) of intentional type. They can be considered as vectors that are designated as need → potential goal engrams of memory. A set of such vectors has internal energy or inducing forces that depend on the intensity of needs. Similarly to psychophysics, vector need → potential goal engrams may be of higher or lower level than the existing absolute threshold according to their energetic characteristics. If need → potential goal engrams are above the existing energetic threshold then a person is aware of their feelings or desires to act toward achieving a potential goal. There is a possibility that not one but several potential vectors may exceed the threshold level. Which of the vectors will dominate depends on the difference in an engram's potential of specific vectors. It is possible to evaluate engrams' potential of such vectors utilizing psychophysical methods. Without going into the psychophysical analysis of this issue, we

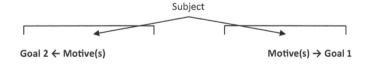

Subject

Goal 2 ← Motive(s)   Motive(s) → Goal 1

FIGURE 6.4   Formation of two opposing goals.

just want to state that if a person has two divergent goals, need → potential goal at the same time and both of them reach sufficiently high activation level, a subject can be well aware of them. In such cases we can use the term vectors motive → potential goal.

Of course, such difference in terminology have some arbitrary distinction. Nevertheless, this difference can be useful. In the first case, the goal might be unconscious or not sufficiently conscious. In the second case, we speak about conscious goals or even well-realized motives. If two motive → potential goal vectors have approximately the same intensity, occur at the same time, and have potential high intensity engrams, this results in a conflict of motives. When one vector begins to dominate over the other, a person starts to act in accordance with the chosen vector. The selected vector need → potential goal is transformed into motive → potential goal, and at the final steps the predominant vector is transformed into the vector motive → goal and a subject starts to act according to this vector.

The misinterpretation of goal and its relation to the motive when the goal is mixed with motives can be observed in the following hypothetical example. If a thief spots a hundred dollar bill in somebody's pocket his goal is to pull it out as gently as possible so the victim doesn't realize it. Obtaining this money is his motive but not a goal when he is in action. The more money in the victim's pocket and the more the thief needs the money, the more the thief is motivated to steal. "Obtaining the money" is simply the verbalized description of one possible motive in this example. Another motive may be, for example, the desire to demonstrate his "perfect skills to another craft-brother." Refocusing attention of the thief on the fact that he'll get money, but not on the goal of the performed actions, can lead to inaccuracies in performance of actions, and therefore in undesirable consequences.

When a computer programmer works with an issue that requires fixing a production database, their goal is to make the corrections without damaging sensitive information while the motive is to resolve the issue at hand. Such a task may be accompanied by great emotional tension, because an error can lead to severe undesirable consequences. The more significant the task is, the higher the subject's motivation to reach the goal.

The indistinctness of the term goal in cognitive psychology results in suggestions to eliminate this concept altogether. For example, Diaper used an example to prove his position that the human goal is a redundant concept. He wrote (Diaper, 2004, p.17):

> A hierarchy of goals, as used in HTA, consists of multiple related goals, but a person can also perform an action on the basis of unrelated goals. Furthermore, unrelated goals that nonetheless motivate the same behavior cannot be simply prioritized in a list, because different goals have more or less motivational potency depending on their specific context.

For example, a chemical plant operator's unrelated goals for closing a valve might be 1) to stop the vat temperature rising; 2) to earn a salary, and 3) to avoid criticism from the plant manager. The first might concerns the safety of large numbers of people, the second is socio-psychological and might concern the operator's family responsibilities, and the third is personal and might concern the operator's self esteem. These three goals correspond to different analysis perspectives, the sociological, the socio-psychological, and the personal psychological; and there are other possible perspectives as well. Furthermore, people might have different goals within a single perspective.

The above statement demonstrates that the concept of human goal is confusing for HCI practitioners. What is an unrelated goal, for instance? Such a notion does not exist in AT. It is not Diaper's fault that he confuses the goal with motives because the concept of goal is not precisely defined in psychology and ergonomics, and goal and motives are not distinguished. Our behavior is poly-motivated and can include a number of motives. If the goal includes various motives, according to cognitive psychology, a person pursues not one but multiple goals → motives at any given time during task performance.

Per SSAT, the goal is an informational or cognitive component of human activity. In contrast, motives or motivation is an energetic, inducing aspect of activity. Therefore, in the above example the goal of human activity is "closing a valve." In contrast, the motives, which push the operator to close the valve, may be 1) to stop the vat temperature rising, 2) to earn the salary, and 3) to avoid criticism by the plant manager. The listed above wishes or desires (which are described verbally) are transformed into motives through their connections with the goal. They create motivational inducing forces to reach one single goal: closing a valve.

The example above demonstrates that including motivational components into a goal of a task or action makes it difficult to perform task analysis in the specific settings. Hence, Diaper and Stanton (2004) suggest eliminating the concept of human goal from the task analysis altogether. The authors suggest substituting the concept of goal, and connected motives, with their method called forward scenario simulation (FSS) approach. As a result, the critically important area of psychology that is involved in studying anticipation, forecasting, formation of hypothesis, goal formation, etc., would be entirely ignored. The traditional area of ergonomics that utilizes cognitive approach does not consider the concept of goal at all.

Analysis of the material presented here brings us to the following conclusion. The vector motive → goal is the basic concept of AAT and SSAT. However, according to SSAT, a goal is an informational and motivation an energetic component of activity. Therefore, motives cannot move to the goal, because they are different mechanisms of activity regulation. If we accept the statement that this vector disappears when motives move to the goal then the goal directness of activity is also eliminated. Integration of motives with goal as done in cognitive psychology results in one task having multiple goals. The task analysis is then impossible to conduct. Therefore, experimental cognitive psychologists do not use the concept of goal in task analysis. This term has some common meanings in their opinion. The traditional interpretation of goal eliminates the concept of a goal-directed process of activity self-regulation. Self-regulation can be considered only as a homeostatic process, the approach that has been criticized by various authors in personal and social psychology.

6.5 MOTIVATIONAL ASPECTS OF SELF-REGULATION

The study of motivation helps to understand how a subject accepts or formulates goals and pursues achieving them. Here we demonstrate how analysis of self-regulation helps to understand the motivational process. The notion of self-regulation, which is the foundation of functional analysis, links motivation, cognition, and behavior into a unitary system.

Traditionally, motivation is considered an important concept in the personnel/organizational area of industrial/organizational psychology, but motivational factors are practically ignored in ergonomic studies. However, from the self-regulation point of view, motivation influences interpretation of the meaning of events. Therefore, motivation is critically impotent for accurate understanding of the human informational processing. Cognitive processes should be considered in unity with motivation. Motive and goal form goal-directed tendencies of the self-regulation process until the self-regulative cycle is completed.

Functional analysis that considers activity as a self-regulative, goal-directed system pays attention to situation specific aspects of motivation where conscious and unconscious motivational components interact with each other. General needs, conscious, and unconscious motives influence human activity through their integration with the situation-specific conscious goal, or unconscious set and reflection of the situational requirements.

In SSAT, motivation is considered a goal-directed process with various cognitive mechanisms that have feedforward and feedback connections. An advanced model of activity self-regulation that depicts these connections can be found in G. Bedny, I. Bedny (2018). This model includes goal, set, assessment of task difficulty, etc. In this model a goal performs the integrative function.

For example, according to this model the essential purpose of the mechanism, assessment of task difficulty, is to evaluate how difficult the task would be for a person who has to perform it. An individual may under- or overestimate the objective complexity of the task. If the difficulty of the task is overestimated, the task can be rejected by an individual while objectively they would be able to complete it. On the other hand, an individual can underestimate the difficulty of the task, and as a result select inadequate or inappropriate strategies for the task performance, thereby failing to solve the problem.

The concept of difficulty can be approached from two different points of view. It can be studied as a characteristic of the task or as a functional mechanism of self-regulation. In the first case, the concept of difficulty becomes important for evaluation of a task complexity, which is the objective characteristic of the tasks. The more complex the task is, the higher the probability is that it is going to be difficult for the subject to perform. The concept of difficulty can also be considered as the functional mechanism of self-regulation that influences the strategies of task performance. Then it is important to evaluate when a person believes that they possesses the necessary abilities and experience to accomplish the goal of the task.

Therefore, self-concepts of abilities, self-efficacy, etc., are imperative for functional analysis. Bandura (1982) considered self-efficacy to be an important mechanism of motivation. This means that self-concept of abilities, self-efficacy etc., and their relation to task are important for the mechanism, assessment of task difficulty. At the same time, evaluation of task difficulty or difficulty of the goal attainment does not predetermine the motivational processes. Considering that the mechanism, assessment of task difficulty, has complex relations with other mechanisms, no simple derivation about the motivation level can be made. Incorrect assessment of difficulty can result in inadequate personal sense or motivation to sustain the efforts for completing the task.

6.6   MOTIVATIONAL STAGES OF SELF-REGULATION

The model of self-regulation allows identification of various stages of the motivation process. Each stage has its own specific features. Our model outlines five stages that are listed below.

Motivation is important not only at the conscious, but also at unconscious levels of self-regulation. At the unconscious level information about external situations can interact with current needs that govern motivation as well as with set, particularly with those aspects of set that depend on instructions (goal-directed set). A mental set can be formed based on the analysis and integration of the information. An important aspect of this mechanism is the formation of motivational tendency that makes activity a goal-directed process. This motivational tendency, which is also called the preconscious motivational stage, can trigger executive aspects of self-regulation or initiate the formation of the conscious goal.

The second stage of motivation is involved in the formation or acceptance of the conscious goal. This stage can be developed after the preconscious stage of motivation is formed, and the set is transformed into a conscious goal. The goal formation or the goal acceptance are accomplished via the feedback process. This stage of motivation is also called the goal-related stage. The preconscious stage of motivation cannot just precede other motivational stages, but also function in parallel with them. Furthermore, the goal-related stage of motivation could be transformed into a preconscious motivational stage. This stage involves transformation of the conscious goal into an unconscious set.

The third motivational stage is associated with the evaluation of difficulty and significance of a task. This motivational stage is called task-evaluative aspects of motivation.

The fourth stage is related to the executive aspects of motivation and is associated with the goal attainment process. This stage is called the executive or process related stage of motivation. It is associated with task performance. At this stage of motivation, interaction between difficulty, sense of significance, and formation of a program of performance, making decisions about corrections, and program of performance are the most important ones.

The fifth stage of motivation is related to the evaluation of the activity result (result-evaluative stage of motivation). At this motivational stage, analyses of the relationship between negative and positive evaluation of a result and their connection with motivation play a leading role. Analysis of this stage of motivation also involves studying the connection between motivation and the subjective standard of a successful result.

All stages of motivation are intimately connected and can be in agreement or in conflict with each other. For example, positive or negative evaluation of the activity result has meaning for a person only if they are motivated to achieve a goal. Similarly, motivation to achieve a goal may be unrelated to the executive or process related stage of motivation. In this case positive motivation to achieve a goal may be combined with a negative motivation for task performance, which creates a conflict or contradiction between different motivational stages. One often encounters this in both work and learning activity. Positive motivation for goal attainment can be combined with negative motivation associated with the assessment of the task difficulty. This can lead to a rejection of a desired goal. Overestimation of one's own result can lead to ignoring external evaluation, thereby losing valuable information.

There is a complex relationship between difficulty and motivation. An increase in difficulty of a goal attunement does not always increase the level of motivation as stated in the goal-setting theory (Lee, Locke, Latham, 1989). The level of motivation depends on a complex relationship between the assessment of difficulty and the assessment of the sense of task (task significance).

The mechanisms assessment of difficulty and assessment of the sense of task facilitate explanation of the strategies of task performance. Complexity is the objective characteristic of task, and difficulty its subjective characteristic. These function blocks demonstrate that in general, the more complex the tasks are, the more difficult they are to perform.

The relationship between difficulty and significance is presented in Figure 6.5.

Task difficulty and significance can change from a very low level to a very high level, hence the relationship between difficulty and sense of significance can vary. For example, task difficulty can be evaluated as very high and significance as very low, as depicted on Figure 6.5a.

Then a person would not be motivated to perform the task because they would not have any reason to waste a great deal of effort on a task that has no significance for them.

If the difficulty is very low and significance is very high this can produce a sufficient level of motivation for task performance (see Figure 6.5b).

If difficulty and significance are low but not extremely low, a person is still motivated to perform a task but with very low motivation (see Figure 6.5c). This kind of work is usually perceived as monotonous and boring. So various correlations between difficulty and sense of significance can produce different motivational states.

Bandura (1977) developed the construct of self-efficacy, according to which all motivation manipulations are effected through self-efficacy. According to this author, the higher the self-efficacy the higher the motivation.

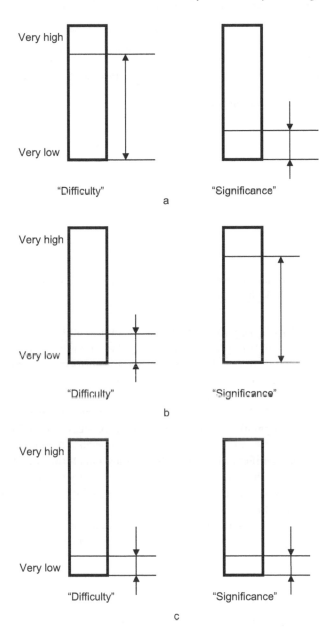

FIGURE 6.5   Relationship between difficulty and significance.

From a self-regulation point of view, if a person evaluates a goal as being not significant and very difficult due to their low self-efficacy, the resulting negative influence on motivation increases the probability that the goal will be avoided. On the other hand, if the goal is significant or highly desirable, those with low self-efficacy can nevertheless be motivated to reach such a goal. In certain situations, high

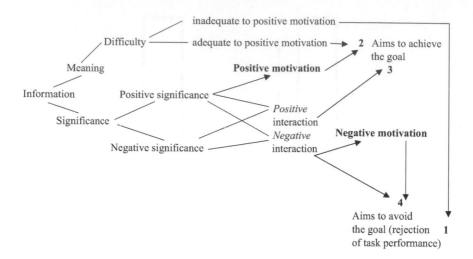

FIGURE 6.6 The role of relationship between positive and negative significance, and adequate and inadequate difficulty of task in motivation.

self-efficacy can have a negative effect. For example, athletes that regard their self-efficacy as very high might underestimate the strengths of their opponents, diminishing their motivation and consequently their performance. We can see the notion of self-efficacy does not fully explains motivational processes. Similarly, it can't be explained just by specificity of the goal formation process. Motivation is a dynamic process that depends on the complex relationship between various mechanisms of self-regulation. Finally, motivation process cannot be understood or analyzed outside of the social context in which it occurs. Accepted by a subject, goals become more significant to him or her if they match the socially developed norms and standards of the community in which the activity takes place. Therefore, goals that are acceptable by the social norms and standards are accompanied by strong motivational factors.

The figure above depicts the relationship between positive and negative significance, and adequate and inadequate difficulty of task in the motivational process.

This figure depicts four situations marked by bold numbers that occur with various probabilities:

1. Difficulty of task is inadequate for a subject; it results in *negative motivation;*
2. Difficulty of task is adequate for a subject and it has only positive significance; result—*positive motivation;*
3. There is interaction between positive significance and negative significance, and this interaction is positive when difficulty is adequate for a subject; result—*positive motivation;*
4. There is interaction between positive and negative significance, and this interaction is negative, *even if difficulty is adequate for a subject;* result—*motivation is negative.*

The above model demonstrate that presented to a subject, information has meaning and significance. These two factors influence interpretation of information and should be considered in unity.

CONCLUSION

There are various concepts of motivation that are not considered here in detail because the main topic of this discussion has been to study motivation from the systemic structural activity theory perspective.

The main components of motivation are energetic function that determines the intensity of motivation; directing function that guides the behavior in a specific direction; and maintaining function that sustains behavior until the activity is completed.

Systemic-structural activity theory sees difficulty and the significance of tasks as key factors of the subject's motivation.

The notion of motivation is tritely connected with the concept of needs that derive from motivation but should be distinguished from motives. Needs are the source of motivation but they are still not motives. Only when needs are connected with a goal are they transformed into motives. These concepts are clearly defined and differentiated only in activity theory.

This paper describes the relationship between goal and motives, stages of motivation and its role in the activity self-regulation, regulatory aspects of motivation, and stages of work motivation.

REFERENCES

Bandura, A. (1982). Self-efficacy mechanism in human agency. *American Psychologist,* 37(2), 122–147.

Bandura (1977). *Social learning theory.* Englewood Cliffs, NJ: Prentice Hall.

Bedny, G. Z., Bedny, I.S. (2018). *Work Activity Studies Within the Framework of Ergonomics, Psychology and Economics.* Boca Raton: CRC, Taylor and Francis.

Boekaerts, M., Pintrich, P. R., Zeidner, M. (Eds.). (2005). *Handbook of Self-Regulation.* San Diego, California: Academic Press, pp. 303–341.

Carver C. S., Scheier, M. F. (1998). *On the self-regulation of behavior.* New York: Cambridge University Press.

Carver C. S., Scheier, M. F. (2005). On the structure of behavioral self-regulation. In (Eds.) M. Boekaerts, P. R. Pintrich, M. Zeidner. *Handbook of self-regulation*: Academic Press, pp. 42–84.

Diaper, D. (2004). Understanding task analysis for human-computer interaction. In D. Diaper, N. Stanton (Eds.). *The handbook of task analysis for human-computer interaction.* Lawrence Erlbaum Associates, Publishers, Mahwah, New Jersey, pp. 5–47.

Kahneman, D. (1973). *Attention and effort.* Englewood Cliffs, NJ: Prentice-Hall.

Kleinback, U., & Schmidt, K. H. (1990). The translation of work motivation into performance. In V. Kleinback, Quast H. H., Thierry, H., H. Hacker (Eds.), *Work Motivation* (pp. 27–40). Hillsdale, NJ: Lawrence Erlbaum Associates, Inc.

Lee, T. W., Locke, E. A., & Latham, G. P. (1989). Goal setting, theory and job performance. In A. Pervin (Ed.), *Goal concepts in personality and social psychology* (pp. 291–326). Hillsdale, NJ: Lawrence Erlbaum Associates.

Leont'ev, A. N. (1978). *Activity, consciousness and personality.* Englewood Cliffs. NJ: Prentice Hall.

Locke and Lathman (1984). *Goal setting: A motivational technique that works*. Englewood Cliffs, NJ: Prentice Hall.

Locke, E. A. & Latham, G. P. (1990). Work motivation: The high performance cycle. In V. Kleinbeck, et al. (Eds.), *Work Motivation* (pp. 3–26). Hillsdale, NJ: Lawrence Erlbaum Associates.

Lomov, B. F. (1981). Toward the problem of activity in psychology. *Psychological Journal*. 2, 5, 3–22.

Lomov, B. F. Editor, (1986). *Engineering Psychology Textbook*. Moscow, Higher Education. p.173.

McClelland D.C., Atkinson J.W., Clark R.A., Lowell E.L. (1953). *The achievement motive*. New York: Appleton-Century-Crofts.

Pervin, L. A. (Ed.) (1989). *Goal concepts in personality and social psychology. Lawrence*, Erlbaum Associates, Publishers. Mahwah, New Jersey.

Pintrich, P. R. (2003). A motivational science perspective on the role of student motivation in learning and teaching contexts. *Journal of Educational Psychology*, 95, 667–686.

Simonov (1982). Need-informational theory of emotions. *Problems of Psychology*, 6, 44–56.

Tomashevski, T. (1971). People in the work environment. In "*Ergonomics*," pp. 106–121.

Vekker, L.M., and Paley, I. M. (1971). Information and energy in psychological reflection. In B. G. Anan'ev, (Ed.), *Experimental Psychology*, 3, 61–66. Leningrad: Leningrad University.

Zarakovsky, G. M., & Pavlov, V.V. (1987). *Laws of functioning man-machine systems*. Moscow: Soviet Radio.

Section II

Efficiency of Performance
and Quantitative Methods
of Its Assessment

7 Evaluation of the Factor of Significance in Human Performance (Applied Activity Theory Approach)

Alexander M. Yemelyanov
Georgia Southwestern State University Georgia, US

Alina A. Yemelyanov
Georgia State University, Georgia, US

In loving memory of our father and grandfather Mikhail Kotik

CONTENTS

7.1   INTRODUCTION

This current work presents an evaluation of the factor of significance in human performance with the aim of subsequent analysis of people's attitudes toward emotionally-driven events. The described psychological experiments have been performed by Tartu University professor Mikhail Kotik (1974–1995). These experiments, which possess a number of theoretical and practical applications, were primarily published in the Russian language, and therefore remained largely inaccessible to English-speaking audiences. The purpose of this current work is to try to remedy this situation. It is worth noting that these experiments were initially designed to study human-operator performance (based on the actual practices of combat pilots) in situations that may pose a threat towards life or health (physical risks). Subsequently, along with physical risks, social and material risks began to be analyzed, as well. In addition to the evaluation of the factor of significance of anxiety-inducing events, further experiments evaluated the factor of significance of attractive or appealing events in social, material, and physical aspects. The final series of experiments linked both of these factors of significance for determining an individual's attitude toward anticipated events, as well as the decisions that are made in light of them. This was possible due to the application of "soft" verbal characteristics, such as weak-strong and seldom-often, in experiments. It turned out that, depending on which soft characteristics the individual uses to describe an event, it is possible to make conclusions regarding how this event is perceived. Overall, these studies have experimentally identified and graphically presented the general principles of people's attitudes toward emotional events regardless of their physical, social, or material nature. To establish validity and illustrate reliability, further experimental studies were conducted.

Mikhail Arkadyevich Kotik was one of the leading engineering psychologists in his field who stood at the forefront of the creation and development of applied activity theory (Bedny and Meister, 1997). His primary research avenues were associated with the psychological aspects of reliability and safety of human activities, in which he created a new direction of research: the psychology of activity safety.

It should be noted that Kotik's research was carried out as part of an applied activity theory, which preceded the emergence of the systemic-structural activity theory approach (Bedny, Karwowski, and Bedny, 2015) for studying activity theory. Therefore, the meaning that Kotik applied to understanding significance differs from the one adopted in SSAT. Thus, in his studies, the factor of significance is considered to be a source of emotional tension that creates motives for corresponding activity. This means that such significance already includes the factor of difficulty, and according to SSAT, determines the level of motivation needed to achieve the goal of the activity. Therefore, when studying the works of Kotik, it is necessary to bear in mind such an expanded interpretation of the factor of significance, as well as the significance-as-anxiety and significance-as-value it incorporates, which are connected to SSAT's negative and positive motivation, respectively.

The advantages offered by this approach include its ease of implementation in practice and of the resulting data interpretation, as well as its ability to assess not only people's conscious emotional manifestations, but their subconscious motives, as well. In the compilation, systematization, and interpretation of these experimental findings, which have been presented in numerous publications up to this date, their distinct language, examples, and logic style have been preserved as much as possible to maintain his original format and style. The list of cited literature has been left without considerable changes, as well.

7.2   SIGNIFICANCE AS AN EMOTIONAL ASPECT OF SENSE

Kotik (1994a) elaborated that in order to determine people's attitudes toward their surroundings, standard surveys, questionnaires, interviews, etc., turn out to be of little value because they do not consider the emotional factor involved. This is related to a host of reasons. First of all, people are largely conditioned to the point that they are often wary of surveys and often do not answer what they actually think, but what they think is safe in order to look better than they really are. Second of all, and when they are actually being truly genuine and answer surveys with complete sincerity, their answers nevertheless do not fully reflect their inherent feelings, since they are only based on the premises of the conclusions of their own rationales. Furthermore, an individual's feelings toward something or someone, as is known, are constructed in large part on the basis of his subconscious perceptions, hidden from his own awareness; naturally, this fact in such answers is not considered. But an individual's attitude toward anything in the outside world, as is known, is substantively established upon the basis of his subconscious motives; this circumstance, of course, is typically not reflected in surveys. If researchers somehow try to wrestle with the first issue by means of using a variety of interpretations to deduce the truth from misleading answers, then in light of the existing methods of surveying, they are simply powerless with the second issue. On the other hand, for the analysis of attitudes of individuals in some risky professions (pilots, submarine operators, etc.) toward dangerous situations that occur in their lines of duty, which does include evaluations of emotional factors, some physiological indicators may sometimes be used, which include heart rate, respiratory frequency, galvanic skin response, brain waves, etc. However, the implementation of this physiological method is quite problematic—it is a considerable challenge to effectively simulate dangerous situations in experiments, just as it is not an easy nor reliable task to measure individuals' physiological reactions. Essentially, people's emotional attitudes toward events with a positive valence can only be measured via surveys, questionnaires, and other similar self-evaluations, whereas people's emotional attitudes toward events with a negative valence (those that may pose a threat to life) are best measured by means of physiological indicators. With all this in mind, it becomes evident that it is necessary to develop more convenient, accurate, and effective methods that would allow determining people's inherent attitudes toward emotional events by means of indirect questions on both the conscious and subconscious levels.

Below we will look more specifically at some basic principles of the proposed approach. We can speak of two polar categories of emotions that can arise in different life situations: significance-as-value and significance-as-anxiety. These concepts, coined by Kotik & Yemelyanov (1992), originated from activity theory as the established notions of "significance" and "meaning" (Vygotsky, 1956; Leont'ev, 1975), based on the following hierarchical scheme (Figure 7.1):

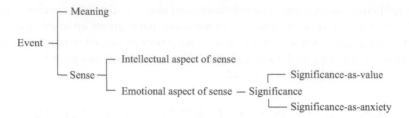

FIGURE 7.1 Significance-as-anxiety and significance-as-value in forming the emotional aspect of an event.

The significance of a certain situation, according to Leont'ev (1975), forms in a given social sphere and reflects a situation's more essential connections and relations. Sense provides a reflection of a given meaning in an individual's conscience, which is why it can be called subjective. Vygotsky (1956) pointed out that sense is not only perceived by the subject, but is also empathized by him or her, which is the reason he or she represents the unity of affective and intellectual processes. Bassin (1973) pointed out that sense without feeling is a logical construction, while feeling without sense is more likely to be a physiological category. Therefore, the emotional (affective) side of sense, which Bassin tightly linked with the subconscious behavior of the subject, Kotik (1995) identified as a significance of the event for the subject, and refers to two categories of significance, significance-as-value and significance-as-anxiety. Essentially, he views *significance-as-value* as a level of feeling toward a subject's success, expectations for that success, and eager anticipation of its positive outcomes, while *significance-as-anxiety* as a level of apprehension, originating from difficulties, risks in achieving a goal, and negative outcomes from not achieving it.

The concept of significance leads directly to K. Lewin's well-known field theory (Lewin, 1939), according to which this indicator can be considered as a source of emotional tension that creates motives for corresponding behavior and/or action. This motivating force—the significance of the event (S)—can be considered to be a function of two basic indicators of the theory: the valence (V) of the event (the strength of its attractiveness or repulsion) and the expected probability (P) of occurrence of an event with this valence: $S = f(V, P)$. This field theory position (Lewin at al., 1944) later became widespread in various models devised by supporters of the theory (Heckhausen, 1991): psychological interface (Tolman, 1951), social learning (Rotter, 1954), expectancy (Vroom, 1964), motivation (Atkinson, 1957), and various others. Meyer (1956) introduced the concept of emotions of expectation, a force directing a subject's behavior and depending as much upon the severity of attractiveness as on the expectations of its achievement. These models formed the basis for developing new branches of field theory, e.g., the theory of expected value, decision-making, resulting valence, achievement motivation, etc., that found various practical applications. All this related to attractive goals with a positive valence and to aversive goals with a negative valence; in other words, this terminology relates to objects that are significant-as-value and significant-as-anxiety. It must be noted, however, that neither Lewin nor his successors could clarify specific functional dependences that existed among the indicators in the formula $S = f(V, P)$. They considered motivation to be a simple product of valence(s) and probability(s) without concrete evidence. The attempt has been made to establish this dependence experimentally, using several special characteristics of its likelihood and intensity.

7.3 EXPERIMENTAL RESEARCH

To establish a functional dependency, M. Kotik conducted a series of experiments, the results of which are described below. In the first series of experiments (Section 7.3.1), the relation between the indicated parameters (C, V, and P) associated with negative events that are alarming for the individual—different types of losses of a social, material, and physical nature—was studied. In the second series of experiments (Section 7.3.2), such a relation was studied in terms of attractive events, now related to the gains of a social, material, and physical nature. The resulting dependencies in both series of experiments turned out to be rather similar, which allowed the inference that they are relatively universal and reliable in demonstrating the relation among the parameters C, V, and P of situations of different natures.

It is worth noting that in reality, people encounter situations in their everyday lives that contain within them both elements of significance-as-value and significance-as-anxiety (Kotik, 1993). Therefore, for example, information regarding a danger that might have come up on the path to a goal is simultaneously significant-as-value and significant-as-anxiety (because it allows for finding ways in which it can be avoided). It is also the case that valuable information regarding a success usually contains within itself some elements of worry—in other words, the success may be large, but only temporary and short-lived. Speaking more abstractly, people are forced to face many bittersweet situations in their lives. This is why there was a practical interest for researchers to obtain an integrated characteristic of significant situations that take into account both attractiveness and anxiety. With this goal in mind, another series of experiments has been conducted (Section 7.3.3).

7.3.1 EXPERIMENTAL STUDY OF SIGNIFICANCE-AS-ANXIETY

7.3.1.1 Evaluation of Significance-as-Anxiety with the Use of "Hard" Characteristics

Here we will only be discussing events, tasks, and situations in terms of significance-as-anxiety. In order to more accurately present a problem, we will use a specific example in which the anticipated losses are related to *physical dangers*. Let us imagine that an individual is required to walk over a durable but narrow board that is five meters long. This board is placed at a height approximately 1 meter above the ground. Such a task, naturally, would be relatively easy and of little worry to him. Now let's suggest that the same board is placed over a deep ravine. The new task will become much more worrying just because it will be far more dangerous—the cost of an error would become considerably greater. If an ordinary robot were to walk across the same board, this sort of characteristic would be meaningless.

From the aforementioned, it follows that the cost of failure (in other words, the severity of its consequences) will obviously prejudge the anxiety of this problem. However, if the cost of failure in the problem will be high, while the likelihood of such errors occurring is very insignificant, the anxiety arising from such a problem, as can be expected, is quite small. At the same time, a task in which mistakes are less severe, but may arise quite frequently, is sometimes more alarming than the first. Thus, on the basis of these considerations, it can be assumed that the anxiety caused by a task must be a function of two main arguments: the *probability (likelihood)* of the occurrence of errors, and the *severity* of the consequences resulting from these errors. In order to test this hypothesis, a special study was conducted (Kotik, 1974a; 1978b). It consisted of 50 participants: psychology major students, safety engineers, and economists.

In the first part of this experiment, the participants were given six error consequence (severity) options: microtrauma (1), light trauma (2), moderate trauma (3), severe trauma (4), disability (5), and death (6). Each participant was given a sheet of graph paper on which two coordinate axes were drawn. The horizontal x-axis measured the indicator of consequence severity (S). At the beginning of this axis, point 1 was plotted, and at the end, point 6, representing, respectively, microtrauma and fatal trauma. Each participant was supposed to plot points 2, 3, 4, and 5 between these two points on the horizontal axis, identifying intermediate degrees of injury severity: light, moderate, severe, and disability. Moreover, it was necessary to arrange these points in such a way that the distance between different consequences corresponded to participants' differences in the extent of severity. In other words, the participants were required to build a psychological scale of the severity of an error.

After the horizontal x-axis was scaled based on the severity of consequences, participants were asked to perform the second part of the study. Here, it was necessary for them to evaluate the potential effects of six different errors with the following standpoints. First, specify the chances at which the possibility of a microtrauma occurring already makes the task *alarming*. These chances needed to be plotted on the vertical y-axis of the graph P, scaled in percent format from 0 to 100%. Then, regarding the next point 2 (light trauma), the participants were asked to indicate the likelihood (chances) with which the occurrence of such consequences would seem alarming to them. Likewise, they were asked to rate the other four categories of error consequences in the same way, i.e., to build a y-coordinate of the appropriate height for the remaining four points. Following this, the participants connected their marked points, and each one obtained his own curve $P(S)$. This curve defined the boundary between tasks on the graph plane that still could not be considered *sufficiently alarming* (under the curve) and those that can be (especially the higher a point is above the curve). The graphs created in this way were statistically processed, and from the resulting dataset, an averaged function $P(S)$ was obtained, which is represented in Figure 7.2.

This graph indicates confidence intervals of statistical data at a level $C = 99\%$ (i.e., $\alpha = 0.01$). In this way, an averaged curve was obtained, which describes the relationship between the degree of severity of the error (S) and the chances of its occurrence (P), starting from which the situation in question already began to appear alarming (T) to the subjects: $T = f(S, P)$. The curve turned out to be decreasing

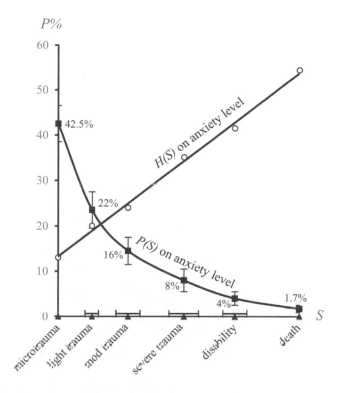

FIGURE 7.2 Dependence of the anxiety level on the severity of physical danger and likelihood of its occurrence ($\alpha = 0.01$).

and changing approximately according to the exponential law. Although individual participants' curves were located at different levels (for more anxious people, their curves were placed below average, while for more self-confident, above average), all of them, as the statistics show, changed according to approximately one law and were close to the averaged curve.

The third installment of this experiment was analogous to the second part and only differed from it in terms of the fact that here, the participants were given the task of identifying the likelihood and creating curves based on situations when a task isn't alarming, but only *slightly alarming*, and in situations when that same problem becomes *very alarming* and *extremely alarming*. After statistical processing of the data obtained at this stage, three more averaged curves were created, in addition to the first. In this way, in Figure 7.3 it was possible to present four curves, determining pairs of parameters (S, P) based on which the task was, on average, evaluated by participants as *slightly alarming (T$_1$)*, *alarming (T$_0$)*, *very alarming (T$_2$)*, and *extremely alarming (T$_3$)*. On the horizontal axis of this graph, averaged points from the scale of physical danger were marked, along with their confidence intervals at a level $C = 95\%$. The statistical difference among the obtained curves, according to the sign test (Dixon and Massey, 1969), was also at a confidence level of $C = 95\%$ ($\alpha = 0.05$).

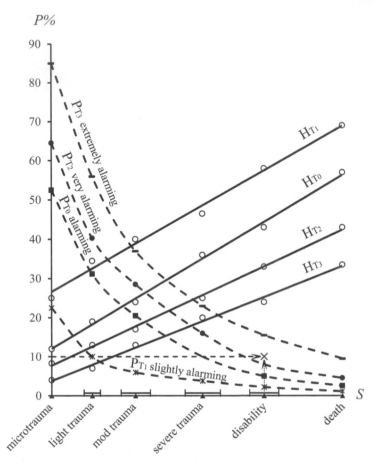

FIGURE 7.3 Dependence of different anxiety levels on the severity of physical danger and likelihood/uncertainty of its occurrence.

Let us now return to the task mentioned above regarding walking across a narrow board to analyze it from the position of the obtained curves. Let's assume that in both cases, the likelihood of a person falling from the board will be 10%. Furthermore, it can be assumed that if a person falls from a height of 1 meter, in a worst-case scenario, they would get a microtrauma, but if they fell into a deep ravine, they would become handicapped. After drawing a horizontal line at a level of 10% at point 1 (microtrauma) on the graph in Figure 7.3, we find that passing over the board at a height of 1 meter is associated with the task as being less than "slightly alarming." At the same time, the task of walking across the same board over a ravine for point 5, which leads to disability, is seen as more than "very alarming."

The experimental curves $P(S)$ for different levels of anxiety (T_1, T_0, T_2, and T_3), as seen in Figure 7.3, are approximately exponential. If the likelihood (P), indicated by the subjects, is to be approximately expressed in terms of probability (p), and then, based on the formula $H = -log_2 p$, in terms of entropy, then for the selected levels of severity, we will obtain the corresponding four dependences H (S). This dependence, as can be seen from Figure 7.3, turns out to be close to linear, which indicates that the factors that make up the given level of anxiety (S, H) are connected among themselves by a straight proportionality. In other words, the tasks will maintain a given constant level of anxiety, if there will be a proportional increase in the entropy of the occurrence of these consequences with increasing severity of the potential consequences.

The experiment described above pertains to the anxiety caused by physical danger that may pose a threat to an individual's life. Naturally, the question arises: what are the patterns of anxiety occurrence in relation to other categories of danger—*social* and *material*? To resolve this, two analogous series of experiments were conducted in exactly the same sequence as the one related to physical danger (Kotik and Yemelyanov, 1983).

In one of them, the considered task was studied in relation to *social danger*. Here, the participants, 40 university students, were presented with five levels of social dangers: an unsatisfactory grade in a current course (1), a failing grade on an exam (2), expulsion from a university (3), the loss of a loved one (4), and personal death (5). With the other group of 40 students, anxiety related to *material danger* was studied. Participants rated the following in terms of an increasing level of material danger: loss of 3 rubles (1), loss of a scholarship (2), theft of all personal belongings (3), house fire (4), and personal death (5). It turned out that the characteristics for social and material dangers were very similar in nature to the analogous characteristics for physical danger. This fact allows the conclusions that, regardless of the nature of the existing dangers (*physical, social,* or *material*) and the anxiety caused by them, a function of the level of danger severity and the likelihood of its realization is formed by roughly the same laws.

From the analysis of the obtained dependences, it is possible to ascertain that for the observable levels of anxiety, the following approximate functional relationships exist:

$$H_{T_1} = 2.5 + 0.4S, \quad H_{T_0} = 1.1 + 0.4S, \quad H_{T_2} = 0.7 + 0.3S, \quad H_{T_3} = 0.3 + 0.3S,$$

where $S = 0.5 \div 10$.

If we consider $A_T = S/H_T = $ const. as an indicator of the level of anxiety, then for the considered four levels, we obtain the following figures: $A_1 \approx 1.5$, $A_0 \approx 2$, $A_2 \approx 2.5$, and $A_3 \approx 3$.

On the basis of these criteria, it is possible for two estimates, which are used by the participant to identify a risk (its severity S and uncertainty H_T) to calculate the level of anxiety for participant. For example, if a person is afraid of the physical effects of severity $S \approx 6$ (according to Figure 7.3, this is something like a severe trauma) and determines the likelihood of its occurrence with chances $P = 20\%$, which equals: $H_T = 2.32$ bites, then the level of his or her anxiety based on the calculation will turn out to be $A = 2.58$, i.e., a little more than "very alarming."

Thus, concluding the description of the first series of experiments, we can establish that the significance-as-anxiety of events in a goal-directed activity is determined by the severity of a situation and its likelihood of occurring. In order to determine the degree of anxiety for the participant solving a specific task, it is necessary to establish the level of severity of the expected danger in this task and mark it as an x-coordinate on the horizontal axis. Next, it is necessary to establish the likelihood with which the participant expects this danger to occur, and then plot this figure as an y-coordinate on the vertical axis for this point. Assessing the position of a point with coordinates (x, y) relative to curves of different anxiety levels, we can determine a participant's level of anxiety when solving the task.

7.3.1.2   Evaluation of Significance-as-Anxiety with the Use of "Soft" Characteristics

In the first approach, study participants provided quantitative assessments of the likelihood (chances) of a variety of different degrees of danger occurring. Nevertheless, it is generally known that people tend to experience difficulties when it is necessary to quantify a circumstance, and especially when they must provide a concrete prognosis in numerical format. When an individual is interrogated regarding the severity of consequences and their likelihood, it is easier for him or her to characterize them by using soft (vague or fuzzy) descriptors, such as strong and often, as opposed to numerically identifying their chances of occurrence. According to Zadeh (1973), "Elements in human thinking are not numbers, but labels of fuzzy sets, that is, classes of objects in which the transition from membership to non-membership is gradual rather than abrupt." Furthermore, when choosing a descriptive word that conveys this type of vague understanding, an individual cannot help but express their own perceptions toward the situation in question. This can be demonstrated with an example. Let's consider the following situation: "Peter received an award of 30 rubles," which quite accurately describes the size of the award. However, someone could point out that "Peter received a large award," while another observer might remark that "Peter received a small award." In these two examples, a specific amount describing the size of the award was replaced by an unspecified and vague descriptor; in other words, from a data-specific point of view, the new descriptors became less informative. Regardless, the use of soft characteristics such as "large" or "small" allowed observers to express their opinions toward the situation. Essentially, by choosing an indefinite quantity to specify the situation, an individual expresses his perception of it; in other words, replacing a specified amount with an undefined, or fuzzy, descriptor leads to a loss in terms of the amount of information communicated, yet more fully conveys their emotional perceptions toward the issue.

It is noteworthy that by evaluating events using such words, people usually do not think in terms of specifics: their judgements seem to arise quite spontaneously. For example, if an individual is asked to judge certain events by using words establishing their likelihood and intensity, they generally will not acknowledge which of the events is considered most significant. By judging one event, for example, as "strong – very seldom", but another as "weak – very often", people may not realize which of these events they regard as more alarming, and therefore they cannot adequately articulate their attitude toward the particular event.

All this led us to conduct an additional experiment on the given issue, in which the participants would be able to assess dangerous events, relating them to certain categories of soft characteristics (Kotik and Yemelyanov, 1983). As can be deduced from the previous series of experiments, significance-as-anxiety of events is evaluated based on two primary criteria: the degree of severity of consequences and the likelihood of their occurrence. Both of these criteria can be expressed through soft characteristics. Thus, the criteria of severity can be determined based on which one of the following characteristics of intensity fits the observable situation: zero, extremely weak, very weak, weak, not weak – not strong, strong, very strong, extremely strong, and max. Likelihood can be assessed based on the criterion of which of the following sets it will be assigned to: never, extremely seldom, very seldom, seldom, not seldom-not often, often, very often, extremely often, and always. Thus, the basis for determining dangerous events' levels of significance that are of interest to us were two sets of soft characteristics—according to the severity of danger of a situation and its likelihood. It was possible to build coordinate axes and then mark on the horizontal axis the abovementioned quantities based on the criteria of intensity, while on the vertical axis, quantities based on the criteria of likelihood. In light of this, it was assumed that both groups of values steadily grow and can be viewed as equally distributed across the axes. To obtain the significance characteristics that interest us, we used a method of constructing indifference curves (Larichev, 1979). Five different levels of significance-as-anxiety were considered, according to the following pairs of characteristics: very weak – very seldom (1), weak – seldom (2), not weak-not strong – not seldom-not often (3), strong – often (4), and very strong – very often (5).

Study participants were given sheets of graph paper with scaled coordinate axes on a plane marked with thick circles to denote the aforementioned initial points: R_2, R_3, R_4, R_5, and R_6. Next, they were provided with the following instructions: "Imagine that you are expecting to receive some sort of punishment (an electric shock; criticism; etc.) with a certain level of intensity and some likelihood of its occurrence. You are presented with a graph on which verbal ratings of intensity are plotted on the x-axis, and the likelihood of the punishment to occur is plotted on the y-axis. The graph is marked with five points: $R_2(x_2, y_2)$, $R_3(x_3, y_3)$, $R_4(x_4, y_4)$, $R_5(x_5, y_5)$, and $R_6(x_6, y_6)$. Select one of these points, consider the significance of such an impact, and then find all other points $R(x, y)$ with the same level of significance to the right and left of it. Mark these points and connect them with a curve. Similarly, build curves representing other points' significance levels by using the rest of the marked points".

In this study, 80 upperclassmen physics and psychology university students participated. Based on the results of the research and after a thorough statistical analysis, five averaged graphs of different levels of significance-as-anxiety were constructed (Figure 7.4). Next to each graph, deviations from the mean around each curve are presented by confidence intervals at a level of $C = 95\%$ ($\alpha = 0.05$). The lower curves of smaller levels of anxiety turned out to be relatively close in terms of form to the curves obtained in the first series of experiments.

The characteristics established in Figure 7.4 can be used to determine the level of anxiety of a wide variety of events or situations. For this, all that is required is that a person indicate what he thinks the likelihood of an adverse influence and its intensity in terms of these sets would be. According to these two criteria, we find a point on the graph, and based on its position relative to the graph's curves, we establish a subject's level of anxiety in a given situation.

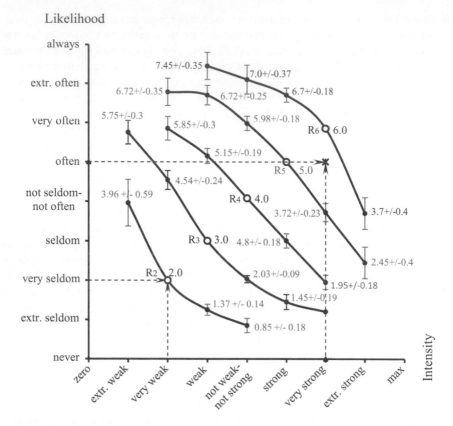

FIGURE 7.4 Levels of significance-as-anxiety as a function of intensity and likelihood presented as soft characteristics.

Let's return to some examples previously described, related to the task of walking across a narrow board. It's possible to consider that when the board is at a height of 1 meter above the ground, the cost of the error in such a scenario is "very weak" in terms of intensity and "very seldom" in terms of the likelihood of its occurrence. In this case, we arrive at point R_2 which is located at the very *lowest* level of anxiety. However, when it is required to walk across that same board while over a deep ravine, the severity of the trauma would be "very strong," and the likelihood of falling here would be rated as "often." In the given example, the significance of a task, as can be seen in Figure 7.4, will turn out to be closest to its *highest* level in terms of anxiety. But let us imagine that the second task will be performed by a circus tightrope-walker. More than likely, they too would rate the intensity of danger as "very strong", but would nevertheless rate the likelihood of their fall as "extremely seldom"—in this case, the level of their anxiety would turn out to be as low as in the first simple task.

7.3.2 EXPERIMENTAL STUDY OF SIGNIFICANCE-AS-VALUE

Measurements of significance-as-value occurrences are studied within the framework of information theory (Harkevich, 1960), although they do not determine an individual's emotional attitudes toward these situations, but rather these situations' utility for him. During the study of anxiety-inducing events, we used indicators of likelihood and intensity. Similarly, we can assume that people's attitudes toward attractive events are presented with the same indicators; in other words, an attractive event must contain a value factor of an anticipated event, as well as a prognosis of the probability (likelihood) of its practical realization. To test this hypothesis, another series of experiments was conducted (Kotik, 1984).

7.3.2.1   Evaluation of Significance-as-Value with the Use of "Hard" Characteristics

In the first study, 30 student-psychologist upperclassmen participated. This study was conducted in the following manner. The participants were distributed sheets of graph paper, which had axes 10 cm in length drawn on. At the same time, they were informed that there are six types of situations for them to consider, presented in order of increasing importance of the *social value*.

1. Receiving an excellent grade in a course.
2. Having your research published in the Scientific Notes of Tartu University.
3. Having your research published in a prominent international journal.
4. Receiving an Estonian state award for your research.
5. Receiving a national (USSR) award for your research.
6. Receiving the Nobel Prize for your research.

After scaling the horizontal x-axis, which determines the degree of attractiveness of these situations, participants were prompted to complete the second part of the experiment. First, they were required to rate the likelihood at which receiving an excellent grade in a course would make this event significant-as-valuable for them. These likelihoods were to be graphed on the vertical y-axis in percent format. Then, for both the second and all other subsequent points, participants were required to plot corresponding y-coordinates in the same way on the y-axis, which would determine the likelihood at which each of these events would acquire significance-as-value for them. Then the marked y-coordinates were to be connected, which gave each subject a curve that determined the value (C) of the event as a function of its attractiveness (S) and likelihood (P) of occurrence.

During the third phase of the experiment, which was conducted analogous to the second one, study participants were given the task of specifying likelihood and creating graphs based on two different guidelines: when the given information is presented to them as being slightly valuable, and then separately for when the given information is presented as being very valuable. In this way, each participant built three curves on his or her graph, determining pairs of parameters (S, P) based on which information was perceived by him or her as *slightly valuable* (C_0), *valuable* (C_1), and *very valuable* (C_2). After statistical processing of all the participant's graphs, an averaged graph was built, which presented three curves based on three abovementioned levels of values $P_{C_0}(S), P_{C_1}(S)$, and $P_{C_2}(S)$ (Figure 7.5).

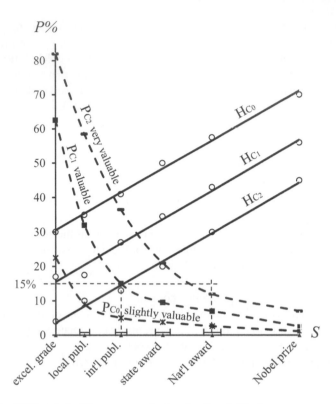

FIGURE 7.5   Different significance-as-value levels of social factors.

Marked on the horizontal axis of this graph are averaged points of the scale of social values and their confidence intervals at a level of C = 95%. The statistical difference among the obtained curves, according to the sign test, were also at a confidence level of 95%. If chances (p), indicated by the participants, can be approximately expressed through probability (p), and in turn, probability is transferred into entropy using the formula $H = -log_2 p$, then for the three highlighted levels of value, we will obtain three corresponding dependences $H_{C_0}(S), H_{C_1}(S)$, and $H_{C_2}(S)$. These dependencies, as can be observed from Figure 7.5, turned out to be approximately linear, and this signifies that the importance of the information will remain unchanged, if with the increase of an event's emotional level, its entropy grows linearly with it. To interpret the obtained results it should be noted, as an example, that a 15% likelihood of international publication presented an average level of value for participants, while a national award for research presented a high level of value for them.

Significance-as-value is not only possessed by information regarding *social events*, but also by information involving *material rewards* and *physical pleasures*. Two analogous series of experiments related to material and physical values were conducted (Kotik, 1984). Their results demonstrated that the basic principles of forming evaluations for the significance of things of material and physical values turned out to be similar to principles of creating significance-as-value for social events.

7.3.2.2 Evaluation of Significance-as-Value with the Use of "Soft" Characteristics

It is important to note that when this method was applied in the process of analyzing significance-as-anxiety, it was established on the assumption that soft characteristics for likelihood (never, extremely seldom, …, always) and intensity (zero, extremely weak, …, extremely strong) can be viewed as equally distributed. In order to eliminate this restriction, it was deemed necessary to experimentally examine the association existing between the soft characteristics of both categories.

To address this question, a special study on the scaling of soft characteristics was performed, which was carried out with different methods and in different groups of participants. In one group (30 student philologists), the likelihood was scaled (from never to always)—here participants assigned a corresponding score to different categories (based on a 10-point scale). In the other group (30 physics students), scaling was performed with another method. Participants were presented with sheets of graph paper, which had segments of 10 cm marked. At the beginning of the segment, a point marked "never" was plotted, and at the end was "always." Between these two extreme points, the participants had to mark points of an intermediate value for a given set, and in such a way that the distances between them corresponded to the differences between the indicated categories.

Similarly, both of these methods were used to scale intensity. After statistical processing of this study's results, it was found that both the scales turned out to be uneven. At the same time, the results from different methods of scaling on both sets of data turned out to be quite similar, which may serve as evidence of the validity of the scales.

Next, the participants (physicist students and philologist students, 40 total) were given sheets of graph paper with coordinate axes 10 cm long. The horizontal x-axis represented the intensity scale that was already previously obtained, while the vertical y-axis represented the likelihood scale. A row of points was marked on the graph plane. Initially, participants were asked to identify the point with coordinates (strong, often). The students were told to imagine any pleasant event that can be described as both strong and frequent, immersing themselves in the situation and consider its significance-as-value, and then plot all the other points (for pairs of intensity-likelihood) to the right and left of this point that had the same level of significance-as-value. By connecting all these points, participants obtained a curve corresponding to the given level of significance. Following this, a (weak, seldom) point was plotted on the graph, and participants then obtained another curve of a different (understandably smaller) level of significance by a similar method. Using this approach, three more curves were created, which corresponded with their significance-as-value to the points (not weak-not strong, not seldom-not often), (very strong, very often), and (very weak, very seldom). With this method of constructing indifference curves, each participant plotted five curves of different levels of significance-as-value on the graph. After statistical processing of individual study participant' graphs, averaged curves were obtained, which defined different levels of significance-as-value events or the information about them as a function of these events' intensity and the likelihood of their occurrence (Figure 7.6). The statistical difference among the obtained curves, according to the sign test, was at a confidence level of $C = 90\%$ ($\alpha = 0.1$).

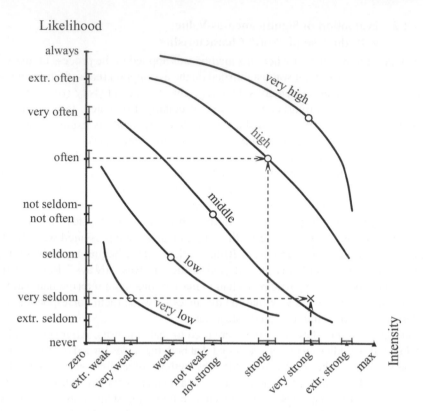

FIGURE 7.6 Levels of significance-as-value as a function of intensity and likelihood presented as soft characteristics.

A comparison of these curves with analogous curves constructed to determine the significance-as-anxiety of an event (Figure 7.4) shows that they are actually quite similar. Due to the unevenness of the scales, our last curves (especially the lower and upper ones) turned out flatter. If the curves of significance-as-anxiety would also have been built based on uneven scales of likelihood and intensity as the last ones were for significance-as-value, then it can be assumed that their similarity would have been even greater. With the curves depicted in Figure 7.6, comparisons of significance-as-value for events of a diverse nature can be conducted. This requires that the participant using these fuzzy characteristics determines the intensity of a given event and the likelihood of its occurrence. Based on these two estimates, on the graph plane (Figure 7.6) exists a point whose location among the depicted curves can be used to conclude the extent of significance-as-value for a subject of a given event.

In this way, the described approaches allow to conduct an analysis of conflicting situations that arise from the necessity of selecting the more preferable choice out of two valuable options. Section 7.3.1 listed the methods of analysis of a similar option, only in the situation when it is necessary to select the lesser evil out of several. Thus, the suggested approach, as we can see, could serve as a convenient tool for conducting an analysis of the process of decision-making.

Regarding the practical application of the aforementioned approach, one important reservation should be made. It is well known that people tend to use soft characteristics of evaluation only in a relative sense and with regard to specific situations. For example, when one person states that he was "very lucky" with a ticket purchase to a theater show, and another says that he was "very lucky" in receiving a job promotion, this in no way indicates the same positive effect nor the same attitudes toward the given events for these two people. It should therefore be expressly agreed that the application of the described approach in comparing different events according to their levels of significance is possible only in those cases, when such comparisons are made by one individual, in the same situation, in the solving of the same problem, and at the same time a list of the same soft characteristics is continually used. In the example above, these conditions had been met.

7.3.3 EXPERIMENTAL STUDY OF SIGNIFICANCE: EVALUATION WITH THE USE OF "SOFT" CHARACTERISTICS

This section was compiled from materials in (Kotik and Yemelyanov, 1986, 1992).

In practice, people have to deal with situations that are fraught with elements of both significance-as-value and significance-as-anxiety. For example, information about the dangers that arise on the way to a goal is both significant-as-anxiety and significant-as-value (since it enables us to search for ways to avoid these dangers). Likewise, valuable information regarding success is usually fraught with elements of worry or anxiety based on the idea that success could be greater and/or that it is temporary. Figuratively speaking, people usually have to deal with bittersweet situations. It was therefore of practical interest to obtain some integrated characteristics that synthesize the significance of the situation while taking into account both its value and anxiety. For this reason, another experiment was conducted, in which the participants performed comparisons of different events characterized by certain levels of significance-as-value and significance-as-anxiety and then separated these events into groups, each of which needed to be equivalent. Estimates of significance-as-value and significance-as-anxiety could be carried out on the previously established intensity scale, ranging in values from zero to extremely strong. The experiment involved 50 university students from the departments of psychology and physics.

Each participant was presented with a sheet of graph paper on which axes were marked, each 10 cm in length. The horizontal axis represented the intensity scale in terms of significance-as-anxiety, while the vertical axis had the same scale, but in terms of significance-as-value (zero, very weak, weak,..., extremely strong). We continue our analysis based on an example of an everyday problem, which is provided in the following form.

"You have taken up heavy physical labor at a construction site. Plot a number on the horizontal axis against each point, representing the number of hours of such work in a day that you would consider to be of *very weak* intensity, *weak,..., strong, ..., extremely strong.* Then plot a number on the vertical axis (against each point) to represent the pay for a day's worth of such work—what payment you would consider to be *very weak* (in intensity), *weak,..., strong,..., extremely strong.* After having determined the numerical scales of the graph, move to the main portion of the experiment. For each scale on the graph, plot the point of the coordinates that would represent not weak-not strong. This point essentially links the intensity of the work with the intensity of its payment on a level that you think would be about average. Now, to the right and left of this point, you must plot a row of other points on the graph (linking the intensity of labor and its payment) that are equivalent to the starting point. After connecting the marked points, you obtain a type of equipotential curve, an indifference curve that defines the connection between the labor and the amount of compensation, which you believe to be of a *middle* level. Afterwards, at the level of labor intensity not weak-not strong, you must plot on the graph four points that correspond to a payment of very weak, weak, strong, and very strong, and then create separate equipotential curves for each of these points. If you related an earlier curve to some average level of payment for the labor, then these curves, apparently, should be related correspondently to *very low, low, high,* and *very high* levels of labor stimulation."

Thus, each participant built five equipotential curves on the graph. A preliminary study of the obtained curves showed a great similarity among them, despite the fact that the participants all rated the difficulty of their work differently and had different expectations of how much they would be paid for it. The majority of students assigned the average labor intensity to 8 hours of work with an average payment of 10 rubles per day. However, there were also participants who rated their work considerably higher: they linked an average workload of 3 hours per day with a payment of 30 rubles. It is worth noting that despite the differences in the perceptions of difficulty of the work and its compensation, the participants created very similar curves, relating soft characteristics of labor intensity to soft characteristics of its remuneration; the first was viewed as a *significant-as-anxiety* aspect of the work, while the second was seen as a *significant-as-value* aspect of it.

After statistical processing of the experimental results, five averaged curves were created, linking the soft characteristics of both aspects of labor based on the respective levels of satisfaction with this work (Figure 7.7). The statistical difference among the obtained curves, according to the sign test, was at a confidence level of $C = 95\%$ ($\alpha = 0.05$).

In this way, a new, more complex functional dependency was established allowing event evaluation and comparison in terms of the integral significance, and carrying in it elements of value and anxiety.

The following is an example of the practical use of the curves presented in Figure 7.7. Suppose an individual is to select a workplace. In one case, he is offered a job with a high salary, but the commute from home to his workplace would take a few hours. In another case, a similar job is available close to home, but with a considerably smaller salary. Let this individual identify the value of the first option with the word strong and his anxiety with the words very strong. In the second option, he determines his concern regarding the lower wages as not strong- not weak, with the value being very strong. Transferring these characteristics to Figure 7.7, we can see that the overall preference for the first option sits *between the middle and low levels,* and the second turns out to be on a *very high level.* Therefore, it is logical to assume that he will most likely choose the job closer to home.

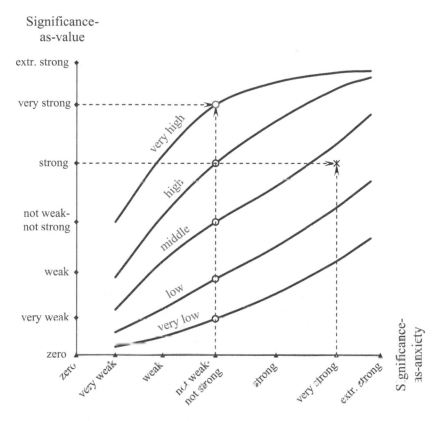

FIGURE 7.7 Levels of significance as a function of significance-as-anxiety and significance-as-value.

7.4 APPLICATIONS OF THE RESEARCH TO STUDYING HUMAN PERFORMANCE

A series of experiments is presented, all of which illustrate the validity and reliability of the suggested method. As was noted above, its advantages include the fact that it is able not only to measure people's conscious emotional states, but also their subconscious motives; another bonus is that it is simple to implement in practice and the results obtained by this method are easy to analyze. Sections 7.4.1 and 7.4.2 describe related experiments on people's attitudes toward dangers in professional work settings and toward events with a positive valence, while Section 7.4.3 demonstrates a use of the suggested approach when both losses (dangers) and gains (values) are present.

7.4.1 STUDY OF PEOPLE'S ATTITUDES TOWARD OCCUPATIONAL HAZARDS

As demonstrated by experiments (Kotik, 1974b), an individual's attitudes toward the difficulties and dangers of the problems that he must solve produce the fundamentals of self-regulation of his activity. The more significant-as-anxiety a problem that needs to be solved seems to the individual, the greater the mobilization of its internal structures it causes, which usually contributes to its more successful resolution (of course, if the mobilization does not turn out to be excessive). Many experimental studies confirm this conclusion (see, e.g., Kitaev-Smyk, 1983; Simonov, 1981; Mckenna, 1982). Self-regulation occurs spontaneously: it only takes an individual to realize the degree of importance of the problem that has arisen, since his own entity/ organism "will be 'fighting' to achieve a result important to him, while actively overcoming obstacles," based on the figurative expression of Bernstein (1962, p. 85). Therefore, it is very important in practice that people adequately assess the significance-as-anxiety problems that they solve. If this indicator (significance-as-anxiety) is underestimated even by well-trained specialists in practice, they often end up making errors, not because they cannot solve the problem, but because they do not use the opportunities available to them to resolve it. Therefore, it is very important that in the training of professionals, during the analysis of the causes of their errors, the method is at hand to identify their attitude toward the problem that needs to be solved. The proposed approach was used in human-operator error and safety analysis (Kotik, Yemelyanov, 1985; Kotik, 1989). It was also used in the study of an individual's attitudes toward occupational hazards in activities such as piloting an aircraft (Kotik, 1978a, 1978b, 1994a), high-voltage network service (Kotik, Öövel, 1980), trolley bus driving (Kotik, Sirts, 1983), law enforcement work, mountain climbing, and taxi driving (Kotik, 1986).

To illustrate the conducted research, a short description of the experiment regarding pilots' attitudes toward flight emergencies is presented. In this study (Kotik, 1994a), 45 pilots of different flight qualifications participated. Each participant was given 18 aviation situations of varying degrees of danger for assessment, all of which were well known from previous flight experience; 9 of those situations were related to technical malfunctioning, and 9 were related to pilot errors. It's apparent that every aviation situation during a flight signifies not only a physical danger for the pilot, but also a social one (a penalty for the error or an inadequate effort to repair any malfunction or a possible loss of reputation, or being prohibited from working on future flights, etc.), in addition to a spiritual danger (being responsible for the loss of life, destroying a costly aircraft, etc.). Therefore, it is necessary to discover not only pilots' general attitudes toward different kinds of aviation accidents or emergencies, but also the various indicators of these attitudes.

Each participating pilot was given the following instructions. "You are to present your attitudes regarding 18 aviation situations, which can all take place during a flight. For each given situation, you are to:

1. express the likelihood of its occurrence, including the point at which this situation becomes alarming for you, conveying its likelihood in percent form and with a word from the list provided below;
2. if you feel that the situation is likely to occur, indicate the extent of the *physical*, *social*, and *spiritual* dangers, starting from the one that already worries you.

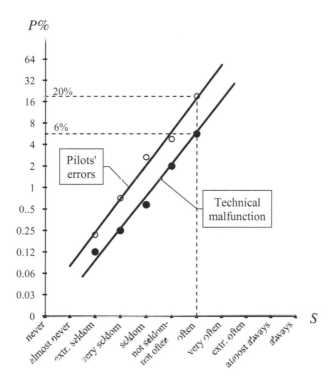

FIGURE 7.8 Comparing pilots' attitudes toward physical danger, occurring in flight during technical malfunctions and personal errors.

In other words, you should describe each aviation emergency with four pairs of characteristics, placing them in the corresponding positions on the form."

The study took place anonymously, only keeping track of each participant's flight qualifications. After statistical analysis of all the aggregated data, averages of certain characteristics of interest to us were examined. We took a closer look at a few of them. Of particular interest was what concerns pilots most: technical malfunctions or their own errors? Based on the physical indicator of danger, the answer to this question is presented in Figure 7.8.

On the horizontal axis, soft verbal characteristics are outlined; on the vertical one, in a logarithmic scale, are the hard numerical characteristics (likelihood presented in percent form) of two situations: one related to technical malfunctions, and the other to pilots' errors. In the establishment of these characteristics, the following assumption could be made: the more alarming a situation is to pilots, the less likely they are to perceive its occurrence as often, very often, etc. As can be observed from Figure 7.8, physical danger, when resulting from pilot errors, was thought of as occurring often with a likelihood of approximately 20%, whereas when resulting from technical malfunctions, with a likelihood of only about 6%. In other words, a very small likelihood (6%) of technical malfunctioning is already enough to make a situation worrisome or even alarming for a pilot, while pilot errors make a situation alarming from a physical danger standpoint with more than three

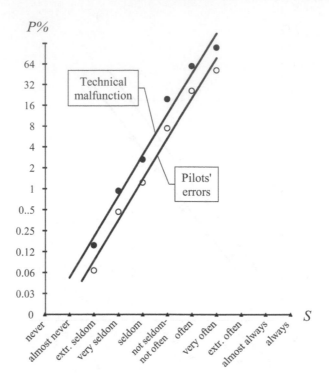

FIGURE 7.9 Comparing pilots' attitudes toward spiritual danger, occurring in flight during technical malfunctions and personal errors.

times higher likelihood (20%). A similar proportion (in a logarithmic scale) can also be found for the descriptor very often. Consequently, pilots' errors, in terms of the appearance of "physical" danger, worry them less than technical malfunctions: confidence based on the sign test at a level of 90% ($\alpha = 0.1$). From the indicated pattern, we can conclude that the lower a line on the graph, the more valence of the event it represents. Figure 7.9 presents descriptors linking pilots' attitudes toward both technical malfunctions and personal errors, but now from the perspective of spiritual danger. Here, it is evident that pilots are more concerned with their own personal errors than with technical malfunctions: confidence based on the sign test at a level of 95% ($\alpha = 0.05$).

By a similar method, it was possible to deduce that the very prospect of an emergency situation occurring during a flight is a more alarming event for first class pilots (more experienced) than for third class pilots. Pilots with a low or medium negative valence of aviation emergencies are most concerned with social danger, while for pilots with a high negative valence of such emergency situations, the primary focus becomes physical danger; in both circumstances, spiritual danger concerns pilots less than physical and social dangers. It is worth noting that all data gathered from this experiment sufficiently reflects reality: the fact that technical malfunctions result in more serious consequences than pilot errors, which are more likely to occur with less experienced pilots who underestimate the potential dangers of a flight, and that social dangers (being prohibited from flying, losing one's reputation, or gaining the status of a coward) are more worrying than are physical dangers for pilots (Lager, 1970).

As a result, on the basis of numerous studies regarding the attitudes of people in different professions toward dangers inherent to their careers, M. Kotik (1986, pp.137–138), came to the conclusion: "It turns out that in occupations that are traditionally considered to be dangerous (pilot, mountain climber, high-voltage electrician, law enforcement officer, taxi driver) because they pose a grave danger to life and health, people are, more often than not, far more concerned with the social dangers than the physical ones! Therefore, the factor of social danger in our time has become paramount. If it is traditionally considered to be relatively important in physically dangerous professions, then what should its role be in professions that are considered to require much responsibility? However, for now this category of danger and its result and impact on the profession are, unfortunately, not being studied."

7.4.2 STUDY OF PEOPLE'S ATTITUDES TOWARD ATTRACTIVE SITUATIONS

Let us consider the role of a significance-as-value occurrence and the subsequent problems in object-oriented activity. If we are to consider an experienced professional, it is possible to demonstrate that such eagerness doesn't really leave a lasting impact (Kotik and Yemelyanov, 1993). In fact, an experienced professional typically manages to successfully solve emerging problems he might encounter, so that such a rate of success becomes the norm and does not cause any particular emotion. Therefore, failures in work and in other types of activities are endured considerably more by experienced specialists than their successes.

However, such a relationship manifests itself not only in work but also throughout an entire human life.

Nevertheless, the experience of valence and significance-as-value presents a scientific and practical interest on somewhat different terms: studying them, we can identify the more attractive objects and phenomena for a subject, and thus derive a judgment concerning his preferences, interests, values, and type of personality, or define these very parameters in specific groups of people. The positive valence and significance-as-value of events can thus be regarded as a kind of characteristic identifying the aspirations of an individual or group of people toward a particular goal or types of activities.

In this study (Kotik, 1994a), the proposed approach was used to study the attitude of people toward events with a positive valence, representing significance-as-value. The subjects were Russian students from the mathematical, medical, and philological departments of the University of Tartu (166 people), who were provided with the following instructions. "You are given a form that lists twenty pleasant events that could theoretically occur in your life. For each event you should indicate the probability of its realization, starting from the moment this event will make you happy. This opportunity must be expressed through a pair of characteristics: in numbers (odds) and words (likelihoods) to indicate the extent of physical, social, and spiritual satisfaction garnered from it."

The greatest interest of this study was in identifying the differences in the attractiveness of these events for students of different specialties and different courses, as well as establishing the influence of their performance, origin, and gender on their preferences.

A comparative analysis of the values of students of different specialties helped identify the following patterns. Philology students physically rated higher than mathematics (at a confidence level of $C = 90\%$ ($\alpha = 0.1$) and medical students ($C = 75\%$) a cruise around Europe; from a social standpoint, they valued long-term pleasure more than doctors ($C = 90\%$) and mathematicians ($C = 75\%$), i.e., being able to attain a good apartment and car after 5–10 years of hands-on work, as well as enjoying a vacation with a loved one at a tropical resort and placing in a top prize-winning spot at student competitions; from a spiritual standpoint, they assessed higher than doctors ($C = 90\%$) and mathematicians ($C = 75\%$) the opportunity to become a prominent expert nationwide by 35 years of age. In general, mathematics students valued more than philologists ($C = 90\%$) the enjoyment attained from a holiday with their girlfriend at a tropical resort, and more than doctors the pleasure attained from 5–10 years of practice to acquire a good apartment and car ($C = 90\%$); spiritually, they valued getting into graduate school more than doctors ($C = 90\%$) and philologists ($C = 75\%$). In general medical students only valued more than philologists ($C = 90\%$) and mathematicians ($C = 75\%$) winning the car "Lada" in a lottery drawing. This comparative analysis allows us to conclude that the philology students, compared with the others, more emotionally perceived the aforementioned pleasant events, while medical students were more restrained in relation to them. On on hand, this could be explained in terms of the philology students being more emotional, probably influenced by a pre-determined choice of their profession, close to more creative fields; on the other hand, it could be explained in terms of the fact that they have less hope for being successful in life in the future (as opposed to the more affluent physicians and mathematicians of our time in the Baltic States), which makes such lavish events seem more valuable to them.

Analysis of the differences in the characteristics of these events by students of various departments shows a significant divergence from students in their first year. In general, first year students appreciate ($C = 90\%$) a vacation with a loved one at a resort, a first place finish in a student competition, and rest in the company of good people more than other students. In a physical sense, freshmen place higher value ($C = 90\%$) on getting a one year internship in the US and attending graduate school. From a social standpoint, they value the pleasure from taking a cruise around Europe more than second year students ($C = 90\%$), and juniors and seniors ($C = 75\%$). In a spiritual sense, freshmen place more value on experiencing happiness from mutual love ($C = 90\%$).

Generally, compared with the rest, upperclassmen more highly value ($C = 90\%$) obtaining a good job after graduating from college and deriving long-term pleasure after 5–10 years of working to amass a good apartment and car. The results obtained from this part of the analysis are so natural and self-explanatory that they require no further justification.

Analysis of the differences in the students' assessments of the levels of their academic performance showed the following. In general, high achieving honors students value the opportunity to earn some extra money during school holidays more than the others ($C = 90\%$). In a physical sense, they value more than the rest a gift of some sort—a fashionable outfit, for example, or an evening with good friends ($C = 90\%$). In a spiritual sense, for excellent students, the value in attending a concert or play is lower than for others ($C = 90\%$). Based on these data, we can say that smart students are more pragmatic than others, which probably contributes to their success, especially in academic fields.

Analysis of the differences between the attitudes of men and women toward the experiences under consideration showed that men, in general, more highly value winning a car in the lottery ($C = 90\%$), while women value having a paying job during a holiday ($C = 90\%$). From a social standpoint, men more highly value winning a top spot in a competition ($C = 90\%$). More than likely, all these differences arise from distinct deeply ingrained characteristics in men—to create and achieve, and in women—to maintain and keep. These are the essential conclusions that we were able to derive from this research, which, in our opinion, are in general accord with the established perceptions of the values possessed by university students in the Baltic States. Therefore, the results of this study can also be regarded as confirmation of the validity of the proposed method.

7.4.3 Study of Students' Attitudes toward Their Future

In this paragraph, the use of the method in the analysis of an individual's future is presented, when both losses (dangers) and gains (opportunities) are likely to occur. It is well known that the future always concerns an individual, with the interplay of highly desired events (values) merging with worries about potential failures (anxieties). Considering this, it was interesting to discover how the suggested method allows assessment and evaluation of people's attitudes toward events that contain both value and anxiety, based on the following example with Tartu University students.

In this experiment (Kotik, 1994b, 1995), 160 second, third, and fourth year Tartu University students participated. The study sample consisted of 40 students each from the schools of mathematics, medicine, philology, and economics, with 20 Estonian and 20 Russian students per group. The students' attitudes toward their future professional careers and family lives were studied. The future was divided into three stages: the beginning, a period of stabilization, and the final stage. Expectations of success and failure were estimated with respect to all stages, and success and failure were expressed in general, material, and spiritual terms.

Each participant was given a list of 12 questions about the three stages of his or her professional career and family life. The first question concerned expected success, the next one, expected failure. Words from two soft categories were to be used in the responses, which assessed the expected intensity of the indicated events and chances of this event occurring with the given intensity. The list of words expressing the intensity of the event included a range from extremely weak, very weak, weak, not weak-not strong to extremely strong. The list expressing the likelihood of occurrence of the events included the same number of words (seven) from extremely seldom, very seldom, seldom, not seldom-not often to extremely often. As a result, in responding to each question, the participants had to answer by choosing three pairs of words (from the categories intensity and likelihood for evaluating the future in general, material, and spiritual terms).

Table 7.1 presents a fragment of the questionnaire the students had to complete. They also had to indicate the department they attended, their academic performance (according to a 5 point gradation), and their family and ethnic background (Estonian or Russian).

The questionnaire was anonymous, and time for its completion was unlimited. The participants were given only general directions to consider as deeply as possible the events under question, and thereafter, without any deliberation, to judge the pairs of respective words separately in terms of general, material, and spiritual life, inserting these words in the questionnaire on the proper line.

After the questionnaires were completed, they were subjected to preliminary processing. By means of the graph in Figure 7.6 (estimation of the intensity and likelihood of an event), the significance of the event for the subject was established. For example, if a student judged the intensity of the expected success at the beginning of his career as strong and the likelihood of such success as not seldom-not often, then, as can be seen from Figure 7.6, the significant value for the participant was between middle and high. Each pair of estimations was assessed by means of the graph indicating the significance of the events. As a result, based on the responses to each question, three new variables of significance of the event were formed in general, material, and spiritual terms.

In the current study, as has already been mentioned, three stages of the participant's professional career (questions 1-6) and family life (7-12) were considered. Therefore, it was interesting to determine not only the significance of success and failure during each separate stage, but also the overall significance of the entire stage for the participant. For this, we used the graph in Figure 7.7, which enabled us to estimate the third indicator on the basis of the first and second ones. For example, if a participant rated the significance-as-value of success at a given stage as very weak, but the significance-as-anxiety associated with failure as not weak-not strong, then, as can be seen from Figure 7.7, the overall significance of this stage for the participant proves to be very low. For each stage of their future professional career and family life, values of overall significance were revealed in *general*, *material*, and *spiritual* terms.

All these variables were statistically analyzed; as a result, the primary indicators of significance of the events and their confidence intervals were established. This procedure and factor analysis both enabled us to compare the responses of Estonian and Russian students (young men and women) attending different university departments, with different academic proficiency and social backgrounds.

Below we analyze some specific results of the study. First of all, the findings point to some differences between Estonian and Russian students in terms of their expectations regarding the future. The Russian students had more hope of success in their professional careers and in material terms, in general, but they were also more apprehensive of failure. These differences (at a confidence level of $C = 99\%$, i.e., $\alpha = 0.01$) characterized their views with regard to both the beginning of their careers and the period of stabilization. The Estonian students were more afraid of failure in spiritual terms at the beginning of their careers (difference at a confidence level

TABLE 7.1

Fragment of the questionnaire completed by Students.

| | | | Assessment of the future in… | | | | | |
| | | | General terms | | Material terms | | Spiritual terms | |
Stage No	Question No	Question	Likelihood	Intensity	Likelihood	Intensity	Likelihood	Intensity
1.		Professional career						
	1.	How do you estimate a chance of remarkable success at the beginning of your future career in the attained occupation, and what do you predict will be the level of that success?						
	2.	How do you estimate a chance of remarkable failure at the beginning of your professional career in the attained occupation, and what do you predict will be the level of that failure?						
3.		Family life						
	11.	What chance do you have of attaining happy family life, and what will be the level of such happiness?						
	12.	What chance will there be for failure in your family life, and what will be the level of such failure?						

of $C = 99\%$). The overall evaluation of the Estonian students' future professional careers (involving all three stages) was more optimistic than that of their Russian counterparts ($C = 95\%$). Essentially, we can conclude that the Russian students are more prepared for future success and failure in general and in material terms; the Estonians, although fearing failure in spiritual terms, are generally more optimistic about their future professional careers.

The comparison of differences between the Estonian and Russian students concerning their expectations of family life demonstrates that Russians are more confident of their success in general terms at all three stages ($C = 99\%$). At the same time, the Russians are more afraid of failure at the middle and later stages in general terms, and in spiritual terms at all stages of family life. The overall evaluations of the expectations of the Estonian students about their family life in general terms at the beginning, and in material terms at all three stages, were higher than those of the Russians ($C = 90\%$). These findings suggest that the Russian students are better prepared for success and for failure in their family life. The Estonians, meanwhile, are oriented toward more of a stable family life and are somewhat more optimistic in this respect.

It can therefore be concluded that the research results adequately reflect, on one hand, the uncertainty of Russian students about their future in present day Estonia, and on the other hand, the greater confidence of the Estonian students. Some differences in the ratings by young men and women should also be mentioned. For example, men are more oriented than women toward a successful professional career in spiritual terms ($C = 90\%$) and are also more prepared for failure ($C = 90\%$). However, young women are generally more optimistic than men about their future career opportunities ($C = 95\%$).

The responses of men and women also display differences with regard to expectations concerning family life. Men believe in the material well being of their future family life during all three of the considered stages more so than women do ($C = 90\%$), but are also more afraid of material failure ($C = 95\%$). Young women are more fearful of failure in family life in general terms ($C = 90\%$).

With regard to a comparison of the research results in general, we can say that they sufficiently reflect reality. Attitudes toward future family life reflect innate differences between men and women: men have a drive to create and advance, whereas women are traditionally more concerned with maintaining and caring for the things that have already been achieved.

For revealing differences in the values of students majoring in different disciplines, we employed factor analysis. This analysis indicated that differences concerned only expectations of success. For example, it was found that philology students believed more in general and material success than did medical students (difference at a confidence level of $C = 95\%$) after both attaining and ending a job career. Mathematics students believed more in the success of their professional career in spiritual terms than did medical students ($C = 95\%$). This suggests that medical students are more modest in their expectations of success in their professional career than are philology students (who are usually more enthusiastic) and mathematics students (who are, as a rule, more self-confident). Hence, the findings seem to be quite valid. Insofar as judgements about future family life were concerned, no significant differences were found in the responses of students majoring in different disciplines.

The analysis also revealed a certain effect of students' academic performance on their view of the future. The best students were the most enthusiastic believers in a successful professional career in spiritual terms. In material terms, students whose academic performance was about average were most convinced of their success. We consider these findings quite valid: the best students are oriented more toward the intellectual aspect of their future career and believe in it, whereas average students give their consideration to the material side of their future career. Students' academic performance seems not to affect their views about future family life.

With regard to students' social background, only one significant difference influences their future family life: those with a professional background (physicians, teachers, engineers, etc.) expect more success in intellectual terms from their future family life than do those who come from employee families ($C = 95\%$). This result suggests that the children of intellectuals are more oriented toward the spiritual aspect of well being in their future family life. This finding seems to be quite valid.

Winding up the discussion of the research results, it can be said that the conclusions are generally in accordance with socially established ideas about the discussed problems. Therefore, the results indicate that the approach we employed is rather reliable.

Concluding the discussion of the primary results of the conducted study, it is possible to determine that they are in accord with existing social standards, conditions, and views on the issues discussed, so they can be regarded as evidence of the validity of the proposed approach and the ensuing method of evaluating people's attitudes toward emotive (anxiety-inducing and attractive) events. Of course, as M. Kotik notes (1995, p. 112), to check the validity of the approach, it would have been better to compare the means of its results with those of the other—proven to be more reliable—approach. However, the other approach, which allows assessment of people's attitudes toward emotive events while considering their conscious and subconscious motives, was not known to him. Thus, when evaluating the validity of the suggested approach, he conducted comparisons of the obtained results with practice.

REFERENCES

1. Atkinson, J. (1957). Motivational determinants of risk-taking behavior. *Psychological Review*, 64, 359–372.
2. Bassin, F. V. (1973). The problem of meaning and sense. *Questions of Psychology*, 6, 13–24.
3. Bedny, G., Meister, D. (1997). The Russian *Theory of Activity: Current Application to Design and Learning*. Mahwah, NJ: Lawrence Erlbaum Associates.
4. Bedny, G., Karwowski, W., and Bedny, I. (2015). *Applying Systemic-Structural Activity Theory to Design of Human-Computer Interaction Systems*, Boca Raton, CRC Press/Taylor & Francis Group.
5. Bernstein, N. A. (1962). New fields in the development of physiology and their relation to cybernetics. *Questions of Philosophy*, 8, 78–87.
6. Dixon W, Massey F. (1969). *Introduction to statistical analysis*. New York: McGraw-Hill.
7. Harkevich A. A. (1960). On the value of information. In A. A. Lyapunov (Ed.), *Problems of Cybernetics*, v.5 (pp. 53–57). Russia, Moscow: Publishing House for Physical and Mathematical Literature.

8. Heckhausen, H (1991). *Motivation and action*. Berlin, Germany: Spring-Verlag.
9. Kitaev-Smyk, L. A. (1983). *Psychology of stress*. Moscow, Russia: Nauka (Science) Publishers.
10. Kotik, M. A. (1974a). On the apprehension of significance as regulators of activity. *Studies in psychology III. Scientific Notes of Tartu University*, v. 335 (pp. 52–65). Tartu, Estonia: Tartu University Press.
11. Kotik, M. A. (1974b). *Self-regulation and reliability of operator*. Tallinn, Estonia: Valgus.
12. Kotik M. A. (1978a). A method of diagnostics of a person's attitude towards an alarming event. *Problems of communication and perception. Studies in psychology VII. Scientific Notes of Tartu University*, v. 474, Tartu, Estonia: Tartu University Press, pp. 162–179.
13. Kotik M. A., (1978b). On the method of evaluating the conscious and subconscious in the significance factor. In *Subconscious: principles, functions, and methods of experimental analysis*, v. 3 (pp. 651–659). Georgia, Tbilisi: Mecniereba.
14. Kotik, M. A., Öövel, L. (1980). Developing a method to measure attitudes towards dangerous or alarming events. *Problems of cognitive psychology. Studies in psychology VIII. Scientific Notes of Tartu University*, v. 522 (pp. 77–85). Tartu, Estonia: Tartu University Press.
15. Kotik, M. A., Sirts T. (1983). The influence of the attitude towards danger on the accident-proneness. *Problems of perception and social interaction. Studies in psychology. Scientific Notes of Tartu University*, v. 638 (pp. I2I–I34). Tartu, Estonia: Tartu University Press.
16. Kotik, M. A., Yemelyanov, A. M. (1983). A method for evaluating the significance-as-anxiety of information. *Studies in artificial intelligence. Scientific Notes of Tartu University*, v. 654 (pp. 111–129). Tartu, Estonia: Tartu University Press.
17. Kotik, M. A. (1984). A method for evaluating the significance-as-value of information. *Studies in artificial intelligence. Scientific Notes of Tartu University*, v. 688 (pp. 86–102). Tartu, Estonia: Tartu University Press.
18. Kotik, M. A., Yemelyanov, A. M. (1985). *Errors in control action. Psychological causes, method of automated analysis*. Estonia, Tallinn: Valgus.
19. Kotik, M. A. (1986). New studies of dispositions toward professional dangers. *Structure of Cognitive Processes. Studies in psychology XIV. Scientific Notes of Tartu University*, v. 753 (pp. 123–139). Tartu, Estonia: Tartu University Press.
20. Kotik, M. A., Yemelyanov, A. M. (1986). A rapid method for evaluating subjective preferences in problem solving. *Structure of Cognitive Processes. Studies in psychology XIV. Scientific Notes of Tartu University*, v. 753 (pp. 140–160). Tartu, Estonia: Tartu University Press.
21. Kotik, M. A. (1989). *Psychology and safety* (3rd ed.). Tallinn, Estonia: Valgus.
22. Kotik, M. A., Yemelyanov, A. M. (1992). Emotions as an indicator of subjective preferences in decision making. *Psychological Journal*, 13 (1), 118–125.
23. Kotik, M. A., Yemelyanov, A. M. (1993). *The origin of human-operator errors*. Russia, Moscow, Transport Publishers.
24. Kotik, M. A. (1994a). New method of evaluating people's attitudes toward anxiety-inducing events. *Questions of Psychology*, 1, 97–104.
25. Kotik, M. A. (1994b). Developing applications of "Field Theory" in mass studies. *Journal of Russian East European Psychology*, 2 (4), July–August, 1994, 38–52.
26. Kotik M. A. (1995). How students of Tartu University view their future. *Questions of Psychology*, 2, 105–112.
27. Lager, K. (1970). Experimental methods and results of measuring stress in the modeling of flight conditions. In *Emotional stress* (pp. 290–295). Russia, Leningrad: Medical Publishers.

28. Larichev, O. I. (1979). *Science and art of decision making.* Moscow, Russia: Science Publisher.
29. Leont'ev, A. N. (1975). *Activity. Consciousness, Personality.* Moscow, Russia: Political Publishers.
30. Lewin, K. (1939). Field Theory and Experiment in Social Psychology. *American Journal of Sociology,* 44 *(6),* 868–896.
31. Lewin, K., Dembo, T., Festinger, L., and Sear, P.S., (1944). Level of aspiration. In J. M. Hunt (Ed.), *Personality and the behavior disorders* (p. 333–378). New York: Roland Press.
32. Mckenna, F. P. (1982). The human factor in driving accidents. *Ergonomics,* 23 (10), 867– 877.
33. Rotter, J. B. (1954). *Social learning and clinical psychology.* New York: Prentice Hall.
34. Meyer, L. B. (1956). *Emotion and Meaning in Music.* Chicago and London: The University of Chicago press.
35. Simonov, P. V. (1981). *The emotional brain.* New work, Plenum Press.
36. Tolman, E.C. (1951). A psychological model. In T. Parsons and E.A. Shils (Eds.), *Toward a general theory of action* (pp. 279–361). New York: Harper Torchbooks.
37. Vroom, V. H. (1964). *Work and motivation.* New York: Wiley.
38. Vygotsky, L. S. (1956). *Selected Psychological Research.* Moscow, Russia: Academy of Pedagogical Sciences.
39. Zadeh, L. A. (1973). Outline of a new approach to the analysis of complex systems and decision processes. *IEEE Transactions on Systems, Man, and Cybernetics,* v. 3, 28–44.

APPENDIX

Mikhail A. Kotik

Mikhail Kotik's research work and teaching were primarily conducted at Tartu University. However, Kotik started his research long before he began his work at the University. As an aeronautical engineer by training, he was responsible for flight safety and worked as chief electrical engineer of military aviation in Tartu, Estonia from 1959 until 1970. After ending up in air traffic accidents numerous times, as well as actively participating in the investigation of the origin of airplane crashes, Kotik was capable not only of determining the causes of the accidents, but also delving deeper into the particular working conditions of pilots in both regular and emergency situations. He probed further into research, while constructing special devices that enabled him to conduct flight experiments and record data on pilot behavior in different emergency situations. The research for all his books and articles was conducted on Tartu's airbase. On the basis of his studies, he published numerous books and handbooks, the first of which was titled "*A Short Discourse on Engineering Psychology*" (1971). Based on all this research, general guidelines in the field of psychology of occupational safety and risk prevention were developed. In 1981, Kotik published "*Psychology and Safety,*" the first work on this topic in Eastern Europe. In the last decade of his life, Kotik studied the psychological background of human operator error causes, as well as developing their mathematical models.

8 Mental Workload vs. Human Capacity Factor: A Way to Quantify Human Performance

Ephraim Suhir
Portland State University, Maseeh College of
Engineering and Computer Science, CA, USA

"A pinch of probability is worth a pound of perhaps."

James G. Thurber,
American writer and cartoonist

CONTENTS

ACRONYMS

ATC   air traffic controller
COV   coefficient of variation
DEPDF   double exponential probability distribution function
EVD   extreme value distribution
FDR   flight data record
FOAT   failure-oriented-accelerated testing
GWB   George Washington bridge
HCF   human capacity factor
HITL   human-in-the-Loop
KCAS   knots calibrated air speed
LGA   La Guardia airport
MWL   mental workload
NTCB   National Transportation Safety Board
PM   predictive modeling
PRM   probabilistic risk management
RAT   ram air turbine
SF   safety factor
SM   safety margin
TRACON   terminal radar approach control

SUMMARY

While **considerable** improvements in various vehicular (aerospace, maritime, automotive, railroad, etc.) technologies and other human-in-the-loop (HITL) related missions and situations can be achieved through better ergonomics, better work environments, and other traditional means that directly affect human behaviors, there is also the potential opportunity for a further reduction in vehicular casualties, as well as in increasing the likelihood of the success and safety of vehicular missions and off-normal situations, through better understanding of the role that various uncertainties play in the designer's and operator's world of work. By employing quantifiable and measurable ways to assess the role of these uncertainties, and by treating a HITL as a part, often the most critical part, of the complex human-instrumentation-equipment-vehicle-environment system, one could improve dramatically the individual's performance, to predict, minimize, and even specify the probability of the occurrence of a mishap that is never completely avoidable.

It is the author's belief that adequate human performance in whatever situation or application cannot be assured if it is not quantified, and since no one is perfect, that such quantification should be done preferably on a probabilistic basis. In effect, the only difference between what is perceived as a failure-free and an unsatisfactory human performance is, in effect, the difference in the levels of the never-zero probability of his or her failure. In the simplest model, such a failure should be attributed to an insufficient human capacity factor (HCF) when they have to cope with a high cognitive (mental) workload (MWL).

Our suggested MWL/HCF models and their possible modifications and generalizations can be helpful, after appropriate sensitivity factors are established and sensitivity analyses (SA) are carried out 1) when developing guidelines for personnel selection and training; 2) when choosing the appropriate simulation conditions;

and/or 3) when there is a need to decide, if the existing levels of automation and of the employed equipment (instrumentation) are adequate in off-normal, but not impossible, situations, and if not, 4) whether additional and/or more advanced and perhaps more expensive equipment or instrumentation should be developed, tested, and installed, so that the requirements and constraints associated with a mission or a situation of importance are met. Our MWL/HCF-based approach is, in effect, an attempt to quantify, on a probabilistic basis, using probabilistic risk management (PRM) techniques, the role that the human plays, in terms of their ability (capacity) to cope with a mental overload. Using an analogy from the reliability engineering field and particularly with the well known stress-strength interference model, the MWL could be viewed as a certain possible demand (stress), while the HCF can be viewed as an available or a required capacity (strength).

The MWL level depends on the operational conditions and the complexity of the mission, i.e., it has to do with the significance of the general task, while the HCF considers, but might not be limited to, the human's professional experience and qualifications, capabilities and skills; level and specifics of their training; performance sustainability; ability to concentrate; mature thinking; ability to operate effectively in a tireless fashion, under pressure, and if needed, for a long period of time (tolerance to stress); team player attitude; swiftness in reaction, if necessary, etc., That is, all the critical qualities that would enable them to cope with the high MWL. It is noteworthy that the ability to evaluate the absolute level of the MWL, important as it might be for numerous existing non-comparative evaluations, is less critical in our approach: it is the comparative levels of the MWL and the HCF, and the comparative assessments and evaluations that are important in our approach.

The author does not intend to come up with an accurate, complete, ready-to-go, off the shelf type of methodology in which all the i's are dotted and all the t's are crossed, but intends to show how the powerful and flexible PRM method could be effectively employed to quantify the role of the human factor by comparing, on the probabilistic basis, the actual and/or possible MWL and the available or required HCF levels, so that an adequate and sufficient safety factor is assured. In this chapter, the famous miracle-on-the-Hudson event is used as a suitable example to illustrate the concept in question. We believe that the approach taken, with the appropriate modifications and generalizations, is applicable to many HITL situations, not necessarily in the vehicular domain, when a human encounters an uncertain environment and/or a hazardous situation, and/or interacts with never perfect hardware and software.

The author realizes that his approach might not be accepted easily by some traditional psychologists. They might feel that the problem is too complex to lend itself to any type of formalized quantification. With this in mind we are suggesting possible future work that could be conducted using, when necessary, flight simulators to correlate the suggested probabilistic models with the existing practice. Testing on a flight simulator is analogous to highly-accelerated life testing (HALT), and particularly failure-oriented-accelerated testing (FOAT) in electronics and photonics reliability engineering.

The famous Hudson event is chosen in this chapter to illustrate the possible application of the MWL-HCF bias in HITL related missions and situations. It is important to emphasize that this is merely an illustration on how these two major aspects of the HITL related situation could be treated, and not to show, in a rather tentative fashion, why indeed Capt. Sullenberger was successful in an extraordinary situation, where other navigators may or may have been. As Gottfried Leibnitz, the famous German mathematician put it, "There are things in this world, far more important than the most splendid discoveries … it is the methods by which they were made."

8.1　INTRODUCTION

Application of the quantitative probabilistic risk management (PRM) concept should complement in various HITL related situations, whenever feasible and possible, the existing vehicular psychology practices, which are typically qualitative a-posteriori statistical assessments. A PRM approach based on the double exponential probability distribution function (DEPDF) of the extreme value distribution (EVD) type is suggested as a suitable quantitative technique for assessing the probability of the human non-failure in an off-normal flight situation. The HCF is introduced in this distribution and considered along with the (elevated) short-term MWL that the human (pilot) has to cope with in an off-normal (emergency) situation. The famous 2009 US Airways miracle successful landing (ditching) and the infamous 1998 Swissair UN shuttle disaster are chosen to illustrate the usefulness and fruitfulness of the approach. It is shown that it was the exceptionally high HCF of the US Airways crew, and especially that of its captain Sullenberger, that made a reality what seemed to be at first glance, a miracle. It is shown also that the highly professional, and in general, highly qualified Swissair crew exhibited inadequate performance (quantified in our analysis as a relatively low HCF level) in the off-normal situation they encountered. The Swissair crew made several fatal errors, and as a result, crashed the aircraft. In addition to the DEPDF-based approach, we show that the probability of a safe landing can be evaluated by comparing the (random) operation time (consisting of the decision-making time and the landing time) with the "available" time needed for landing. We conclude that the developed formalisms, after trustworthy input data are obtained (using flight simulators or applying the Delphi method) might be applicable even beyond the vehicular domain and can be employed in various HITL situations when a short term high human performance is imperative and therefore the ability to quantify it is highly desirable. We also conclude that although the obtained numbers make physical sense, it is the approach, not the numbers, that is, in the author's opinion, the main merit of the paper.

Human error contributes to about 80% of vehicular (avionic, maritime, railroad, automotive) casualties (see, e.g., Reason, 1990, 1997; Kern, 2001; O'Neil, 2001; Foyle, Hooey, 2008; Harris, 2011; Hollnagel, 1993). Ability to understand their nature and minimize their likelihood is of obvious and significant importance. Considerable safety improvements in various off-normal vehicular situations can be achieved through better training, better ergonomics, better work environment, and other human psychology related means and efforts that directly affect human behavior and performance: psychological analysis of casualties, computer aided simulations (including attempts to mimic the actual situation in an aircraft cockpit or in a space shuttle cabin), and a-posteriori statistical analyses of the occurred casualties and accidents. There is also an opportunity for casualty reduction through better understanding of the role that different uncertainties play in the operator's world of work: environmental conditions; dependability and availability of instrumentation and equipment; trustworthiness, consistency and user friendliness of the obtained information; predictability and timeliness of the response of the object of control (aircraft, spacecraft, boat) to the navigator's actions; performance of the interfaces of these factors, etc. By employing quantifiable and measurable ways of assessing the role of various critical uncertainties and by treating a HITL as a part (often as the most crucial part) of the complex human-instrumentation-equipment-vehicle-environment system, one could dramatically improve an individual's performance, and predict and minimize the probability of a mission failure (Suhir, 1997; Suhir, 2010; Suhir and Mogford, 2011; Suhir, 2011 and 2012).

PRM-based concepts, methods, approaches, and algorithms could and should be widely used, in addition to the psychological activities and efforts, when there is a need to evaluate, quantify, optimize, and when possible and appropriate, even specify the human capacity to cope with an elevated MWL. The following ten factors that affect mission success and safety in various HITL situations should be considered:

- human performance (capacity) factor;
- navigation, information, and control instrumentation (equipment) factor;
- vehicle (object of control) factor;
- environmental factor; and
- six interfaces between (interactions of) the above factors.

All these factors and their interfaces are associated with uncertainties that contribute to the cumulative probability that a certain established safety criterion for a particular anticipated casualty or mishap is violated. These uncertainties are characterized by their probability distributions, safety criteria, consequences of possible failure, and the levels of the acceptable risk.

When adequate human performance in a particular critical HITL situation is imperative, the ability to quantify the human factor is highly desirable. Such quantification could be done particularly by comparing the actual or anticipated MWL with the likely (available) human capacity factor (HCF). The MWL vs. HCF based models and their modifications and generalizations can be helpful, particularly after appropriate algorithms are developed and extensive sensitivity analyses are carried out to

- evaluate the role that the human plays, in terms of their ability to cope with a MWL in various situations when human factors, equipment/instrumentation performances and uncertain and often harsh environments contribute jointly to the success and safety of a task or a mission;
- assess the risk of a particular mission success and safety, with consideration of the human-in-the-loop performance;
- develop guidelines for personnel selection and training;
- choose the appropriate simulation conditions; and/or decide if existing levels of automation and the employed equipment (instrumentation) are adequate in possible off-normal situations (if not, additional and/or more advanced and perhaps more expensive equipment or instrumentation should be developed, tested, and installed).

In the analysis that follows the DEPDF-based model is applied for the evaluation of the likelihood of a human non-failure in an emergency vehicular mission success and safety situation. The famous 2009 miracle on the Hudson event and the infamous 1998 UN shuttle disaster are used to illustrate the substance and fruitfulness of the approach. We try to shed probabilistic light on these two well-known events. As far as the the Hudson incident is concerned, we intend to provide quantitative assessments of why such a "miracle" could have actually occurred, and what had been and had not been indeed a miracle in this incident: a divine intervention, a perceptible interruption of the laws of nature, or simply a wonderful and rare occurrence that was due to a heroic act of the aircraft crew and especially of its captain Sullenberger, the lead "miracle worker" in the incident. As to the UN shuttle crash, we are going to demonstrate that the crash occurred because of the low HCF of the aircraft crew in an off-normal situation that they had encountered and that was, in effect, much less demanding than the Hudson situation. Some other reported water landings (ditchings) of passenger airplanes are listed in Appendix A. Some of them have ended successfully.

8.2   PRM-BASED HCF VS. MWL APPROACH: TEN COMMANDMENTS

Here are the major principles (i.e., ten commandments) of our PRM-based approach:

1. HCF is viewed in this approach as an appropriate quantitative measure (not necessarily and not always probabilistic though) of the human ability to cope with an elevated short term MWL.
2. It is the relative levels of the MWL and HCF (whether deterministic or random) that determine the probability of human non-failure in a particular HITL situation.
3. Such a probability cannot be low, but need not be higher than necessary either: it has to be adequate for a particular anticipated application and situation.
4. When adequate human performance is imperative, ability to quantify it is highly desirable, especially if one intends to optimize and assure adequate HITL performance.
5. One cannot assure adequate human performance by just conducting routine today's human psychology-based efforts (which might provide appreciable improvements, but do not quantify human behavior and performance; in addition, these efforts might be too and unnecessarily costly), and/or by just following the existing "best practices" that are not aimed at a particular situation or an application; the events of interest are certainly rare events, and best practices might or might not be applicable.
6. MWLs and HCFs should consider, to the extent possible, the most likely anticipated situations; obviously, the MWLs are and HCFs should be different for a jet fighter pilot, for a pilot of a commercial aircraft, or for a helicopter pilot, and should be assessed and specified differently.

7. PRM is an effective means for improving the state-of-the-art in the HITL field: nobody and nothing is perfect, and the difference between a failed human performance and a successful one is the result of the level of non-failure probability.

8. Failure-oriented-accelerated testing (FOAT) on a flight simulator is viewed as an important constituent part of the PRM concept in various HITL situations. It is aimed at better understanding the factors underlying possible failures; it might be complemented by the Delphi effort (see the first link in the References);

9. Extensive predictive modeling (PM) is another important constituent of the PRM based effort, and in combination with highly focused and highly cost effective FOAT, is a powerful and effective means of quantifying and perhaps nearly eliminating human failures.

10. Consistent, comprehensive, and psychologically meaningful PRM assessments can lead to the most feasible HITL qualification (certification) methodologies, practices, and specifications.

8.3   MOST LIKELY (NORMAL) MENTAL WORKLOAD (MWL)

Our HCF vs. MWL approach considers elevated (off-normal) random relative HCF and MWL levels with respect to the ordinary (normal, established) deterministic HCF and MWL values. These values could and should be established based on the existing human psychology practices.

The interrelated concepts of situation awareness and MWL ("demand") are central to today's aviation psychology. Cognitive overload has been recognized as a significant cause of aviation error. The MWL is directly affected by the challenges that navigators face when controlling the vehicle in a complex, heterogeneous, multitask, and often uncertain and harsh environment. Such an environment includes numerous and interrelated concepts of situation awareness: spatial awareness for instrument displays; system awareness for keeping the pilot informed about actions that have been taken by automated systems; and task awareness that has to do with the attention and task management. The time lags between critical variables require predictions and actions in an uncertain world. The MWL depends on the operational conditions and on the complexity of the mission. MWL has to do therefore with the significance of the long- or short-term task. The long-term MWL is illustrated in Figure 8.1.

Task management is directly related to the level of MWL, as the competing demands of the tasks for attention might exceed the operator's resources—their capacity to adequately cope with the demands imposed by the MWL.

Measuring the MWL has become a key method of improving aviation safety. There is an extensive published work in the psychological literature devoted to the measurement of MWL in aviation, both military and commercial. Pilot's MWL can be measured using subjective ratings and/or objective measures. The subjective ratings during FOAT (simulation tests) can be, e.g., after the expected failure is defined, in the form of periodic inputs to some kind of data collection device that prompts the pilot to enter a number between 1 and 10 (for example) to estimate MWL every few minutes. There are some objective MWL measures, such as heart rate variability for

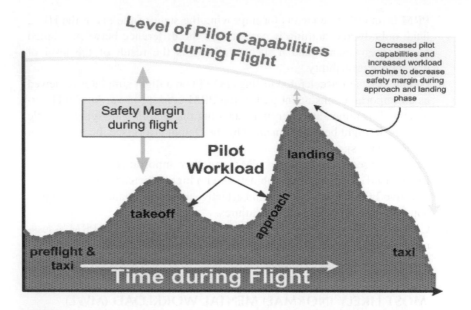

FIGURE 8.1   Long-term (pilot capabilities) HCF vs. MWL (pilot workload).

example. Another possible approach uses post-flight paper questionnaires. It is easier to measure MWL on a flight simulator than in actual flight conditions. In a real aircraft, one would probably be restricted to using post-flight subjective (questionnaire) measurements, since you would not want to interfere with the pilot's work.

Given the multidimensional nature of MWL, no single measurement technique can be expected to account for all the important aspects of it. In modern military aircraft, the complexity of information, combined with time stress, creates difficulties for the pilot under combat conditions, and the first step in mitigating this problem is to measure and manage the MWL. Current research efforts in measuring MWL use psycho-physiological techniques, such as electroencephalographic, cardiac, ocular, and respiratory measures in an attempt to identify and predict MWL levels. Measuring cardiac activity has been a useful physiological technique employed in the assessment of MWL, both from tonic variations in heart rate and after treatment of the cardiac signal.

8.4 MOST LIKELY (NORMAL) HUMAN CAPACITY FACTOR (HCF)

HCF includes, but might not be limited to, the following major qualities that would enable a professional to successfully cope with an elevated off-normal MWL:

- psychological suitability for a particular task
- professional experience and qualifications
- education, both special and general
- relevant capabilities and skills
- level, quality, and timeliness of training
- performance sustainability (consistency, predictability)
- independent thinking and independent acting, when necessary

- ability to concentrate
- ability to anticipate
- self control and ability to act in a rational and methodical manner in haz-ardous and even life threatening situations
- mature (realistic) thinking
- ability to operate effectively under pressure, and particularly under time pressure
- ability to operate effectively, when necessary, in a tireless fashion, for a long period of time (tolerance to stress)
- ability to act effectively under time pressure and make well substantiated decisions in a short period of time
- team player attitude, when necessary
- swiftness of reaction, when necessary

These and other qualities are certainly of different importance in different HITL situations. It is also clear that different individuals possess these qualities in different degrees. Long-term HCF could be time dependent. To come up with a suitable figures of merit (FOM) for the HCF, one could rank the above list and perhaps other qualities on a scale from, for example one to four, and calculate the average FOM for each individual and particular task and/or a mission or a situation (see, e.g., Tables 8.5, 8.6 and 8.8 below).

8.5 DOUBLE-EXPONENTIAL PROBABILITY DISTRIBUTION FUNCTION (DEPDF)

Different PRM approaches can be used in the analysis and optimization of the interaction of the MWL and HCF. When the MWL and HCF characteristics are treated as deterministic ones, a high enough safety factor $SF = \dfrac{HCF}{MWL}$ can be used. When both MWL and HCF are random variables, the safety factor can be determined as the ratio $SF = \dfrac{\prec SM \succ}{S_{SM}}$ of the mean value $\prec SM \succ$ of the random safety margin $SM = HCF - MWL$ to its standard deviation S_{SM}. When the capacity demand (strength-stress) interference model is used (Figure 8.2) the HCF can be

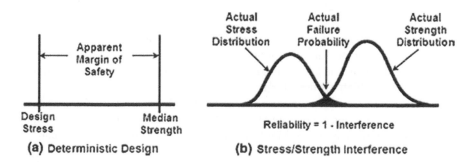

FIGURE 8.2 Capacity demand (strength-stress) interference model.

viewed as the capacity (strength) and the MWL as the demand (stress), and their overlap area could be considered as the potential (probability) of possible human failure. The capacity and the demand distributions can be steady state or transient, i.e., their mean values can move toward each other when time progresses, and/or the MWL and HCF curves can be spread over larger areas. Yet another PRM approach is to use a single distribution that accounts for the roles of the HCF and MWL, when these (random) characteristics deviate from (are higher than) their (deterministic) most likely (regular) values. This approach is used in the analysis below.

A double-exponential probability distribution function (DEPDF).

$$P_h(G,F) = P_0 \exp\left[\left(1 - \frac{G^2}{G_0^2}\right)\exp\left(1 - \frac{F^2}{F_0^2}\right)\right], \quad G \geq G_0, \quad F \geq F_0 \qquad (8.1)$$

of the extreme value distribution (EVD) type (see, e.g., [8]) can be used to characterize the likelihood of a human non-failure to perform his/her duties, when operating a vehicle [10, 11]. Here $P_h(G,F)$ is the probability of non-failure of the human performance as a function of the off normal mental workload (MWL) G and outstanding human capacity factor (HCF) F, P_0 is the probability of non-failure of the human performance for the specified (normal) MWL $G = G_0$ and the specified (ordinary) HCF $F = F_0$. The specified (most likely, nominal, normal) MWL and HCF can be established by conducting testing and measurements on a flight simulator. The calculated probabilities

$$P = \frac{P_h(G,F)}{P_0} = \exp\left[\left(1 - \frac{G^2}{G_0^2}\right)\exp\left(1 - \frac{F^2}{F_0^2}\right)\right], \quad G \geq G_0, \quad F \geq F_0 \qquad (8.2)$$

(that are, in effect, ratios of the probability of non-failure in the off-normal conditions to the probability of non-failure in the normal situation) are shown in Table 8.1. The following conclusions can be drawn from the table data:

- At a normal (specified, most likely) MWL level ($G = G_0$) and/or at an extraordinary (exceptionally) high HCF level ($F \mapsto \infty$) the probability of human non-failure is close to 100%.
- The probabilities of human non-failure in off-normal situations are always lower than the probabilities of non-failure in normal (specified) conditions.
- When the MWL is extraordinarily high, the human will definitely fail, no matter how high his/her HCF is.
- When the HCF is high, even a significant MWL has a small effect on the probability of non-failure, unless the MWL is exceptionally high. For high HCFs, the increase in the MWL has a much smaller effect on the probabilities of failure than for relatively low HCFs.
- The probability of human non-failure decreases with an increase in the MWL, especially at low MWL levels, and increases with an increase in the HCF, especially at low HCF levels.

TABLE 8.1
Calculated probability ratios of human non-failure.

G/G_0	1	2	3	4	5	8	10	∞
F/F_0	x	x	x	x	x	x	x	x
1	1	4.979 E-2	3.355 E-4	3.059 E-7	3.775 E-11	4.360 E-28	1.011 E-43	0
2	1	0.8613	0.6715	0.4739	0.3027	0.0434	0.007234	0
3	1	0.9990	0.9973	0.9950	0.9920	0.9791	0.9673	0
4								0
5								0
8					1.0000			0
10								0
∞								1.0000

These intuitively more or less obvious conclusions are quantified by the Table 8.1 data. These data show also that the increase in the probability ratio above 3.0 ("three is a charm") has a minor effect on the probability of non-failure. This means particularly that the navigator (pilot) does not have to be trained for an unrealistically high MWL, i.e., does not have to be trained by a factor higher than 3.0 compared to a navigator of ordinary capacity (skills, qualification). In other words, a pilot does not have to be a superman to successfully cope with a high level MWL, but still has to be trained in such a way that, when there is a need, he or she would be able to cope with a MWL by a factor of 3.0 higher than the normal level, and their HCF should be by a factor of 3.0 higher than what is expected of the same person in ordinary, normal conditions. Of course, some outstanding individuals (like Capt. Sullenberger, for instance) might be characterized by the HCF that corresponds to MWL's somewhat higher than 3.0 (see Table 8.5).

8.6   PHYSICAL MEANING OF THE DEPDF

From (2) we find, by differentiation:

$$\frac{dp}{dG} = -2\frac{H(p)}{G}\frac{1}{1-\frac{G_0^2}{G^2}} \tag{8.3}$$

where $H(p) = -p\ln p$ is the entropy of the distribution of the relative probability of the human non-failure in extraordinary (off-normal) operation conditions. When the MWL G is significant, the formula (3) can be simplified:

$$\frac{dp}{dG} = -2\frac{H(p)}{G}. \tag{8.4}$$

This result explains the physical meaning of the distribution: the change in the probability of human non-failure (provided that the probability of non-failure in normal conditions is simply 100%) when the change in the MWL is, for large MWL levels, proportional to the uncertainty level that is defined by the entropy of the distribution in question and is inversely proportional to the MWL level. The right part of the formula (8.4) can be viewed as a kind of coefficient of variation (COV), where the role of the uncertainty level in the numerator is played by the entropy, rather than by the standard deviation, and the role of the stress (loading) level in the denominator is played by the MWL rather than by the mean value of the random characteristic of interest.

From (2) one could find also:

$$\frac{dp}{dF} = 2\frac{H(p)F}{F_0^2} \tag{8.5}$$

When the random HCF F is equal to its nominal value F_0, this formula yields:

$$\frac{dp}{dF} = 2\frac{H(p)}{F_0} \tag{8.6}$$

This result can also be used to interpret the physics underlying the DEPDF (2): the change in the probability of human non-failure with the change in the HCF at its nominal (normal) level is proportional to the entropy of the distribution (2) and is inversely proportional to the nominal HCF.

8.7   HCF NEEDED TO SATISFACTORILY COPE WITH A HIGH MWL

From (2) we obtain:

$$\frac{F}{F_0} = \sqrt{1 - \ln\left(\frac{\ln(p)}{1 - \frac{G^2}{G_0^2}}\right)} \tag{8.7}$$

This relationship is tabulated in Table 8.2. The following conclusion can be drawn from the computed data:

- The HCF level needed to cope with an elevated MWL increases rather slowly with an increase in the probability-of-non-failure, especially for high MWL levels, unless this probability is very low (below 0.1) or very high (above 0.9).
- In the region $p = 0.1 \rightarrow 0.9$, the required high HCF level increases with an increase in the MWL level, but this increase is rather moderate, especially for high MWL levels.
- Even for significant MWLs that exceed the normal MWL by orders of magnitude, the level of the HCF does not have to be very much higher than the HCF of a person of ordinary HCF level. When the MWL ratio is as high as 100, the HCF ratio does not have to exceed four to assure the probability of non-failure of as high as 0.999.

TABLE 8.2
Relative HCF F/F_0 vs. relative probability of non-failure and relative MWL.

p	E-12	E-3	E-2	0.1	0.5	0.9	0.99	0.9999
G/G_0	x	x	x	x	x	x	x	x
5	1.0681	1.4985	1.6282	1.8287	2.1318	2.5354	2.9628	3.6590
10	1.5087	1.9138	2.0169	2.1820	2.4416	2.8010	3.1930	3.8478
100	2.6251	2.8771	2.9467	3.0621	3.2522	3.5300	3.8484	4.4069
1000	3.3907	3.5893	3.6453	3.7392	3.8964	4.1311	4.4063	4.9016
10000	4.0127	4.1819	4.2301	4.3112	4.4483	4.6552	4.9011	5.3508

8.8 DIFFERENT APPROACH: OPERATION TIME VS. "AVAILABLE" LANDING TIME

The above time independent DEPDF-based approach enables one to compare, on a probabilistic basis, the relative roles of the MWL and HCF in a particular off-normal HUTL situation. The role of time (e.g., reaction swiftness) is accounted for in an indirect fashion, through the NCF level. In the analysis that follows we assess the likelihood of safe landing by considering the roles of different times directly, by comparing the operation time, which consists of the decision-making time and actual landing time, with the "available" landing time (i.e., the time from the moment when an emergency was determined to the moment of landing). Particularly, we address item 10 of Table 8.4, i.e., the ability of the pilot to anticipate and make a substantiated and valid decision in a short period of time ("We are going to be in the Hudson"). It is assumed, for the sake of simplicity, that both the decision-making and the landing times could be approximated by Rayleigh's law, while the available time, considering (in the case of the miracle-on-the-Hudson flight) the glider conditions of the aircraft, follows the normal law with a high ratio of the mean value to the standard deviation. Safe landing could be expected if the probability that it occurs during the available landing time is sufficiently high. The formalism of such a model is similar to the helicopter-landing-ship (HLS) formalism developed earlier [9].

8.8.1 PROBABILITY THAT THE OPERATION TIME EXCEEDS A CERTAIN LEVEL

If the (random) sum, $T = t + \theta$, of the (random) decision making time, t, and the (random) time, θ, needed to actually land the aircraft is lower, with a high enough probability, than the (random) duration, L, of the available time, then safe landing becomes possible. In the analysis that follows, we assume the simplest probability distributions for the random times of interest. We use Rayleigh's law

$$f_t(t) = \frac{t}{t_0^2} \exp\left(-\frac{t^2}{2t_0^2}\right) \quad f_\theta(t) = \frac{\theta}{\theta_0^2} \exp\left(-\frac{\theta^2}{2\theta_0^2}\right) \tag{8.8}$$

as a suitable approximation for the random times t and θ of decision-making and actual landing, and the normal law

$$f_l(l) = \frac{1}{\sqrt{2\pi}\sigma} \exp\left(-\frac{(l-l_0)^2}{2\sigma^2}\right) \quad \frac{l_0}{\sigma} \geq 4.0 \tag{8.9}$$

as an acceptable approximation for the available time, L. In the formulas (8) and (9), t_0 and θ_0 are the most likely times of decision-making and landing, respectively (in the case of Rayleigh law these times coincide with the standard deviations of the random variables in question), l_0 is the most likely (mean) value of the available time, and σ is the standard deviation of this time. The ratio $\frac{l_0}{\sigma}$ (safety factor) of the mean value of the available time to its standard deviation should be large enough (larger than 4), so that the normal law could be used as an acceptable approximation for a random variable that, in principle, cannot be negative, as it is the case when this variable is time.

The probability, P_*, that the sum $T = t + \theta$ of the random variables t and θ exceeds a certain time level, $\hat{T}$, can be found on the basis of the convolution of two random times distributed in accordance with Rayleigh law as follows:

$$P_* = 1 - \int_0^{\hat{T}} \frac{t}{t_0^2} \exp\left[-\frac{t^2}{2t_0^2}\right]\left[1 - \exp\left(-\frac{(T-t)^2}{2\theta_0^2}\right)\right]dt = \exp\left(-\frac{\hat{T}^2}{2t_0^2}\right) +$$

$$+ \exp\left[-\frac{\hat{T}^2}{2(t_0^2 + \theta_0^2)}\right]\left\{\frac{\theta_0^2}{t_0^2 + \theta_0^2}\left[\exp\left[-\frac{t_0^2\hat{T}^2}{2\theta_0^2(t_0^2 + \theta_0^2)}\right]\right] - \exp\left[-\frac{\theta_0^2\hat{T}^2}{2t_0^2(t_0^2 + \theta_0^2)}\right]\right\} +$$

$$+ \sqrt{\frac{\pi}{2}} \frac{\hat{T}t_0\theta_0}{(t_0^2 + \theta_0^2)^{3/2}}\exp\left[-\frac{\hat{T}^2}{2(t_0^2 + \theta_0^2)}\right]\left\{\left[erf\left[\frac{t_0\hat{T}}{\theta_0\sqrt{2(t_0^2 + \theta_0^2)}}\right]\right] + erf\left[\frac{\theta_0\hat{T}}{t_0\sqrt{2(t_0^2 + \theta_0^2)}}\right]\right\}$$

$$\tag{8.10}$$

where

$$erf(x) = \frac{2}{\sqrt{\pi}}\int_0^x e^{-z^2}dz \tag{8.11}$$

is the error function. When the most likely duration of landing, θ_0, is very small compared to the most likely decision-making time, t_0, the expression (10) yields:

$$P_* = \exp\left(-\frac{\hat{T}^2}{2t_0^2}\right) \tag{8.12}$$

i.e., the probability that the total time of operation exceeds a certain time duration, $\hat{T}$, depends only on the most likely decision-making time, t_0. From (12) we obtain:

$$\frac{t_0}{\hat{T}} = \frac{1}{\sqrt{-2\ln P_*}}. \tag{8.13}$$

If the acceptable probability, P_*, of exceeding the time, $\hat{T}$ (e.g., the available time, if this time is treated as a non-random variable of the level $\hat{T}$), is, say, $P = 10^{-4} = 0.01\%$, then the time of making the decision should not exceed $0.2330 = 23.3\%$ of the time, $\hat{T}$ (expected available time), otherwise the requirement $P \le 10^{-4} = 0.01\%$ will be compromised. If the available time is for example 2 min, then the decision-making time should not exceed 28 sec, which is in good agreement with Capt. Sullenberger's actual decision-making time. Similarly, when the most likely time, t_0, of decision making is very small compared to the most likely time, θ_0, of actual landing, the formula (10) yields:

$$P_* = \exp\left(-\frac{\hat{T}^2}{2\theta_0^2}\right) \tag{8.14}$$

i.e., the probability of exceeding a certain time level, $\hat{T}$, depends only on the most likely time, θ_0, of landing.

As follows from the formulas in this chapter, the probability that the actual time of decision-making or the time of landing exceed the correspondingly most likely times is expressed by the formulas of the types (12) and (14), and is as high as $P_* = \dfrac{1}{\sqrt{e}} = 0.6065 = 60.6\%$. In this connection we would like to mention that the one-parametric Rayleigh law is characterized by a rather large standard deviation, and therefore might not be the best approximation for the probability density functions for the decision-making time and the time of landing. A more powerful and flexible two-parametric law, e.g., the Weibull law, might be more suitable as an appropriate probability distribution of the random times t and θ. Its use, however, will make our analysis unnecessarily more complicated. Our goal is not so much to "dot all the i's and cross all the t's," as far as modeling the human factor role in the problem in question is concerned, but rather to demonstrate that the attempt to use PRM methods to quantify the role of the human factor in avionics safety and similar problems might be quite fruitful. When developing practical guidelines and recommendations, a particular law of the probability distribution should be established based on the actual statistical data, and employment of various goodness-of-fit criteria (Pierson's, Kolmogorov's, etc.) might be needed in detailed statistical analyses.

When the most likely times t_0 and θ_0, required for making the go-ahead decision and for the actual landing, are equal, the formula (10) yields:

$$P_* = P_*\left(\frac{t_0}{\hat{T}}, \frac{\theta_0}{\hat{T}}\right) = \exp\left(-\frac{\hat{T}^2}{2t_0^2}\right)\left[1 + \sqrt{\pi}\frac{\hat{T}}{2t_0}\exp\left(\left(\frac{\hat{T}}{2t_0}\right)^2\right)erf\left(\frac{\hat{T}}{2t_0}\right)\right] \tag{8.15}$$

TABLE 8.3

The probability P_* that the operation time exceeds a certain time level $\hat{T}$ vs the ratio $\hat{T}/t_0$ of this time level to the most likely time t_0 of decision making for the case when the time t_0 and the most likely time θ_0 time of landing are the same. For the sake of comparison, the probability $P°$ of exceeding the time level $\hat{T}$, when either the time t_0 or the time θ_0 is zero, is also indicated.

$\hat{T}/t_0$	6	5	4	3	2
P_*	6.562E-4	8.553E-3	6.495E-2	1.914E-1	6.837E-1
$P°$	1.523E-8	0.373E-5	0.335E-3	1.111E-2	1.353E-1
$P_*/P°$	4.309E4	2.293E3	1.939E2	1.723E1	5.053

For large enough $\dfrac{\hat{T}}{t_0}$ ratios $\left(\dfrac{\hat{T}}{t_0} \geq 3\right)$ of the critical time $\hat{T}$ to the most likely decision-making or landing time, the second term in the brackets becomes large compared to unity. The calculated probabilities of exceeding a certain time level, $\hat{T}$, based on the formula (15), are shown in Table 8.3. In the third row of this table we indicate, for the sake of comparison, the probabilities, $P°$, of exceeding the given time, $\hat{T}$, when only the time t_0 or only the time θ_0 is different from zero, i.e., for the special case that is mostly remote from the case $t_0 = \theta_0$ of equal most likely times. Clearly, the probabilities computed for other possible combinations of the times t_0 and θ_0 could be found between the calculated probabilities P_* and $P°$. The following conclusions can be drawn from the Table 8.3 data:

- The probability that the total time of operation (the time of decision-making and the time of landing) exceeds the given time level $\hat{T}$, thereby leading to a casualty, rapidly increases with an increase in the total time of operation.
- The probability of exceeding the time level $\hat{T}$ is considerably higher when the most likely times of decision-making and of landing are finite, and especially when they are close to each other, in comparison with the situation when one of these times is significantly shorter than the other, i.e., zero or next-to-zero. This is particularly true for short operation times, like in the situation in question: the ratio $P_*/P°$ of the probability P_* of exceeding the time level $\hat{T}$ in the case of $t_0 = \theta_0$ to the probability $P°$ of exceeding this level in the case $t_0 = 0$ or in the case $\theta_0 = 0$ decreases rapidly with an increase in the time of operation. There exists therefore a significant incentive for reducing the operation time. The importance of this intuitively obvious fact is quantified by the table data.
- Another useful bit of information that could be drawn from the data of the type shown in Table 8.3 is whether it is possible at all to train a human to make a decision in just a couple of seconds. It took Capt. Sullenberger about 30 sec to make the right decision, and he is an exceptionally highly qualified pilot, with an outstanding HCF. If a very short-term decision could not be expected, and a low probability of human failure is still required, then

one should decide on a broader involvement of more sophisticated, more powerful, and more expensive equipment and instrumentation to do the job. If pursuing such an effort is decided upon, then probabilistic sensitivity analyses of the type developed above will be needed to determine the most promising way to go. It is advisable, of course, that the analytical predictions are confirmed by computer-aided simulations, and verified by highly focused and highly cost effective FOAT conducted on flight simulators.

8.8.2 PROBABILITY THAT THE LANDING TIME EXCEEDS THE AVAILABLE TIME

Since the available time L is assumed to be a random normally distributed variable, the probability that this time is found below a certain level $\hat{L}$ is

$$P_l = P_l\left(\frac{\sigma}{\hat{L}},\frac{l_0}{\hat{L}}\right) = \int_{-\infty}^{\hat{L}} f_l(l)dl = \frac{1}{2}\left[1+erf\left(\frac{\hat{L}-l_0}{\sqrt{2}\sigma}\right)\right] = \frac{1}{2}\left[1+erf\left(\frac{1-\frac{l_0}{\hat{L}}}{\sqrt{2}\frac{\sigma}{\hat{L}}}\right)\right]. \quad (8.16)$$

The probability that the available time is exceeded can be determined by equating the times $\hat{T} = \hat{L} = T$ and computing the product

$$P_A = P_a\left(\frac{t_0}{T},\frac{\theta_0}{T}\right)P_l\left(\frac{\sigma}{T},\frac{l_0}{T}\right) \quad (8.17)$$

of the probability, $P_a\left(\frac{t_0}{T},\frac{\theta_0}{T}\right)$, that the time of operation exceeds a certain level, T, and the probability, $P_l\left(\frac{\sigma}{T},\frac{l_0}{T}\right)$, that the available time is shorter than the time T. The formula (8.17) considers the roles of the most likely available time, the human factor, t_0 (the most likely time required for the pilot to make their go-ahead decision), and the most likely time, θ_0, of actual landing (which characterizes both the qualification and skills of the pilot and the qualities and behavior of the flying machine) a safe landing.

Carrying out detailed computations based on the formulas (8.10), (8.16), and (8.17) is however beyond the scope of the present article.

8.9 "MIRACLE-ON-THE-HUDSON": INCIDENT

US Airways Flight 1549 was a domestic passenger flight from LaGuardia Airport (LGA) in New York City to Charlotte/Douglas International Airport, Charlotte, North Carolina. On January 15, 2009, the Airbus A320-214 flying this route struck a flock of Canada geese during its initial climb out, lost engine power, and ditched in the Hudson River off midtown Manhattan. Since all 155 occupants survived and safely evacuated the airliner, the incident became known as the "Miracle on the Hudson" [13, 14].

The bird strike occurred just northeast of the George Washington Bridge (GWB) about three minutes into the flight and resulted in an immediate and complete loss of thrust from both engines. When the crew determined that they would be unable to reliably reach any airfield, they turned southbound and glided over the Hudson, finally ditching the airliner near the USS *Intrepid* museum about three minutes after losing power. The crew was later awarded the Master's Medal of the Guild of Air Pilots and Air Navigators for successful "emergency ditching and evacuation, with the loss of no lives…a heroic and unique aviation achievement…the most successful ditching in aviation history." The pilot in command was 57-year-old Capt. Chesley B. "Sully" Sullenberger, a former fighter pilot who had been an airline pilot since leaving the United States Air Force in 1980. He is also a safety expert and a glider pilot. The first officer was Jeffrey B. Skiles, 49. The flight attendants were Donna Dent, Doreen Welsh, and Sheila Dail.

The aircraft was powered by two GE Aviation/Snecma-designed CFM56-5B4/P turbofan engines manufactured in France and the US. One of seventy-four A320s then in service in the US Airways fleet, it was built by Airbus with final assembly at its facility at Aéroport de Toulouse-Blagnac in France in June 1999, and delivered to the carrier on August 2, 1999. The Airbus is a digital fly-by-wire aircraft: the flight control surfaces are moved by electrical and hydraulic actuators controlled by a digital computer. The computer interprets pilot commands via input from a side stick, making adjustments on its own to keep the plane stable and on course. This is particularly useful after engine failure by allowing the pilots to concentrate on engine restart and landing planning. The mechanical energy of the two engines is the primary source of electrical power and hydraulic pressure for the aircraft flight control systems. The aircraft also has an auxiliary power unit (APU), which can provide backup electrical power for the aircraft, including its electrically powered hydraulic pumps; and a ram air turbine (RAT), a type of wind turbine that can be deployed into the airstream to provide backup hydraulic pressure and electrical power at certain speeds. According to the NTSB [14], both the APU and the RAT were operating as the plane descended into the Hudson, although it was not clear whether the RAT had been deployed manually or automatically. The Airbus A320 has a ditching button that closes valves and openings beneath the aircraft, including the outflow valve, the air inlet for the emergency RAT, the avionics inlet, the extract valve, and the flow control valve. It is meant to slow flooding in a water landing. The flight crew did not activate the ditch switch during the incident. Sullenberger later noted that it probably would not have been effective anyway, since the force of the water impact tore holes in the plane's fuselage much larger than the openings sealed by the switch.

First Officer Skiles was at the controls of the flight when it took off at 3:25 pm, and was the first to notice a formation of birds approaching the aircraft about two minutes later, while passing through an altitude of about 2,700 feet (820 m) on the initial climb out to 15,000 feet (4,600 m). According to flight data recorder (FDR) data, the bird encounter occurred at 3:27:11, when the airplane was at an altitude of 2,818 feet (856 m) above ground level (agl) and at a distance of about 4.5 miles north-northwest of the approach end of runway 22 at LGA. Subsequently, the airplane's altitude continued to increase while the airspeed decreased, until 3:27:30, when the airplane reached its highest altitude of about 3,060 feet (930 m), at an airspeed of about 185 kts calibrated airspeed (KCAS). The altitude then started to decrease as the airspeed started to increase, reaching 210 KCAS at 3:28:10 at an altitude of about 1,650 feet (500 m) The windscreen quickly turned dark brown and several loud thuds were heard. Capt. Sullenberger took the controls, while Skiles began going through the three-page emergency procedures checklist in an attempt to restart the engines.

At 3:27:36 the flight radioed air traffic controllers at New York Terminal Radar Approach Control (TRACON). "Hit birds. We've lost thrust on both engines. We're turning back towards LaGuardia." Responding to the captain's report of a bird strike, controller Patrick Harten, who was working the departure position, told LaGuardia tower to hold all waiting departures on the ground, and gave Flight 1549 a heading to return to LaGuardia. Sullenberger responded that he was unable.

Sullenberger asked if they could attempt an emergency landing in New Jersey, mentioning Teterboro Airport in Bergen County as a possibility; air traffic controllers quickly contacted Teterboro and gained permission for a landing on runway 1. However, Sullenberger told controllers that "We can't do it," and that "We're gonna be in the Hudson," making clear his intention to bring the plane down on the Hudson River due to a lack of altitude. Air traffic control at LaGuardia reported seeing the aircraft pass less than 900 feet (270 m) above GWB. About 90 seconds before touchdown, the captain announced, "Brace for impact," and the flight attendants instructed the passengers how to do so. The plane ended its six-minute flight at 3:31 pm with an unpowered ditching while heading south at about 130 knots (150 mph, 240 km/h) in the middle of the North River section of the Hudson River roughly abeam 50th Street (near the Intrepid Sea Air Space Museum) in Manhattan and Port Imperial in Weehawken, New Jersey. Sullenberger said in an interview on CBS television that his training prompted him to choose a ditching location near operating boats so as to maximize the chance of rescue. After coming to a stop in the river, the plane began drifting southward with the current.

National Transportation Safety Board (NTSB) Member Kitty Higgins, the principal spokesperson for the on scene investigation, said at a press conference the day after the accident that it "has to go down [as] the most successful ditching in aviation history... These people knew what they were supposed to do and they did it, and as a result, nobody lost their life." The flight crew, particularly Captain Sullenberger, was widely praised for their actions during the incident, notably by New York City Mayor Michael Bloomberg and New York State Governor David Paterson, who opined, "We had a miracle on thirty-fourth street. I believe now we have had a miracle on the Hudson." Outgoing U.S. President George W. Bush said he was "...inspired by the skill and heroism of the flight crew," and he also praised the emergency responders and volunteers. Then President-elect Barack Obama said that everyone was proud of Sullenberger's "heroic and graceful job in landing the damaged aircraft," and thanked the A320's crew.

The NTSB ran a series of tests using Airbus simulators in France, to see if Flight 1549 could have returned safely to LaGuardia. The simulation started immediately following the bird strike and "...knowing in advance that they were going to suffer a bird strike and that the engines could not be restarted, four out of four pilots were able to turn the A320 back to LaGuardia and land on Runway 13." When the NTSB later imposed a 30 second delay before they could respond, in recognition that it wasn't reasonable to expect a pilot to assess the situation and react instantly, all four pilots crashed.

On May 4, 2010, the NTSB released a statement that credited the accident outcome to the fact that the aircraft was carrying safety equipment in excess of that mandated for the flight, and excellent cockpit resource management among the flight crew. Contributing factors to the survivability of the accident were good visibility, and fast response from the various ferry operators. Captain Sullenberger's decision to ditch in the Hudson River was validated by the NTSB. On May 28, 2010, the NTSB published its final report into the accident [14]. It determined the cause of the accident to be "the ingestion of large birds into each engine, which resulted in an almost total loss of thrust in both engines."

8.10 "MIRACLE-ON-THE-HUDSON": FLIGHT EVENTS

The US AW Flight 1549 events and durations are summarized in Table 8.4. It took only 40 sec for Capt. Sullenberger to make his route change decision and another 2 min to land the aircraft.

TABLE 8.4
US AW Flight 1549, January 15, 2009 (Wikipedia).

Flight segment	Time (EST)	Duration, sec	Altitude	Speed	Event
1	3:25:00 pm	60.00 (16.6667)	0	279.6 km/h	Aircraft took off from LGA and started climbing up. First officer Skiles runs the aircraft.
2	3:26:00 pm	71.00 (19.7222)	820 m	–	Skiles noticed a flock of birds.
3	3:27:11 pm	19.00 (5.2778)	856 m	322.2 km/h	Bird strike (north-east of GWB, NYC).
4	3:27:30 pm	6.00 (1.6667)	930 m	342.6 km/h	Highest altitude reached.
5	3:27:36 pm	24.00 (6.6667)	–	359.3 km/h	Radioed TRACON traffic controllers: "Hit birds. Lost thrust on both engines. Turning back toward LGA."
6	3:28:00 pm	10.00 (2.7778)	609 m	374.1 km/h	Complete loss of thrust (engine power).
7	3:28:10 pm	30.00 (8.3333)	500 m	388.9 km/h	Sullenberger takes over control.
8	3:28:40 pm	20.00 (5.5555)	500 m	388.9 km/h	Sullenberger makes route change decision and turns southbound.
9	3:29:00 pm	10.00 (2.7778)	396 m	353.7 km/h	Started gliding over Hudson River.
10	3:29:10 pm	90.00 (25.0000)	–	–	"Brace for impact" command.
11	3:30:40 pm	20.00 (5.5555)	0	240 km/h	Touch down (ditching) Hudson River.
12	3:31:00 pm	–	0	0	Full stop, start drifting.

8.11 "MIRACLE-ON-THE-HUDSON": QUANTITATIVE AFTERMATH

In this section we intend to demonstrate how the Hudson event could be quantified using the DEPDF-based evaluations.

8.11.1 SULLENBERGER'S HYPOTHETICAL HCF

Sullenberger's HCF is computed in Table 8.5. The calculations of the probability of the human non-failure are carried out using formula (8.2) and are shown in Table 8.6. We did not try to anticipate and quantify a particular (most likely) MWL level, but rather assumed different MWL deviations from the most likely level. A more detailed

TABLE 8.5
Sullenberger's HCF.

No	Relevant qualities	Relative HCF rating $\left(\dfrac{F^*}{F_0} \right)$	Comments
1	psychological suitability for the given task	3.2	1. 57-year-old former fighter pilot who had been a commercial airline pilot since leaving the US Air Force in 1980. He is also a safety expert and a glider pilot [7]. See also Appendix B. "I was sure I could do it". "The entire life up to this moment was a preparation for this moment... ."I am not just a pilot of that flight. I am also a pilot who has flown for 43 years..."
2	professional qualifications and experience	3.9	
3	level, quality, and timeliness of past and recent training	2.0	
4	mature (realistic) and independent thinking	3.2	
5	performance sustainability (predictability, consistency)	3.2	
6	ability to concentrate and act with a "cool demeanor" in hazardous and even life threatening situations	3.3	2. Probability of human non-failure in normal flight conditions is assumed to be 100%.
7	ability to anticipate ("expecting the unexpected")	3.2	3. The formula
8	ability to operate effectively under pressure	3.4	$p = \exp\left(1 - \dfrac{G^2}{G_0^2} \right)$ would have to be
9	self-control in hazardous situations	3.2	used to evaluate the probability of non-failure in the case of a pilot of ordinary skills. The computed numbers are shown in parentheses. The computed numbers show that such a pilot would definitely fail in the off-normal situation in question
10	ability to make a substantiated decision in a short period of time ("we are going to be in the Hudson")	2.8	
	Average FOM	3.14	

* This is just an example that shows that the approach makes physical sense. Actual numbers should be obtained using FOAT on a simulator and confirmed by an independent approach, such as the, Delphi method: http://en.wikipedia.org/wiki/Delphi_method [12].

TABLE 8.6

Computed probabilities of human non-failure (Capt. Sullenberger).

G/G_0	5	10	50	100	150
p	0.9966	0.9860	0.7013	0.2413	0.0410

MWL analysis can be done using flight simulation FOAT data. The computed data indicate that, as long as the HCF is high (and Capt. Sullenberger's HCF was exceptionally high), even significant relative MWL levels, up to 50 or even higher, still result in a rather high probability of the human non-failure.

Capt. Sullenberger's HCF was extraordinarily, exceptionally high. This was due to his age, old enough to be an experienced performer and young enough to operate effectively under pressure and possess other qualities of a relatively young human. As evident from the computed data, the probability of human non-failure in off-normal flight conditions is still relatively high, provided that the HCF is significantly higher than that of a pilot of normal skills in the profession and that the MWL is not extraordinarily (perhaps, unrealistically) high. Therefore, the actual "miraculous" event was because a person of extraordinary abilities (measured by the level of the HCF) turned out to be in control at the critical moment. Other favorable aspects of the situation were the high HCF of the crew, good weather, and the landing site, perhaps the most favorable one could imagine. Captain Sullenberger knew when to take control of the aircraft, when to abandon his communications with the (generally speaking, excellent) ATCs and to use his outstanding background and skills to ditch the plane: "I was sure I could do it...my entire life up to this moment was a preparation for this moment...I am not just a pilot of that flight. I am also a pilot who has flown for 43 years..." Such a miracle does not happen often, of course, and is perhaps outside any indicative statistics.

8.11.2 Flight Attendants' Hypothetical HCF Estimate: Example

The HCF of a flight-attendant is assessed in Table 8.7, and the probabilities of his/her non-failure are shown in Table 8.8. The qualities expected from a flight attendant are, of course, quite different of those of a pilot. As evident from the obtained data, the probability of the human non-failure of the Airbus A320 flight attendants is rather high up until the MWL ratio of 10 or even slightly higher.

Although we do not try to evaluate first officer Skiles' HCF, we assume that his HCF is also high, although this did not manifest itself during the event. It has been shown elsewhere [10] that it is expected that both pilots have high, and to an extent possible, equal qualifications and skills for a high probability of a mission success, if for one reason or another, the entire MWL is taken by one of the pilots. In this connection we would like to mention that, even regardless of the qualification, it is widely accepted in avionic and maritime practice that it is the captain, not the first officer (first mate) who takes control in dangerous situations, especially life threatening ones. It did not happen, however, in the case of the Swissair UN shuttle's last flight addressed in the next section.

TABLE 8.7
Flight attendant's HCF.

No	Relevant qualities	Relative HCF $\left(\dfrac{F^*}{F_0}\right)$
1	psychological suitability for the task	2.5
2	professional qualifications and experience	2.5
3	level, quality, and timeliness of past and recent training	2.5
4	team-player attitude	3.0
5	performance sustainability (consistency)	3.0
6	ability to perform calmly and methodically in hazardous and even life threatening situations;	3.0
7	ability and willingness to follow orders	3.0
8	ability to operate effectively under pressure	3.4
	Average FOM	2.8625

* *This is just an example. Actual numbers should be obtained using FOAT on a simulator and confirmed by an independent method, such as the Delphi method:* http://en.wikipedia.org/wiki/Delphi_method [12].

TABLE 8.8
Estimated probabilities of non-failure for a flight attendant.

G/G_0	5	10	50	100	150
p	0.9821	0.9283	0.1530	5.47E-4	4.57E-8

8.12 UN SHUTTLE FLIGHT: CRASH

For the sake of comparison of the successful Hudson River case with an emergency situation that ended up in a crash, we have chosen the infamous Swissair September 2, 1998, Flight 111, when a highly trained crew made several bad decisions under considerable time pressure [15] that was not as severe as the Hudson case. Swissair Flight 111 was a McDonnell Douglas MD-11 on a scheduled airline flight from John F. Kennedy (JFK) International Airport in New York City, US to Cointrin International Airport in Geneva, Switzerland. On Wednesday, September 2, 1998, the aircraft crashed into the Atlantic Ocean southwest of Halifax International Airport at the entrance to St. Margaret's Bay, Nova Scotia. The crash site was just 8 km (5.0 nm) from shore. All 229 people on board died, the highest death toll of any aviation accident involving a McDonnell Douglas MD-11. Swissair Flight 111 was known as the "UN shuttle" due to its popularity with United Nations officials; the flight often carried business executives, scientists, and researchers.

The initial search and rescue response, crash recovery operation, and resulting investigation by the Government of Canada took over four years. The Transportation Safety Board (TSB) of Canada's official report stated that flammable material used in the aircraft's structure allowed a fire to spread beyond the control of the crew, resulting in the loss of control and crash of the aircraft. An MD-11 has a standard flight crew consisting of a captain and a first officer, and a cabin crew made up of a maître-de-cabine (M/C - purser) supervising the work of 11 flight attendants. All personnel on board Swissair Flight 111 were qualified, certified, and trained in accordance with Swiss regulations under the Joint Aviation Authorities (JAA).

The flight details are shown in Table 8.9. The flight took off from New York's JFK Airport at 20:18 Eastern Standard Time (EST). Beginning at 20:33 EST and lasting until 20:47, the aircraft experienced an unexplained thirteen minute radio blackout. The cause of the blackout, or if it was related to the crash, is unknown. At 22:10 Atlantic Time (21:10 EST), cruising at FL330 (approximately 33,000 feet or 10,100 meters), Capt. Urs Zimmermann and First Officer Stephan Loew detected an odor in the cockpit and determined it to be smoke from the air conditioning system, a situation easily remedied by closing the air conditioning vent, which a flight attendant did on Zimmermann's request. Four minutes later, the odor returned and now smoke was visible, and the pilots began to consider diverting to a nearby airport for the purpose of a quick landing. At 22:14 AT (21:14 EST) the flight crew made a radio call to air traffic control (ATC) at Moncton (which handles transatlantics air traffic approaching or departing North American air space), indicating that there was an urgent problem with the flight, although not an emergency, which would imply immediate danger to the aircraft. The crew requested a diversion to Boston's Logan International Airport, which was 300 nautical miles (560 km) away. ATC Moncton offered the crew a vector to the closer, 66 nm (104 km) away, Halifax International Airport in Enfield, Nova Scotia, which Loew accepted. The crew then put on their oxygen masks and the aircraft began its descent. Zimmermann put Loew in charge of the descent, while he personally ran through the two Swissair standard checklists for smoke in the cockpit, a process that would take approximately 20 minutes and later became a source of controversy.

At 22:18 AT (21:18 EST), ATC Moncton handed over traffic control of Swissair 111 to ATC Halifax, since the plane was now going to land in Halifax rather than leave North American air space. At 22:19 AT (21:19 EST) the plane was 30 nautical miles (56 km) away from Halifax International Airport, but Loew requested more time to descend the plane from its altitude of 21,000 feet (6,400 m). At 22:20 AT (21:20 EST), Loew informed ATC Halifax that he needed to dump fuel, which ATC Halifax controllers would say later, was a surprise considering that the request came so late; dumping fuel is a fairly standard procedure early on in nearly any "heavy" aircraft urgent landing scenario. ATC Halifax subsequently diverted Swissair 111 toward St. Margaret's Bay, where they could more safely dump fuel, but still be only around 30 nautical miles (56 km) from Halifax.

In accordance with the Swissair checklist entitled "In case of smoke of unknown origin," the crew shut off the power supply in the cabin, which caused the re-circulating fans to shut off. This caused a vacuum, which induced the fire to spread back into the cockpit. This also caused the autopilot to shut down; at 22:24:28 AT (21:24:28 EST), Loew informed ATC Halifax that "we now must fly manually." Seventeen seconds later, at 22:24:45 AT (21:24:45 EST), Loew informed ATC Halifax that "Swissair 111

TABLE 8.9
Swissair Flight 111, September 2, 1998 (Wikipedia).

Flight segment	Time (EST)	Event
1	20:18:00	Aircraft took off from JFK airport. First officer Stephan Loew runs the aircraft.
2	20:33–20:47	Radio blackout
3	21:10	Captain Urs Zimmermann and first officer Stephan Loew detected an odor in the cockpit and determined it to be smoke from the air conditioning system, a situation easily remedied by closing the air conditioning vent, which a flight attendant did on Zimmermann's request.
4	21:14	Odor returned and smoke became visible. The crew called ATC Moncton indicating an urgent, but not emergency problem, and requested a diversion to Boston's Logan Airport, which was 300 nm (560 km) away. ATC Moncton offered a vector to the closer Halifax Airport in Enfield, Nova Scotia, 66 nm (104 km) away, which Loew accepted.
5	21:14–21:34	The crew put on oxygen masks and the aircraft began to descend. Zimmermann put Loew in charge of the descent, while he ran through the Swissair checklists for smoke in the cockpit, a process that become later a source of controversy.
6	21:18	ATC Moncton handed over traffic control of Swissair 111 to ATC Halifax.
7	21:19	The plane was 30 nm (56 km) away from Halifax Airport, but Loew requested more time to descend the plane from its altitude of 6,400 m.
8	21:20	Loew informed ATC Halifax that he needed to dump fuel. ATC Halifax said later it was a surprise, because the request came so late. Dumping fuel was a fairly standard procedure early on in nearly any "heavy" aircraft urgent landing scenario. Subsequently, ATC Halifax diverted aircraft toward St. Margaret's Bay, where they could more safely dump fuel, but still be only around 30 nm (56 km) from Halifax.
9	21:24:28	In accordance with the Swissair "In case of smoke of unknown origin" checklist, the crew shut off the power supply in the cabin. This caused the re-circulating fans to shut off. This caused a vacuum, which induced the fire to spread back into the cockpit. This also caused the autopilot to shut down. Loew informed ATC Halifax that "we now must fly manually."
10	21:24:45	Loew informed ATC Halifax that "Swissair 111 is declaring emergency."
11	21:24:46	Loew repeated the emergency declaration one second later, and over the next 10 seconds stated that they had descended to "between 12,000 and 5,000 feet" and once more declared an emergency.
12	21:25:40	The flight data recorder stopped recording, followed one second later by the cockpit voice recorder.
13	21:25:50–21:26:04	The doomed plane briefly showed up again on radar screens. Its last recorded altitude was 9,700 feet. Shortly after the first emergency declaration, the captain could be heard leaving his seat to fight the fire, which was now spreading to the rear of the cockpit.

heavy is declaring emergency," repeated the emergency declaration one second later, and over the next 10 seconds stated that they had descended to "between 12,000 and 5,000 feet" and once more declared an emergency. The flight data recorder stopped recording at 22:25:40 AT (21:25:40 EST), followed one second later by the cockpit voice recorder. The doomed plane briefly showed up again on radar screens from 22:25:50 AT (21:25:50 EST) until 22:26:04 AT (21:26:04 EST). Its last recorded altitude was 9,700 feet. Shortly after the first emergency declaration, the captain could be heard leaving his seat to fight the fire, which was now spreading to the rear of the cockpit. The Swissair volume of checklists was later found fused together, as if someone had been trying to use them to fan back flames. The captain did not return to his seat, and whether he was killed from the fire or asphyxiated by the smoke is not known. However, physical evidence provides a strong indication that First Officer Loew may have survived the inferno only to die in the eventual crash; instruments show that Loew continued trying to fly the now crippled aircraft, and gauges later indicated that he shut down engine two approximately one minute before impact, implying he was still alive and at the controls until the aircraft struck the ocean at 22:31 AT (21:31 EST). The aircraft disintegrated on impact, killing all on board instantly.

The search and rescue operations were launched immediately by Joint Rescue Coordination Centre Halifax (JRCC Halifax), which tasked the Canadian Forces Air Command, Maritime Command and Land Force Command, as well as Canadian Coast Guard (CCG) and Canadian Coast Guard Auxiliary (CCGA) resources. The first rescue resources to approach the crash site were Canadian Coast Guard Auxiliary volunteer units—mostly privately-owned fishing boats—sailing from Peggy's Cove, Bayswater, and other harbors on St. Margaret's Bay and the Aspotogan Peninsula. They were soon joined by the dedicated Canadian Coast Guard SAR vessel CCGS *Sambro* and CH-113 Labrador SAR helicopters flown by 413 Squadron from CFB Greenwood.

The investigation identified eleven causes and contributing factors of the crash in its final report. The first and most important was, "Aircraft certification standards for material flammability were inadequate in that they allowed the use of materials that could be ignited and sustain or propagate fire. Consequently, flammable material propagated a fire that started above the ceiling on the right side of the cockpit near the cockpit rear wall. The fire spread and intensified rapidly to the extent that it degraded aircraft systems and the cockpit environment, and ultimately led to the loss of control of the aircraft."

Arcing from wiring of the in-flight entertainment system network did not trip the circuit breakers. While suggestive, the investigation was unable to confirm if this arc was the "lead event" that ignited the flammable covering on MPET insulation blankets that quickly spread across other flammable materials. The crew did not recognize that a fire had started and was not warned by instruments. Once they became aware of the fire, the uncertainty of the problem made it difficult to address. The rapid spread of the fire led to the failure of key display systems, and the crew were soon rendered unable to control the aircraft. Because he had no light by which to see his controls after the displays failed, the pilot was forced to steer the plane blindly; intentionally or not, the plane swerved off course and headed back out into the Atlantic. Recovered fragments of the plane show that the heat inside the cockpit became so great that the ceiling started to melt.

The recovered standby altitude indicator and airspeed indicator showed that the aircraft struck the water at 300 knots (560 km/h, 348 mph) at 20 degrees nose down and in a 110 degree bank turn, or almost upside down. Less than a second after impact the plane would have been totally crushed, killing all aboard almost instantly. The TSB concluded that even if the crew had been aware of the nature of the problem, the rate at which the fire spread would have precluded a safe landing at Halifax even if an approach had begun as soon as the "pan-pan-pan" was declared. The plane was broken into two million small pieces by the impact, making this process time consuming and tedious. The investigation became the largest and most expensive transport accident investigation in Canadian history.

8.13   SWISSAIR FLIGHT 111: EVENTS AND CREW ERRORS

The Swissair Flight 111 events and durations are summarized in Table 8.9. The following more or less obvious errors were made by the crew:

- At 21:14 EST they used poor judgment and underestimated the danger by indicating to the ATC Moncton that the returned odor and visible smoke in the cockpit was an urgency, but not an emergency problem. They requested a diversion to the 300 nm (560 km) away Boston Logan Airport, and not to the closest 66 nm (104 km) away Halifax Airport.
- Capt. Zimmermann put First Officer Loew in charge of the descent and spent time running through the Swissair checklist for smoke in the cockpit.
- At 21:19 EST Loew requested more time to descend the plane from its altitude of 6,400 m, although the plane was only 30 nm (56 km) away from Halifax Airport.
- At 21:20 EST Loew informed ATC Halifax that he needed to dump fuel. As ATC Halifax indicated later, it was a surprise, because the request came too late. In addition, it was doubtful that such a measure was needed at all.
- At 21:24:28 the crew shut off the power supply in the cabin. That caused the re-circulating fans to shut off and caused a vacuum, which induced the fire to spread back into the cockpit. This also caused the autopilot to shut down, and Loew had to "fly manually." In about a minute or so the plane crashed.
- These errors are reflected in the Table 8.10 score sheet and resulted in a rather low HCF and low probability of the assessed human non-failure.

8.14   FLIGHT 111 PILOT'S HCF

Flight 111 pilot's HCF and the probability of human non-failure are summarized in Table 8.10. The criteria used are the same as in Table 8.5 above. The probabilities of human non-failure are shown in Table 8.11.

The computed probability of non-failure is very low even at non-very high MWL levels. Although the crew's qualification seems to be adequate, the qualities #4, 6, 7, 8 and 10, which were particularly critical in the situation in question, turned out to be extremely low.

TABLE 8.10
Flight 111 pilot's HCF.

No	Relevant qualities	$\text{HCF}\left(\dfrac{F^*}{F_0}\right)$
1	psychological suitability for the given task	3.0
2	professional qualifications and experience	3.0
3	level, quality and timeliness of past and recent training	2.0
4	mature (realistic) and independent thinking	1.0
5	performance sustainability (consistency)	2.0
6	ability to concentrate and to act in calm and methodical manner in hazardous situations	1.5
7	ability to anticipate ("expecting the unexpected")	1.2
8	ability to operate effectively under pressure	1.5
9	self-control in hazardous situations	2.0
10	ability to make a substantiated decision in a short period of time	1.2
	Average FOM	1.84

* This is just an example. Actual numbers should be obtained using FOAT on a simulator and confirmed by an independent method, such as, say, Delphi method: http://en.wikipedia.org/wiki/Delphi_method [12].

TABLE 8.11
Computed probabilities of human non-failure (Swissair pilot).

G/G_0	5	10	50	100
p	0.1098	1.1945E-4	0	0

CONCLUSIONS

- The application of a quantitative probabilistic risk management (PRM) approach should complement, whenever feasible and possible, the existing vehicular psychology practices that are, as a rule, qualitative assessments of the role of the human factor when addressing the likelihood of success and safety of various vehicular missions and situations.
- It was the high human capacity factor (HCF) of the aircraft crew and especially of Capt. Sullenberger, that made a reality seem to be a "miracle." The carried out PRM-based analysis enables one to quantify this fact. In effect, it was "miraculous" that an outstanding individual like Capt. Sullenberger turned out to be in control at the time of the incident and that the weather was highly favorable. Under these circumstances, nothing else should be considered a miracle: the likelihood of a safe landing with an individual like Capt. Sullenberger in the cockpit was rather high.

- The taken PRM-based approach, after the trustworthy input information is obtained using FOAT on a simulator and confirmed by an independent approach, such as, say, the Delphi method, is applicable to many other human-in-the-loop (HITL) situations, well beyond the situation in question and perhaps even beyond the vehicular domain.
- Although the obtained numbers make physical sense, it is the approach, not the numbers, that is in the author's opinion, the merit of the paper.

REFERENCES

A.T. Kern, *Controlling Pilot Error: Culture, Environment, and CRM (Crew Resource Management)*, McGraw-Hill, 2001.

W.A. O'Neil, *"The Human Element in Shipping,"* Keynote Address, Biennial Symp. of the Seafarers International Research Center, Cardiff, Wales, June 29, 2001.

D.C. Foyle, B.L. Hooey, *"Human Performance Modeling in Aviation,"* CRC Press, 2008.

D. Harris, *Human Performance on the Flight Deck*, Bookpoint Ltd., Ashgate Publishing, Oxon, UK, 2011.

E. Hollnagel, *Human Reliability Analysis: Context and Control*, Academic Press, London and San Diego, 1993.

E. Suhir, *Applied Probability for Engineers and Scientists*, McGraw-Hill, New York, 1997.

E. Suhir, *"Helicopter-Landing Ship: Undercarriage Strength and the Role of the Human Factor"*, ASME OMAE Conference, June 1-9, Honolulu, Hawaii, 2009; see also *ASME OMAE Journal*, Feb. 2010.

E. Suhir and R H. Mogford, "Two Men in a Cockpit: Probabilistic Assessment of the Likelihood of a Casualty if One of the Two Navigators Becomes Incapacitated," *Journal of Aircraft*, vol.48, No.4, July-August 2011.

E. Suhir, "Human-in-the-Loop: Likelihood of a Vehicular Mission-Success-and-Safety, and the Role of the Human Factor," Paper ID 1168, 2011 IEEE/AIAA Aerospace Conference, Big Sky, Montana, March 5-12, 2011; see also *Journal of Aircraft*, vol.49, No.1, 2012.

J.T. Reason, *Human Error*, Cambridge University Press, Cambridge, UK, 1990.

J.T. Reason, *Managing the Risks of Organizational Accidents*, Ashgate Publishing Company; 1997.

http://en.wikipedia.org/wiki/Delphi_method
http://en.wikipedia.org/wiki/US_Airways_Flight_1549
http://www.ntsb.gov/doclib/reports/2010/AAR1003.pdf
http://en.wikipedia.org/wiki/Swissair_Flight_111

APPENDIX A: OTHER REPORTED WATER LANDINGS (DITCHINGS) OF PASSENGER AIRPLANES

- On 11 July 2011, Angara Airlines Flight 5007 (an Antonov An-24) ditched in the Ob River near Strezhevoy, Russia, after an engine fire. Upon water contact the tail separated and the burnt port engine became detached from its mounts. Otherwise the plane remained intact, but was written off. Out of 37 people on board, including four crew and 33 passengers, 7 passengers died. Of the survivors, at least 20 were hospitalized with various injuries.
- On 6 June 2011, a Solenta Aviation Antonov An-26 freighter flying for DHL Aviation ditched in the Atlantic Ocean near Libreville, Gabon. Three crew and one passenger were rescued with minor injuries.

- On 22 October 2009, a Divi Divi Air Britten-Norman Islander operating Divi Divi Air Flight 014 ditched in off the coast of Bonaire after its starboard engine failed. The pilot reported that the aircraft was losing 200 feet per minute after choosing to fly to an airport. All 9 passengers survived, but the captain was knocked unconscious and although some passengers attempted to free him, he drowned and was pulled down with the aircraft.
- On 6 August 2005, Tuninter Flight 1153 (an ATR 72) ditched off the Sicilian coast after running out of fuel. Of 39 aboard, 23 survived with injuries. The plane's wreck was found in three pieces.
- On 16 January 2002, Garuda Indonesia Flight 421 (a Boeing 737) successfully ditched into the Bengawan Solo River near Yogyakarta, Java Island after experiencing a twin engine flameout during heavy precipitation and hail. The pilots tried to restart the engines several times before making the decision to ditch the aircraft. Photographs taken shortly after evacuation show that the plane came to rest in knee-deep water. Of the 60 occupants, one flight attendant was killed.
- On 23 November 1996, Ethiopian Airlines Flight 961 (a Boeing 767-260ER), ditched in the Indian Ocean near Comoros after being hijacked and running out of fuel, killing 125 of the 175 passengers and crew on board. Unable to operate flaps, it impacted at high speed, dragging its left wingtip before tumbling and breaking into three pieces. The panicking hijackers were fighting the pilots for the control of the plane at the time of the impact, which caused the plane to roll just before hitting the water, and the subsequent wingtip hitting the water and breakup are a result of this struggle in the cockpit. Some passengers were killed on impact or trapped in the cabin when they inflated their life vests before exiting. Most of the survivors were found hanging onto a section of the fuselage that remained floating.
- On 2 May 1970, ALM Flight 980 (a McDonnell Douglas DC-9-33CF), ditched in mile-deep water after running out of fuel during multiple attempts to land at Princess Juliana International Airport on the island of Saint Maarten in the Netherlands Antilles during low visibility weather. Insufficient warning to the cabin resulted in several passengers and crew still either standing or with unfastened seat belts as the aircraft struck the water. Of 63 occupants, 40 survivors were recovered by US military helicopters.
- On 21 August 1963, an Aeroflot Tupolev Tu-124 ditched into the Neva River in Leningrad (now St. Petersburg) after running out of fuel. The aircraft floated and was towed to shore by a tugboat that it had nearly hit as it came down on the water. The tug rushed to the floating aircraft and pulled it with its passengers near to the shore, where the passengers disembarked onto the tug; all 52 on board escaped without injuries.
- On 23 September 1962, Flying Tiger Line Flight 923, a Lockheed 1049H-82 Super Constellation N6923C passenger aircraft, on a military (MATS) charter flight, with a crew of 8 and 68 US civilian and military (paratrooper)

passengers ditched in the North Atlantic about 500 miles west of Shannon, Ireland after losing three engines on a flight from Gander, Newfoundland to Frankfurt, West Germany. Fort-five of the passengers and 3 crew were rescued, with 23 passengers and 5 crew members being lost in the storm swept seas. All occupants successfully evacuated the airplane. Those who were lost died in the rough seas.

- In October 1956, Pan Am Flight 6 (a Boeing 377) ditched northeast of Hawaii after losing two of its four engines. The aircraft was able to circle around USCGC *Pontchartrain* until daybreak, when it ditched; all 31 on board survived.

- In April 1956, Northwest Orient Airlines Flight 2 (also a Boeing 377) ditched into Puget Sound after a malfunction that was later decided to be caused by failure of the crew to close the cowl flaps on the plane's engines. All aboard escaped the aircraft after a textbook landing, but four passengers and one flight attendant succumbed either to drowning or to hypothermia before being rescued.

- On 26 March 1955, Pan Am Flight 845/26 ditched 35 miles from the Oregon coast after an engine tore loose. Despite the tail section breaking off during the impact the aircraft floated for twenty minutes before sinking. Survivors were rescued after a further 90 minutes in the water.

- On 19 June 1954, Swissair Convair CV-240 HB-IRW ditched into the English Channel because of fuel starvation, which was attributed to pilot error. All three crew and five passengers survived the ditching and could escape the plane. However, three of the passengers could not swim and eventually drowned, because there were no life jackets on board, which was not prescribed at the time.

- On 3 August 1953, Air France Flight 152, a Lockheed L-749A Constellation ditched 6 miles from Fetiye Point, Turkey 1.5 miles offshore into the Mediterranean Sea on a flight between Rome, Italy and Beirut, Lebanon. The propeller had failed due to blade fracture. Due to violent vibrations, engine number three broke away and control of engine number four was lost. The crew of eight and all but four of the 34 passengers were rescued.

- On 16 April 1952, the de Havilland Australia DHA-3 Drover VH-DHA operated by the Australian Department of Civil Aviationwith 3 occupants was ditched in the Bismarck Sea between Wewak and Manus Island. The port propeller failed, a propeller blade penetrated the fuselage, and the single pilot was rendered unconscious; the ditching was performed by a passenger.

- On 11 April 1952, Pan Am Flight 526A ditched 11.3 miles northwest of Puerto Rico due to engine failure after take-off. Many survived the initial ditching but panicked passengers refused to leave the sinking wreck and drowned. Fifty-two passengers were killed, and 17 passengers and crew members were rescued by the USCG. After this accident implementation of pre-flight safety demonstrations for over-water flights was recommended.

APPENDIX B: CAPTAIN SULLENBERGER

Sullenberger was born to a dentist father—a descendant of Swiss immigrants named Sollenberger—and an elementary school teacher mother. He has one sister, Mary Wilson. The street on which he grew up in Denison, Texas, was named after his mother's family, the Hannas. According to his sister, Sullenberger built model planes and aircraft carriers during his childhood, and might have become interested in flying after hearing stories about his father's service in the United States Navy. He went to school in Denison, and was consistently in the 99th percentile in every academic category. At the age of 12, his IQ was deemed high enough to join Mensa International. He also gained a pilot's license at 14. In high school, he was the president of the Latin club, a first chair flute, and an honor student. His high school friends have said that Sullenberger developed a passion for flying from watching jets based out of Perrin Air Force Base.

9 The Study of Production Process in Construction Industry Using Technological Units of Analysis

Gregory Z. Bedny
Essex County College, New Jersey, US

CONTENTS

9.1 INTRODUCTION

Data presented in our previous chapters demonstrates that systemic-structural activity theory (SSAT) distinguishes technological and psychological units of activity analysis.

These concepts and their relationship are important for a description of the structure of work activity. In engineering and ergonomics, elements of human activity are usually described by the use of common language or technical terminology (Bedny, and I. Bedny, 2018).

An example of such descriptions is: "turn two-positioning switch up," "take reading from a digital indicator." In SSAT, these are examples of describing actions by using technological units of analysis. Such a description of activity elements does not have clearly defined psychological characteristics. For a more precise description of these actions, the technological units of analysis should be transferred into psychological units of analysis. An action such as *taking reading from digital indicator* (technological unit of analysis) can be additionally described as *simultaneous perceptual action* with execution time of 0.3 sec. This gives a precise understanding of what action took place.

A combination of such units of analysis allows describing activity at any level of complexity with high precision even when using purely analytical methods.

The use of analytical methods is especially important when the specialist cannot observe the actual performance. Moreover, this kind of documentation can be further employed by other professionals who need it for introducing the task into the production process without observing the real task performance. Using both technological and psychological units of analysis is complicated and labor intensive. In some cases, it is appropriate to use only the technological unit of analysis in the description of the tasks performance. The technological unit of analysis can be larger than the psychological units of analysis (one technological unit can include multiple psychological units). The use of the technological units of analysis is more effective when specialists can observe the performance of a production process, and this process does not include complex cognitive activity.

In this chapter, we will demonstrate the possibility of using only technological units of analysis for the description of human activity. The difficult and dangerous physical work associated with the reconstruction of industrial buildings has been chosen as an object of the study. Specifically, we present the method we developed for the study of the production process in the construction industry. In our study of the existing systems, we utilized the observation. This method is tightly connected with the time study method.

One of the causes that determines the selection of the technological units of analysis in our studies is the fact that in the construction industry quite often it is impossible to use detailed methods of analysis. Thus, the special observation method has been developed specifically for analysis of the considered system. Our object of study was the dangerous and physically demanding operationally related tasks of reconstruction of industrial buildings (R. Chudley, R. Greeno, 2014). Under operationally related tasks, we understand some stage of the entire production process, which is essential in order to accomplish the specific technological purpose. The entire production process is divided into stages that have some technological completeness. The proposed method is used to improve existing methods of work. In this case, the researchers can use direct observation, chronometric analysis, questionnaire, and utilize enlarged technological unit of analysis. The physical components dominate in this type of work. However, if it is necessary, analysis of some cognitive components of work can be studied by utilizing our method that is described in the prior chapters.

The specific feature of the analyzed operationally related tasks is the lack of stable work places, the variety of performed tasks in dangerous conditions, and their limited repeatability. One of the main hazards of the work is that it is performed at heights of 25 feet.

Most of the time, developers and contractors are under pressure to keep a tight schedule due to the project's deadline, which in turn puts workers in the construction industry in risky conditions. But because this kind of job is well compensated, workers are motivated to ignore safety procedures in order to increase productivity and to earn more money. The team performance methods, when workers have to coordinate their performance with others, are important specifics of construction industry jobs. Our study revealed that the method of chronometric study used in the construction industry is not well adapted for specificity of the considered work. The existing method overloads attention of the observers, making it difficult to record data when it is necessary to observe tasks of several workers at the same time.

The gathering of data on individual workers' performance and on the possibility of coordination of their performance in time is not clearly defined by the use of the existing methods of study in the construction industry. Hence, there was a need to develop the new graphical data recording method that includes the chronometric analysis for construction industry. This method allows overcoming the disadvantages of the existing chronometric analysis and presents an opportunity to utilize the quantitative method of analysis of the performance efficiency.

9.2 THE NEW METHOD OF THE CHRONOMETRIC ANALYSIS FOR CONSTRUCTION INDUSTRY

Special paper with a millimeter grid has been prepared as an observation form. For this purpose, horizontal stripes have been drawn on the paper (see Figure 9.1).

Each strip is used for registering the time it takes individual workers to perform various task elements. The performance time of separate elements of work was marked on such stripes along the X-axis. The Y-axis is used for denoting individual workers. As can be seen, there are five people working on high-altitude objects near the edge of a high rise floor (see Figure 9.1). Each stripe was divided into 10 minute intervals of 20 mm. This means that two grids along the X-axis designate one minute. This paper permits registration of chronometric data precision to 0.5 minutes. This is a valid precision of chronometric measurement for the construction industry.

Performance time of the considered element of work is presented on the corresponding stripe for each worker with a specific number as demonstrated on

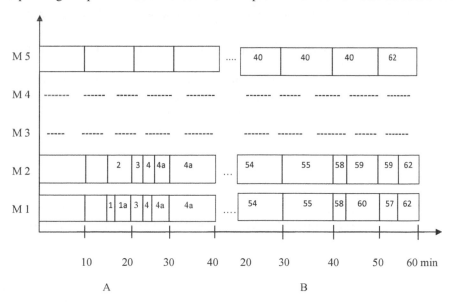

A – the first hour of the shift; B – the third hour of the shift

FIGURE 9.1   Allocation of various elements of work in time.

Figure 9.1. There are numbers and verbal description of the work under Figure 9.1. Several specialists conducted this chronometric study. Some of them measured performance time, while others filled out the chart prepared for chronometric study to register this time. Figure 9.1 depicts a fragment of the chart with chronometric recording during the three hours' interval (the first and the third hours of work, as an example). Personal interviews of individual workers were conducted at the end of the shift. The interviewers paid special attention to the critical situation during the shift.

At the following stage, data obtained in the chronometric study was compared with data collected during the interviews. The engineering department representatives who were responsible for the development of the considered work process were also involved in the process of observation. They performed their own observation, and their input was taken into account when the data gathered by the researchers was further utilized in the analysis of the considered fragment of the work process.

Figure 9.1 demonstrates that some workers were not ready to start the work at the beginning of the shift. Specifically, worker number 5 was not present at his workplace during the first 40 minutes of the shift. Figure 9.1 reflects what each worker really did and the time each operation took. Subsequently, we conducted a standardized classification of all types of work with the following determination of time for each type of work. Conducting chronometry measurements at the first stage and the classification of types of work at the following stage made it much easier to accomplish the data collection in the construction industry during this time study.

We have also developed a special classification of various work types that can be used for qualitative and quantitative analysis of the obtained data. The following work types have been identified:

1. Main work, when a worker is directly involved in the performance of the task
2. Ancillary work involving the preparation stage for performance of the main part of the work
3. Non-productive work when a worker performs unnecessary elements of work that are excluded from the task performance because the worker is not sufficiently skilled or violates technological injunction
4. Unnecessary waiting period or down time, which can be avoided in efficient work conditions
5. Additional main work that depends on the specificity of conditions in which construction is performed in comparison to the conventional forms of construction
6. Additional ancillary work that depends on the specificity of conditions in which construction is performed in comparison to the conventional forms of construction
7. Work involving safety requirement violations

In situations where a group of workers performed the operationally related tasks together, the chronometric data was usually collected separately for each worker. At the next step, performance time of the specific types of work was determined for all workers involved in the team's performance. The performance time of various types of work has been defined for both individual workers and for the whole team. This helps us to define the efficiency of the team performance as well as the efficiency of individual workers' performance.

The construction sector involves high risk due to its production processes and labor intensive characteristics (Y. Fatih, U. B. Celebi, 2015). Reconstruction of industrial buildings is an especially dangerous and difficult type of work in the construction industry. It often requires ancillary work types in comparison to the performance of similar types of work in the new building construction process. This type of work should be designed in such a way that methods of work in reconstruction conditions can approach work conditions in the new building construction process. This is the reason we singled out work types number 6 and 7 listed above. The production process can be considered optimal when ancillary work, unnecessary waiting periods, and non-productive types of work are minimized, and violation of safety requirements is totally eliminated.

Determining the time spent on each type of work allows to convert qualitative data into the quantitative one by using a set of coefficients specially designed for this purpose (G. Bedny, I. Bedny, 2018). It is important to note that for this purpose, we used only technological units of analysis, and not psychological units of analysis. Such units of analysis as members of human algorithm that are classified according to psychological principles, or the concepts of cognitive and behavioral actions and psychological operations are not utilized in this study.

9.3   QUANTITATIVE EVALUATION OF THE PRODUCTION PROCESS

The economic efficiency of the construction industry has been studied by a number of authors (Kantorer, S., 1977). We have developed a new type of coefficients for quantitative evaluation of the production process (operationally related tasks).

The fraction of the main work in comparison to the total amount of work is determined by the following coefficient:

$$\Delta T_{mw} = T_{mw} / T_g, \tag{9.1}$$

where T_{mw} is time spent for performing the main type of work and T_g is the total time of the shift.

The fraction of additional main work during the reconstruction condition in comparison to conventional construction conditions is defined according to the formula

$$\Delta T^{ad}_{mw} = T^{ad}_{mv} / T_{mw}, \tag{9.2}$$

where $T_{ad/mw}$ is the time of additional main work in the reconstruction conditions and T_{mw}, is the time of main work in conventional conditions.

The fraction of the ancillary work in the total shift is:

$$\Delta T_{anc} = T_{anc} / T_g, \tag{9.3}$$

where T_{anc} is the time of ancillary work and T_g is the total time of the shift.

The fraction of ancillary work in the conditions of reconstruction in comparison to the ancillary work in conventional construction conditions is defined according to the formula:

$$\Delta T^{ad}_{anc} = T^{ad}_{anc} / T_{anc},\tag{9.4}$$

where T^{ad}_{anc} is the time of additional ancillary work in reconstruction conditions and T_{anc} is the time of ancillary work in conventional conditions.

The fraction of the non-productive work in the total time of work is:

$$\Delta T_{n\text{-}pr} = T_{n\text{-}pr} / T_g,\tag{9.5}$$

where $T_{n\text{-}pr}$ is the time spent on the non-production work and T_g is the total time of the shift.

The fraction of time when the safety requirements are violated in the total time of the shift is defined as:

$$\Delta T_{sv\text{-}r} = T_{sv\text{-}r} / T_g,\tag{9.6}$$

where $T_{sv\text{-}r}$ is the time of the work performed with the violations of safety requirements and T_g is the total time of the shift.

The coefficient of work safety is defined as:

$$T_{s\text{-}r} = 1 \text{-} T_{sv\text{-}r},\tag{9.7}$$

where $T_{s\text{-}r}$ is the coefficient of work safety and $T_{sv\text{-}r}$ is the time of work that is performed under safety requirement violations.

The fraction of the unnecessary waiting period or down time that can be avoided by utilizing the efficient process planning in the total time of the shift can be determined as:

$$\Delta T_{un\text{-}wait} = T_{un\text{-}wait} / T_g,\tag{9.8}$$

where $T_{un\text{-}wait}$ is the unnecessary down time.

The principles applied for development of the discussed above coefficients are based on the classification of different types of work according to the criteria considered above.

Such analysis helps us to determine how efficiently labor in the construction industry is organized. The use of these coefficients becomes especially effective for a large amount of data. The coefficients can be calculated for each worker individually, or for the whole team of workers. Analysis of the obtained data and their comparison allows getting detailed information about the organization of the work process. For example, low value of ΔT_{anc} (ancillary work) is a positive factor. However, in this situation high values of coefficients such as $\Delta T_{n\text{-}pr}$ (non-productive work) and $\Delta T_{sv\text{-}r}$ (violation of safety requirements) can be used as evidence of the reduction of ancillary work that is accompanied by the violation of technological and safety requirements. In our studies, we have discovered that in reconstruction conditions ΔT^{ad}_{mw} (time of additional main work in reconstruction conditions) and ΔT^{ad}_{anc} (time of additional ancillary work in reconstruction conditions) were equal to zero. However, comparison of these coefficients with ΔT_{sv-r}, T_{s-r} and ΔT_{n-pr} demonstrates that this led to violations of safety requirements and technological requirements.

Analysis of this coefficient for individual workers allowed evaluating their effectiveness. Below we present an example of quantitative measures of assessment of the efficiency of team performance in the production process:

$$\Delta T_{mw} = 0.32; \Delta T^{ad}_{mw} = 0; \Delta T_{anc} = 0.26; \Delta T^{ad}_{anc} = 0;$$

$$\Delta T_{n-pr} = 0.37; \Delta T_{sv-r} = 0.51; T_{s-r} = 0.49; \Delta T_{un-wait} = 0.05$$

The data obtained in this study shows that the fraction of T_{mw} (main work) is only 0.32. This is a very low percentage of the time spent on the main productive work. The workers also spent relatively little time doing ancillary work (ΔT_{anc}). These two facts resulted in an increase in non-productive work (ΔT_{n-pr}), and reduced the fraction of the main work time (ΔT_{mw}). It is easy to understand the reason for such values. Ancillary work should be at the optimal level according to the specificity of tasks. All necessary preparations for main tasks should be completed during this type of work.

Both ΔT^{ad}_{mw} and ΔT^{ad}_{anc} are equal to zero and could be considered a positive factor. However, comparison of this data with safety requirements demonstrates that $\Delta T_{sv-r} = 0.51$ and $T_{s-r} = 0.49$. This means that 51% of the time safety requirements have been violated. It can be explained by the fact that ancillary work is necessary in order to prepare the workplace for the main tasks in reconstruction conditions. Scaffoldings and other safety equipment were practically absent. Their use in the production process is time consuming, especially when workers perform their tasks at high altitudes under reconstruction conditions. Productivity and compensation are in conflict with safety requirements, causing workers to be motivated by compensation and subsequently ignore safety requirements. All team members that were involved in reconstruction received good wages. ΔT^{ad}_{mw} and ΔT^{ad}_{anc} required time for work preparation in order to comply with safety requirements. Despite the fact that all the work tasks should be performed at a height of 8 meters, the workers made little use of safety equipment.

It is understandable that described types of work can overlap in time. For example, the main types of work can be accompanied by violations of safety requirements. Therefore, the sum of coefficients' value cannot be equal to one. The presented material shows that to obtain information about the organization of the work process and its efficiency, all coefficients that are described above should be considered not only in isolation but also compared and related to each other.

In this study, only technological units of analysis were utilized. When it is necessary, psychological units of analysis can be used for a more detailed study of the separate critical elements of the task.

For example, in our studies we selected the safety belts for workers who worked at heights of 8 meters. For this purpose, workers tried on different safety belts by opening and closing the locks. All their actions were videotaped. At the next stage of our study, motor actions were described by using the MTM-1 system (Bedny, 2015). Based on the findings of our study we were able to recommend the best safety belt.

CONCLUSION

The material presented here allows us to conclude that our method of analysis of work processes in the construction industry is efficient. The method is based on the graphic recording of the chronometric data and its classification according to the suggested criteria. The graphical data recording method does not overload specialists' attention, and in some cases allows them to conduct observation of several workers' performance simultaneously. The suggested method can be used for analysis of efficiency of performance of the individual workers and of the team performance.

REFERENCES

Bedny, G.Z., Bedny, I., (2018). *Work Activity Studies Within the Framework of Ergonomics, Psychology, and Economics*. CRC, Taylor and Francis, Boca Raton, FL.

Bedny, G. Z. (2015). *Application of Systemic-Structural Activity Theory to Design and Training*. CRC, Taylor and Francis, Boca Raton, FL.

Chudley, R., Greeno, R., (2014). *Building construction handbook*. Routledge. New York, NY.

Fatih, Y., Celebi, U.B. (2015). The Importance of Safety in Construction Sector: Costs of Occupational Accidents in Construction Sites. *International Journal of Economics and Business Research*, 6(2):25–37.

Kantorer, S. (1977). Calculation of the economic efficiency in the construction industry. In S. Kantorer (Ed.). *Using machine in building construction*. Building Construction Publishers, Moscow, Russia. pp.273–311.

10 Heart Rate as an Indication of the Psychophysiological Strain

Iuriy G. Solonin

Institute of Physiology, Komi Science Centre,
Urals Branch, Russian Academy of Sciences
and Syktyvkar State University, Russia

CONTENTS

10.1 INTRODUCTION

Blood circulation plays a vital role in providing all main physical functions of the human body. The high reactivity of the circulatory system helps the body to adapt to changing environmental conditions, especially related to muscle activity.

Heart rate (HR) plays the leading role in the mechanism of blood circulation amplification. Currently, an increasing number of researchers in the field of applied physiology concentrate on studying it. A number of publications (Scherrer, 1973; Rosenblat and Solonin, 1975; Bedny and Seglin, 1997; Bedny et al., 2001) are devoted to the application of this index in the evaluation of physical and neuro-emotional strain in laboratory studies and in industrial and sports settings studies.

Physicians, labor physiologists, and coaches widely use a traditional method of determining HR called palpation. This method gives fragmentary and not always representative information about the lability of HR. However, existing modern techniques such as radio telemetry, magnetic recording, and others produce more accurate and reliable data on HR for a freely moving person for a long period. Due to these newly available techniques, new data that has been accumulated in the area of human physiology, the importance of such simple indicators as HR has increased, and its interpretation has received further development.

The purpose of this work is to demonstrate the role of HR as an indicator of the physiological strain of the body when assessing blood circulation and level of stress in work and other activities, and to suggest a new method of utilizing HR in work physiology and ergonomics.

10.2 CURRENT STATUS

It's not an exaggeration to say that today there is no other physiological parameter that is used as widely as HR. It is applied in the fields of human physiology and pathology, and is repeatedly tested for its informative and diagnostic value. There are excessive attempts to praise the "enthusiasm pulse," as well as examples of the undeserved "denigration" of its role. Therefore, this indicator should be given the place that it rightfully deserves. There is a need for a clear understanding of the theoretical foundations, possibilities, and limitations of using HR and methods of its application.

Modern views on the mechanisms of regulating the heartbeat and HR are described in "Physiology in the Space Environment" (S. N. Swisher, J. V. Warren, T. G. Coleman, H. T. Milborn, 1968). The functioning of the human heart is influenced by a complex of neural, humoral, and purely mechanical factors that cause changes in the frequency and intensity of its contraction to maintain an optimum level systemic blood pressure and blood flow to the organs and tissues. When the body is at rest, the heart is autoregulated due to sustainable mechanisms such as the intra-cardiac nervous system or a cardiac pacemaker. For activities requiring a coordinated response of the whole body, the central regulatory system is at play including spinal cord, the medulla oblongata, hypothalamus, cerebellum, and cerebral cortex.

The level of HR because of the activity of the sinuauricular node depends on the relationship between the vagus and simpatikusa, coronary blood flow, complex metabolites, and stretching of the heart cavities. Simpatikusa and vagal activity, in turn, is determined by the blood pressure and by the number of metabolites. Sympathetic effects on myocardial function are carried out in the end due to the change in the concentration of catecholamines in heart. It is known that the rise of blood temperature and the excitation of simpatikusa increases toughness and accelerates membrane potential depolarization. There is a direct relationship between the total duration of the action potential and the rate of myocardial contractions.

There is a prevailing opinion that one of the main mechanisms that regulates the functioning of the heart is the ability to change the frequency of contractions. The response of the heart to physical activity is a complex phenomenon involving the interaction of changes in HR, ventricular end-diastolic volume, and neuro-humoral regulation. The increase in HR during physical activity results in the improvement of the contractile state of the myocardium.

The normal reaction of the heart to physical activity is presented as an integrative impact of tachycardia, sympathetic stimulation, and the mechanism of Frank-Starling on the myocardium.

It is believed that when the body is at rest the heart is under homeostatic regulation, but during the muscular activity, homeostasis mechanisms are inhibited, and cardiac activity is regulated by the disturbing influence. The mechanisms of self-regulation include heterometric and homeometric regulation provided by the contractile elements of the myocardium. The Frank-Starling mechanism is the basis of heterometric self-regulation. Homeometric self-regulation ties the strengthening of heart contractions with accelerated HR (Bowditch effect), and myocardial contractility increases with afterload (Anrep effect). Signals from sin-carotid and aortic receptors not just provide an economic mode of the heartbeat, but can also put a reflexive limit on the activity of skeletal muscles, protecting the heart from overloading. In cases where long term adaptation of the heart to high workloads occurs, significant structural biochemical and functional changes in the myocardium occur (Meyerson, Pshennikova, 1988).

The average human HR and its limits (from 30 to 275 bpm) are the result of human evolution. It is known that HR of warm blooded animals is inversely proportional to their body weight (Schmidt-Nielsen, 1987). There are many publications dedicated to the relationship between the body weight, heart size, HR, speed, and duration of excitation to the heart. There is even an established connection between the average HR and longevity of the individual species.

It is reasonable to assume that the range of possible HRs for different species, including humans, is largely confined to the mechanical (geometrical) conditions, which is related to the size of the species. Since the ratio of the heart weight to the body weight of mammals is constant, the metabolic rate decreases with increasing size of the body and its relative surface, the HR also naturally reduces. It is natural to assume that the myocardium of the large heart is unable to contract as fast as the myocardium of the small one. For example, the HR of horses is 35–50 bpm, while mice weighing 1.5–2.5 g, found on the island of Sardinia, have a HR of up to 1200 bpm. Perhaps for the same reasons, children, adolescents, and women have a higher HR than adult males. Bradycardia among long-distance runners and some other athletes, where physiological cardiac hypertrophy along with other factors play a role can be explained similarly.

Heart rate is not directly dependent on the environment, and its influence is manifested through internal regulation mechanisms. The universal nature of the diverse external and internal factors puts the load on various physiological systems and on the body as a whole; they especially affect the heart and its chronotropic response in particular.

Dynamic muscular work and the associated significant increase in oxygen intake is strong and adequate incentive for the increase of blood flow and HR. Motor-visceral impulses, the release of catecholamines in the bloodstream, and sympathetic arousal contribute to the activation of the heart. The "muscle pump" or "peripheral heart" provides a great deal of help to heart functions (N. Arinchin).

The conditions for the heart are worse during static or local dynamic *local* muscle work. Diminished oxygen supply, limited motor-visceral impulses, and a weak "muscle pump" complicates regulation of the heart functions. The reaction in this case is less adequate, since the heart functions are tuned to the global response of the body.

With simultaneous demands from muscle and thermoregulation loads, the heart and circulation are "between the devil and the deep sea" because it is necessary to supply oxygen to working muscles and compensate for heat escaping the body surface at the same time. In such circumstances, the most representative indicators are the integrative properties of the cardiovascular system and its main parameter, namely HR.

Multiple studies demonstrate that HR accurately reflects the intensity of mental strain and information overload (Gorbunov, 1997). However, the main cause of changes in the HR in the described experiments is emotional and muscular components. Some studies demonstrate a weak link between HR and mental strain.

The strength of the emotional impact on the heart can be compared to the impact of muscular activity. According to Pavlov (1932), the emotions of our ancestors were always accompanied by motor acts. Modern psychophysiology considers the cardiac component to be the most reliable indicator of an emotional response (Fedorov, 1997). Acceleration of HR occurs more often than deceleration for various emotional responses.

When emotions are transient in nature, it is difficult to quantify the increase of pulse arrhythmia due to the peculiar reactions of the heart. It is obvious that emotions have a rather negative effect on the circulatory aspect while muscle activity has a positive one. There is reason to believe that along with the rise in blood pressure, stroke volume of the heart is reduced, which may violate the metabolic processes in the myocardium during emotional displays. Many work physiologists described significant changes in HR because of neuro-emotional strain in the workplace (Physiology of Work, 1993).

The integral nature of changes in HR as a response to various stimuli, along with a positive aspect has certain disadvantages, the same as other physiological parameters, because almost all of them are in one way or another integral and non-specific. In many cases, it is impossible to determine "why" and "how much" HR or other indicators change. It is appropriate to assume that acceleration of HR by the same amount is not equivalent with respect to circulatory impact, and the consequences for the body as a whole when the causes for acceleration are different.

As has been mentioned above, the most significant factors for increases in HR and blood circulation are muscle activity and associated increases in metabolic demands. Multiple studies demonstrate the close correlation of HR with energy consumption, the consumption of oxygen, and with the bioelectric activity of skeletal muscles. All other causes of strain on the cardiovascular system such as emotions, thermal load, etc., are not associated with metabolism, and should be viewed as less desirable. Thus, the acceleration of HR due to muscle activity is definitely more favorable than the same increase in HR due to heat stress. For example, subjects refuse to continue participation in an experiment that requires them to sit in a heat chamber at a much lower HR than when they are involved in an intense physical workload.

The complexity of the "gear" regulation of HR suggests that there are non-linear changes in HR in relation to physical activity in the lower and upper levels of regulation. There is evidence of a linear function between acceleration of HR and increasing workload in the range from 100 to 170 bpm.

For muscular work it is very important to identify what group of muscles are involved. It is known that with the same HR the absorption of oxygen is less when the working muscles' weight is greater. In general, the synergetic relationship of HR and oxygen consumption (or body energy) varies depending on the kind of workload involved: general or local, dynamic or static; using flexor or extensor muscles, upper or lower extremities. When the large muscles are in use, venous return increases, thereby increasing heart stroke volume and decreasing HR. This is also evident by the increase of "oxygen pulse" when large muscle ensembles are at work. For the static load, the extent of the muscle strain plays a much greater role than the weight of the strained muscles in acceleration of the HR (Thorevsky et al., 1986).

The main reasons for the distrust some researchers express for the HR as an indicator is due to the lack of standard methods of measuring it, and also because some researchers ignore the characteristics of this indicator and mechanisms that determine its behavior in the body as a whole, and because of the inadequate attention paid to a number of factors that affect HR. Imperfections of the widely used method of pulse palpation measuring, which is quite a labile functional index, imposes a large imprint on its value and evaluation, reducing the value of the considered criterion.

These shortcomings are not an obstacle for the widespread use of HR in experiments, in medicine, and in studying other human activities in order to solve many problems of ecology and physiology of labor and sports. The use of standardized methods of measurement of HR, and a quantitative study of the various factors that influence HR can significantly increase the importance of HR as an informative indicator of physiological strain on the body.

10.3 METHODS OF MEASUREMENT AND EVALUATION OF THE HR

The average HR during normal functioning is used to assess the physiological strain of human activity. It can be measured utilizing the following methods: 1. radio telemetry or other detecting equipment (Rosenblat, 1967) and 2. palpation immediately after the activity (assessment based on the comparison of HR during the active work and during the period of recovery using HR radio measurement for the significant number of people of different professions (Solonin, 1969). The second method is less accurate, but it serves practical needs.

Assessment of any activity (work, exercise, and so on.) in terms of HR may be based on studies involving at least 10 people. HR monitoring should be combined with observation of subjects' activity and timing of its elements. Various devices that enable recording signals such as the electrocardiogram (ECG) on a conventional electrocardiograph, and audio pulse counting utilized for the calculation of radio telemetry of HR for freely moving persons, significantly widens practical application of radio-pulse measurement (Rosenblat, 1967).

HR indication is based on the heart bio potentials (R wave of the electrocardiogram). "The patient's device," the small transmitter weighing 100 g, is placed in the pocket of a person who is instructed to move freely. The receiver, "the researcher's device," weighing about 0.6 kg, decrypts the received radio signal into a changing tone of the sound delivered to the researcher headphones or to the single-channel electrocardiograph called the registrar. The measurements are carried out in 10-second intervals with 5-second pauses, occurring four times every minute. It's also possible to use the totalizer that provides both the control current values for HR every 10 seconds and the cumulative total data to monitor HR. In recent years, compact pulse counters have become available, and are now widely used in sports and other settings.

It should be noted that the auditory calculation method of HR is not less accurate than the graphic one (the interval between the teeth of the ECG). Furthermore, the first method has the highest auditory ratio of sound/noise (S/N) that allows information reception even in a production environment with a high level of interference. Simplicity, accessibility, accuracy, and efficiency of small auditory counting HR allows extensive information collecting on the chronotropic function of the heart in the laboratory and production settings, and during sports activities.

When conducting special research, radio pulse measurement is carried out continuously throughout the operation (shift, sports training, etc.). For practical purposes, the measurements can be conducted in a fragmented manner several times in the course of operations covering typical activities and breaks in the course of the shift. If the workload is relatively even throughout the activity, the measurements are taken in the beginning, in the middle, and at the end of it. The total duration of the measurements for one subject should be about two hours per day. The data is analyzed in two stages: for each subject and for a group of subjects.

HR during various activities (labor, exercises in a continuous loop) and rest are considered separately When analyzing the data for individual subjects. To do this, all digital data on HR during each cycle of activity is grouped in the variation series and the average value is calculated. Such data allows to compare the intensity of workload cycles for the individual, and then to estimate the parameters of the workload as a whole for the work shift or training.

The following three indicators are listed below:

1. The working level of heart rate (HRwork).
   HRwork is calculated using the following formula based on the average data for each cycle:

$$HRwork = (HRwork1 * T1 + HRwork2 * T2 + \cdots + HRwork\, n * Tn)/WT, \quad (10.1)$$

where HRwork1 is HR during the first cycle, HRwork2 is HR during the second cycle, etc., and T1, T2, etc., is the total duration of each of these cycles in minutes. WT is the total duration of the work, which is calculated as:

$$WT = T1 + T2 + \cdots + Tn \quad (10.2)$$

2. The average heart rate at rest (HRrest).
   HRrest. is calculated as the average value of HR for the individual cycles of all of the discrete values.
3. The average heart rate per working day—a shift, training, etc., (HRsh) is calculated as follows:

$$HRsh = (HRwork * WT + HRrest * RT)/(WT + RT) \quad (10.3)$$

where RT is the total rest time (pauses in the work cycle).

It is recommended that all calculations should use values of HR for 10-second intervals, and only then multiply the final results (HRwork, HRrest, HRsh) by 6, translating them into data per minute.

For the derivation of the workload of a group of subjects, individual HR and timing information for individual cycles, and the entire length of the daily activities are averaged. Averages (X), standard deviation indicators (SD), coefficient of variation (CV), and SEM (m) are determined.

When it is not possible to measure HR continuously during activity, HR is measured by palpation immediately after the end of a cycle of activity (working operation, exercise) in the working position (posture). It is preferable to collect data during the first minute of recovery. If the data cannot be collected throughout the first minute, it is possible to restrict the data collection to the range from 15–45 sec or from 20–40 sec (in this case, the obtained value is close to the average value per minute in general). If the data from the first minutes is incomplete or missing, the data from the second minute of recovery should be utilized; if that is not possible the data for the third minute is used. The working level of HR in the previous cycle of activity is covered in the author's publication (Solonin, 1969).

When repeated studies of the same cycle are conducted during various periods of the shift, the average levels are calculated for the respective cycles, and for the periods of rest (the pulse rate measurement is conducted for every minute of relaxation in the range from 20 to 40 seconds; the data for individual pauses is not averaged but summarized for the entire day into a single variation number sequence, and HRrest is calculated based on this data).

The general method of observation is quite similar to that radio pulse measurement method described above. The data is also processed in the same manner.

After collecting the average pulse rate for a group of people, the following parameters of the operational strain of the activity are evaluated: 1. overall intensity of the workload (HRwork and the individual cycles, average levels HRwork1, HRwork2, etc.); 2. total scope of the workload (HRsh).

In terms of total intensity, which represents the main indicator in assessing the strain of the workload and setting the norms for the productivity, production operations are generally classified using Table 10.1. In evaluating the total workload, two groups of professions should be allocated: with mean HRsh of 100 bpm or more that require it with a mean of HRsh less than 100 bpm that does not.

TABLE 10.1

Classification of strenuousness of work based on HR (bpm).

	Quantitative criterion of the type of work (HR)			
	1	2	3	4
Types of muscular work	Easy	Moderate	Heavy	Very heavy
Totally (globally) dynamic	Below 90	Below 100	Below 120 (110)	Above 120 (110)
Regionally dynamic	Below 80	Below 90	Below 110 (100)	Above 110 (100)
Locally dynamic	Below 80	Below 85	Below 95 (90)	Above 95 (90)
Work with static components	Below 85	Below 90	Below 100	Above 100

Note: Average working level of HR_{work} and average shift level HR_{sh} value of HR. Where they do not match, HR_{sh} is given in parentheses.

TABLE 10.2

Heart rate (bpm) for individuals of different sexes 20–29 years old for various workloads when exercising on the bicycle ergometer (X ± SD).

Load, W	Men (n = 13)	Women (n = 12)	P
0	77 ± 9.0	80 ± 7.1	NS
10 (hand)	107 ± 9.3	117 ± 5.7	<0.01
20 (hand)	111 ± 8.2	124 ± 6.8	<0.01
20 (foot)	98 ± 6.4	108 ± 4.4	<0.01
45 (foot)	111 ± 8.2	127 ± 8.8	<0.01
90 (foot)	140 ± 10.0	163 ± 5.4	<0.01
$VO_{2\,max}$, ml/kg*min	40 ± 6.1	38 ± 4.0	NS

Changes in heart rate under the influence of various factors (research results) shows that HR response to the same physical stress differs sharply among people of different sexes ($P < 0.01$), and with increasing strain of work, this difference increases (Table 10.2).

For the subjects in the age range of 20 to 50 years old there is a tendency for HR to decrease with age increase for the same workload (Table 10.3). A significant age difference is identified for moderate workloads (from 20 to 45 W*) in particular. This effect on the HR for different age groups decreases with an increase in the intensity of the workload.

Differences between age groups are also noticeable in the speed of the HR restitution. With 5 minutes of rest, HR decreased after a workload of 45 W in the 20 to 29 year old by 24 bpm, and for the group of subjects in the 40 to 49 year old group

TABLE 10.3

Heart rate (bpm) for men of all ages with different workloads while exercising on the bicycle ergometer (X ± SD).

Load, W	Age, years			P
	20–29 (n = 14) (1)	30–39 (n = 21) (2)	40–49 (n = 17) (3)	
0	85 ± 9.6	81 ± 13.9	79 ± 9.4	NS
20	107 ± 8.8	103 ± 13.0	97 ± 7.3	1,3 < 0.01
45	115 ± 11.8	109 ± 13.5	106 ± 8.2	1,3 < 0.05
90	136 ± 11.4	132 ± 11.7	130 ± 8.6	NS
100	146 ± 12.5	144 ± 15.7	139 ± 10.6	NS
135	165 ± 11.8	166 ± 17.1	163 ± 10.6	NS
VO_{2max}, ml/kg*min	43 ± 6.6	36 ± 7.6	30 ± 4.5	1,2; 1,3; 2,3 < 0.01

TABLE 10.4

Influence of age on heart rate (bpm) for workers of the same professions (X ± SD).

Professions and indicators of HR	Age, years		
	20–29	30–39	40–49
Blacksmith:			P < 0.05
HR$_{rest}$	72 ± 8.6 (n = 12)	72 ± 6.3 (n = 24)	68 ± 5.8 (n = 10)
HR$_{work}$	106 ± 7.8 (n = 12)	106 ± 6.7 (n = 24)	98 ± 9.2 (n = 10)
Metallurgist:			P < 0.05
HR$_{rest}$	80 ± 10.2 (n = 14)	79 ± 7.2 (n = 16)	79 ± 9.9 (n =16)
HR$_{work}$	135 ± 11.0 (n = 14)	130 ± 8.9 (n = 16)	123 ± 11.0 (n = 16)

Note: There was a statistically significant difference for each age group.

by 16 bpm. For the workload of 90 W for the same two age groups it decreases by 42 and 35 bpm, and for the workload of 135 W, by 63 and 54 bpm respectively.

Reported age peculiarities of HR are manifested in a production environment (Table 10.4).

The effect of age is also confirmed by the correlation analysis. For example, the correlation coefficient between age and HR during the shift for the miners working in the copper mine was 0.38 (P < 0.05), and for the metallurgists was 0.43 (P < 0.05).

Tables 10.2 and 10.3 demonstrate the relationship between the level of HR and the intensity of physical activity for individuals of different sex and age groups. It is also worth noticing the significant difference in the changes of HR to the same power load (20 W), performed by hands and by feet when different muscle mass is involved. "Pulse cost" for regional muscle workload (work involving more than one-third of muscle mass) is higher (P < 0.01), than for the global workload (work involving more than two-thirds of muscle mass).

The high correlation of HR with the intensity of the physical workload is also demonstrated by the correlation analysis of the collected data. The correlation coefficient of HR with the efforts and the work intensity for bicycle ergo-metric exercise varies from 0.83 to 0.93 (P < 0.001). The formula for the relationship between physical work capacity and HR is as follows:

$$Y = 0.57 * X + 81, \tag{10.4}$$

where X is the physical work capacity (from 0 to 135 W*) and Y is the HR (from 77 to 160 bpm**).

*Watt (W)—a unit of power in the SI system; equals one joule per second, approximately 0.014 kilocalories per minute, or 0.0013 horsepower.

**The numbers in the notes to the formulas are the minimum and maximum values of correlated variation series. They mean that the regression equation is valid only in the values of these ranges.

In the industrial settings (for 57 different occupational groups), mean for HR for the whole shift also correlates with physical effort ($r = 0.57$; $P < 0.001$) and workload capacity ($r = 0.76$; $P < 0.001$). The formulas for regression are:

$$Y = 0.37 * X1 + 106.2 \tag{10.5}$$

where Y is HRwork (93–149 bpm) and X1 is effort (3–42 kg); and

$$Y = 0.54 * X2 + 80.1 \tag{10.6}$$

where Y is HRwork (93–149 bpm) and X2 is workload (7–100 W).

Analysis of variance for 57 professions showed that the average working level of HR (HRwork) depends on the length of time involved in the severity of physical work 83.6% ($P < 0.001$), and on the sum of such factors as nervous tension, air temperature, noise and dust 13.0% of the time.

Body position or posture affects HR not only during rest, but also during the active work. For example, for welders, in comparison to the standing position, HR slows down an average of 4 bpm for "the slope of" position, by 10 bpm for "the knee" position, by 12 bpm for the "the sitting" position, and by 14 bpm for the "squatting" position.

Body position while working is also important. For example, 13 workers who were plastering walls had HRwork of 109 ± 2.4 bpm, while when they were plastering the ceiling it was 120 ± 2.8 bpm ($P < 0.05$).

Air temperature also has a significant impact on the level of HR when it is outside of comfortable and hygienically acceptable levels. For instance, 18 workers who were unloading fire bricks from a kiln had HRwork of 112 ± 4.9 bpm at a temperature below 20°C, at a temperature 20–30°C it was 123 ± 3.5 bpm ($P < 0.01$) and at a temperature above 30°C it was 136 ± 4.0 bpm ($P < 0.001$). For 10 blacksmiths when doing cold forming of the metal (temperature 22–27°C) HRwork was at 93 ± 4.3 bpm, and when they were doing hot stamping (temperature 38–50°C) HRwork was 118 ± 7.5 bpm ($P < 0.01$).

For 57 occupational groups, the correlation coefficient between the mean-HR and air temperature was +0.39 ($P < 0.01$). The regression formula is:

$$Y = 0.46 * X + 104.3 \tag{10.7}$$

where Y is HRwork (93–150 bpm) and X is air temperature (0–47°C).

Increased HR in the external heat load is associated with the processes of heat loss. HR changes in the cold are likely due to the process of strengthening the body's heat production. Special experiments showed that when working in extreme cold, HR has a tendency to decrease in frequency alone, which is mentioned in some publications (Koshcheev, 1981). Oxygen consumption, as it's measured by Douglas-Holden, increases. Oxygen consumption and HR significantly increase when subjects' workload takes place in cold temperature environments. Obviously, the extreme cold activates gas exchange and heat production, which results in the chronotropic response of the heart.

TABLE 10.5

Heart rate (bpm) of 17 drillers in the North in different seasons under various workloads exercising on bicycle ergometer in the room with a comfortable climate (X ± SD).

Load, W	Season		P
	Summer	Winter	
0	76 ± 9.6	83 ± 10.4	NS
50	94 ± 12.0	103 ± 10.0	<0.05
100	112 ± 12.0	122 ± 10.8	<0.05
150	135 ± 14.4	149 ± 13.6	<0.01
VO_{2max}, ml/kg*min	42 ± 11.6	40 ± 8.0	NS

The working level of HR in production settings is also subject to the influence of the biorhythms. The effect of circadian rhythm is especially pronounced. For instance, HRsh for 14 miners was 94 ± 8.4 bpm for the night shift (12 am to 6 am), 99 ± 9.4 bpm for the day shift (8 am to 2 pm) and 103 ± 6.2 bpm for the evening shift (4 pm to 10 pm) (P < 0.001).

For 11 metallurgists HRwork at night (1 am to 7 am) was 119 ± 6.6 bpm, in the morning (7 am to 1 pm) it was 127 ± 8.7 bpm, in the afternoon (1 pm to 7 pm) it was 143 ± 8.8 bpm, and in the evening (7 pm to 1am) it was 138 ± 9.7 bpm.

Seasonal fluctuations in HR also occurs both during rest and normal operation (Table 10.5 and Table 10.6), which are apparently associated not only with temperature effects. In the cold season, HR increases for the construction workers who work outdoors. Laboratory experiments clearly showed that in the winter "pulse value" of muscle activity increases dramatically even in the comfortable indoor conditions (Table 10.5). Thus, HR responds not only to direct environmental

TABLE 10.6

Heart rate (bpm) depending on the season of the year for different groups of people (X ± SD).

Groups of people	Season		P
	Summer	Winter	
Building construction workers (n = 26):			
HR_{rest}	81 ± 11.0	87 ± 13.5	<0.05
HR_{work}	97 ± 5.5	104 ± 9.0	<0.01
Inhabitants of North (village of Izma, Komi Republic):			
men (n = 69)	65 ± 9.6	69 ± 9.0	<0.05
women (n = 125)	68 ± 13.2	73 ± 12.1	<0.01

conditions but also to the external cold background, related to an increase of body heat loss in winter.

The level of HRrest also differs depending on the latitude of the person's residence. There is a tendency for HR to slow in men living in the latitudes from 40 to 67°N. We measured BP moving north:

Latitude	40−49°N	50−59°N	60−67°N
HR in summer	86±7.4 bpm	78±6.4 bpm	72±8.2 bpm
HR in winter	86±8.7 bpm	80±7.6 bpm	75±9.4 bpm

The same pattern is observed in women. HRrest slows by 11 bpm in winter and summer when residence location changes from the latitude of 50–59°N to the latitude of 60–67°N (P < 0.01).

Integral features of HR are confirmed by the poly-modality effect on HR and by its correlation with other physiological indicators. When exercising on the ergomrtric bicycle, the correlation coefficient of the HRwork and respiration frequency (RF) ranges between 0.7–0.8 (P < 0.01), and with the volume of respiration per minute (MVR), oxygen consumption and energy expenditure (EE) is 0.70–0.85 (P < 0.01).

For various professions, working HR correlates with RF (0.38–0.52), with EE (0.49–0.73), with skin and lung moisture loss (0.39–0.80), etc. Twenty-nine different occupational groups have closely associated HRwork with EE (r = 0.58; P < 0.01) that can be defined by the following formula:

$$Y = 0.054 * X - 2.36 \tag{10.8}$$

where Y is EEwork (2.2–6.6 kcal/min), and X is HRwork (100–149 bpm).

It has been found that for thirty-four occupations HR correlates with MVR (r = 0.58; P < 0.01), and for fifty-seven professions it correlates with the percentage change of the static muscle endurance toward the end of the shift (r = 0.64; P < 0.01).

Another interesting fact is that the level of HR as an indicator of muscular load and physical strain on the body may be also an indicator of the adverse effects of this stress on the workers' health. According to our data, for fifteen occupational groups consisting of men involved in physical labor, mean HRsh is correlated with diseases of the circulatory system (r = 0.54; P < 0.05), of the musculoskeletal system (r = 0.56; P < 0.05), and with temporary disability (r = 0.52; P < 0.05). We derived the following formulas:

$$Y1 = 0.233 * X - 18.96 \tag{10.9}$$

$$Y2 = 0.4 * X - 32.9 \tag{10.10}$$

where X is HRsh (83–112 bpm), Y1 represents the number of diseases of the circulatory system (0–15.1 cases per 100 workers), and Y2 is the number of diseases of the musculoskeletal system (0.5–25.3 cases per 100 working).

In short, the relationship between the number of cases of illnesses and the level of HR on average for the shift is obvious. Therefore, it is possible to predict the number of such cases based the HRsh.

The observed significant correlation between physical strain (for HR) and morbidity of workers tends to increase (r = 0.68–0.76) with the elimination of the factor of work intensity.

Key indicators of HR for subjects in 57 occupational groups vary widely. Preoperational HR (70 to 95 bpm) reflects the initial or "prelaunch" state. Such HR is far from "the basal metabolic rate (BMR)." The other indicators are integral characteristics and hemodynamic status of the body—its functional strain levels of the HR regulation. The maximum level of HR (from 109 to 180 bpm) shows a peak (one-off) voltage, and in some professions, it is close to the extreme value (as in sports). Average HRwork (from 85 to 149 bpm) reflects work (operational) strain, the intensity of labor, and the working activity of the body. HRrest during pauses in the work (from 76 to 106 bpm) is also an average statistical indicator of interoperable "activity," and is associated with the absence of tension during the recovery state. Mean value of the HRsh (from 85 to 125 bpm) generally reflects the physiological "cost" of the professional activity.

The percentage of the use of "the pulse reserve" that is equal to the ratio of the increase of the HRwork to the maximum possible HR (HR reserve) can be used as an indicator of the degree of mobilization of physiological reserve. According to our data, it is equal to 59% (33–92) in average for maximum HRmax, 35% (13–66) for average working level (HRwork), and 29% (12–46) for the average shift level (HRsh).

Naturally, HR limits that have been encountered for some individuals are greater than for the groups. We registered a minimum HR of 48 bpm during rest, and a maximum HR of 214 bpm for the subject who was repairing the open hearth furnace. This data was confirmed electrocardiographically.

In general, our study showed quite substantial changes of HR during work shifts. The average growth of maximum HRmax was 77%, the average working level HRwork increase was 42%, and for the average HRsh it rose 36%. Changes in HR when comparing them with HR before the shift are statistically significant, not to mention their physiological significance. Average HRwork and HRsh levels for specific occupations are given in Table 10.7.

Some statistical parameters of HR (in particular Sigma—SD) are needed to predict and calculate the studied parameters of physiological research. Sigma for different indicators of HR is around 10 bpm in average. The variability of performance indicators has the tendency to decrease, which confirms the presence of "the phenomenon of Barcroft" meaning that the workload makes HR parameters more constant for the group.

Two types of HR variability such as intra-individual (physiological) and inter-individual (biological) are very close in value. The intra-individual coefficient of variation for average HRsh is equal to 10.3%, while the inter-individual one is equal to 9.7%.

TABLE 10.7
Mean group values of heart rate (bpm) collected in production environment using radiotelemetry by the author and his colleagues.

Types of work and profession	HR_{work} $M \pm SD$	HR_{sh} $M \pm SD$
Easy workload (1st class):		
Engineer-designer	$82 \pm 8,7$	$82 \pm 7,9$
Operator involved in cold rolling of metal	$86 \pm 9,8$	$83 \pm 11,2$
Crane operator (female)	$86 \pm 8,5$	$85 \pm 8,3$
Moderate workload (2nd class):		
Driller of gas well	$91 \pm 11,1$	$91 \pm 10,8$
Blacksmith doing cold punching	$94 \pm 9,0$	$91 \pm 6,6$
Operator of gantry crane	$93 \pm 12,3$	$92 \pm 12,8$
Operator doing cold stamping of small parts (female)	$93 \pm 8,4$	$92 \pm 8,4$
Operator drilling rock in the open pit	$98 \pm 13,4$	$92 \pm 14,8$
Operator of the gas compression station	$99 \pm 12,8$	$93 \pm 11,7$
Operator of semiautomatic sawing (cutting) switch in lumbering	$99 \pm 14,1$	$93 \pm 13,8$
Operator of blacksmith's control panel	$96 \pm 6,7$	$96 \pm 9,3$
Mechanical shop foreman	$96 \pm 13,7$	$96 \pm 13,7$
Building construction worker layering bricks	$100 \pm 12,1$	$97 \pm 10,9$
Machine operator	$99 \pm 8,2$	$99 \pm 9,2$
Automobile assembly line fitter	$99 \pm 7,7$	$99 \pm 9,8$
Heavy workload (3th class):		
Sewing machine operator (female)	$100 \pm 13,1$	$100 \pm 11,8$
Operator forming steel at a steel mill shop (female)	$101 \pm 8,4$	$100 \pm 7,7$
Assembler (fitter) involved in building machines	$103 \pm 8,9$	$100 \pm 9,2$
Operator involved in dressing wood	$105 \pm 9,9$	$100 \pm 11,8$
Operator involved in output of steel in electric furnaces	$106 \pm 10,0$	$100 \pm 10,8$
Housings welder	$102 \pm 9,4$	$101 \pm 9,2$
Semi-automatic welding	$102 \pm 8,9$	$101 \pm 9,9$
Bricklayer on building (female)	$104 \pm 11,7$	$101 \pm 12,3$
Replacement of anode in production of aluminum (new technology)	$115 \pm 9,9$	$102 \pm 12,7$
Automobile engine assembler	$103 \pm 6,9$	$103 \pm 10,5$
Smelting silicon in electric furnace	$105 \pm 14,1$	$103 \pm 13,8$
Painter—interior building construction (female)	$106 \pm 12,8$	$104 \pm 15,1$
Smelting of aluminum (new technology)	$109 \pm 8,3$	$104 \pm 9,9$
Machinery on production alumina (female)	$111 \pm 11,2$	$104 \pm 12,8$
Chopping on lumbering	$116 \pm 9,9$	$104 \pm 10,8$
Autobody welder	$105 \pm 12,3$	$105 \pm 11,7$

(Continued)

TABLE 10.7
Mean group values of heart rate (bpm) collected in production environment using radiotelemetry by the author and his colleagues. (Continued)

Types of work and profession	HR_work M ± SD	HR_sh M ± SD
Manual sorting of metal sheets	111 ± 9,5	105 ± 9,9
Bumpers polishing	106 ± 14,1	106 ± 12,9
Regulation of machinery on asbestos preparation (female)	108 ± 5,1	107 ± 9,3
Driver cutting asbestos cardboard	107 ± 8,0	107 ± 6,0
Smelting of nickel	118 ± 11,7	107 ± 13,4
Steeple jack	108 ± 9,9	108 ± 12,1
Lumberjack with gasoline-powered	112 ± 11,8	108 ± 13,4
Blacksmith on large-tonnage press	113 ± 9,5	108 ± 9,6
Driver cutting asbestos cardboard (female)	109 ± 9,5	109 ± 10,8
Face-worker copper mine	111 ± 9,8	109 ± 9,6
Blacksmith on forging hot metal	115 ± 9,5	109 ± 8,6
Plasterer (female)	115 ± 14,2	110 ± 16,1
Sorter of silicon (female)	115 ± 13,8	110 ± 11,9
Helper of driller gas well	119 ± 12,1	110 ± 14,2
Very heavy workload (4th class):		
Smelting of aluminum (old technology)	128 ± 9,9	107 ± 11,2
Replacement of anode in production aluminum (old technology)	123 ± 8,9	108 ± 7,9
Hot rolling of sheet steel metal manually	135 ± 14,5	108 ± 9,2
Tear up block of metal manually	112 ± 9,6	111 ± 11,2
Smelting steel on open-hearth furnace	118 ± 9,8	111 ± 9,6
Tear up block of metal manually (female)	114 ± 8,5	112 ± 9,9
Blacksmith-puncher of hot metal	120 ± 10,0	113 ± 6,0
Smelting of red copper	134 ± 11,9	113 ± 11,7
Unloading backed brick from kiln	121 ± 14,6	114 ± 14,5
Loader of alumina	116 ± 11,0	115 ± 10,6
Cutting of stick hot metal	128 ± 13,9	115 ± 14,0
Hot repairs of open hearth furnace	149 ± 14,1	115 ± 14,4
Chipping steel castings	120 ± 13,8	118 ± 14,5
Smelting of ferro alloy	126 ± 11,9	117 ± 10,7
Loading fire brick manually	120 ± 17,1	119 ± 15,9
Loading in refractories production	123 ± 14,3	119 ± 10,0
Repairs tank apparatus in production alumina	124 ± 10,4	119 ± 14,3
Unloading fire brick from kiln	131 ± 9,1	123 ± 12,5

CONCLUSION

Heart rate (HR) has been recorded for a long time. The collected data provides important information on the status of circulation and strain on the entire body, changes in the physiological state for the individual as they cycles through daily activity in production settings, in physical education, sports, and other human activities.

The level of HR as biological and physiological parameters is determined by multiple factors. All factors influencing HR and that are significant for studies in the physiology of labor and sports, military, and other activities can be divided into two groups: 1. constants related to the individual characteristics of the body (age, sex, body size, metabolism, physical health and fitness, the degree of adaptation to environmental conditions, etc.), 2. temporary factors associated with the short term demands of work and environmental conditions, or short term changes in the body (muscle strain, the weight of clothing and equipment, breathing restrictions when working in a respirator, posture, heat stress, dehydration, cold, vibration, noise, oxygen level in the air and harmful gases in the atmosphere, emotional stress, biorhythm, food intake, the degree of fatigue of the heart and skeletal muscles, etc.). Constant factors determine the overall level of HR and temporary factors define the real ongoing status of this indicator. In order to interpret the levels and changes in HR, the understanding of these effects and their consideration in relation to a particular situation are required.

Thus, when studying and interpreting HR as well as other physiological indicators, one should clearly distinguish between reaction to the change in intensity factor (short term exposure) and response to the average level of prolonged exposure to the constant intensity. In the first case, transitional processes are in play when the high lability (dynamism) of HR, often contradictory reactions, depending on the significance of the stimulus, and phase of the shift, etc. can be observed. This is one of the reasons some researchers are skeptical about the benefits of utilizing HR. In the second case, the process is usually stationary, shifts more unambiguous, reliable, and must first be taken into account when assessing the impact of various factors.

Moreover, a distinction should be made between changes in HR caused by work intensity and by the duration of activity. From this point of view, it is possible to determine primary and secondary reactions. Changes in the internal body environment during continuous operation such as accumulation of metabolites, increased body temperature, blood clots, etc., cause further body reactions in the form of increasing HR and other changes in hemodynamics. Secondary reactions can be associated with fatigue resulting from work or/and sports activity. Circulatory conditions commonly limit human performance in three typical cases: heavy physical labor or exercise, working in the heat load, and static tension of the muscles. In other cases, fatigue is not determined by the circulatory conditions of the body and HR is not an indicator of fatigue.

It can be stated that HR reflects at least four levels of body activity: 1) cellular level (cardiac pacemaker), 2) organ level (heart), 3) system level (blood circulation), 4) the body as a whole. On the one hand, HR is controlled by complex neural (quick) and humoral (strategic) regulation, i.e., it's a controlled parameter. On the other hand, HR is controlling circulatory changes (its cardiac output) and hemodynamics (blood pressure).

Reflecting muscle blood flow and metabolism of all organs, HR at the same time is easily incorporated into the body thermoregulation system (to enhance heat transfer) in the high temperature environment. The fact that HR is highly dynamic and varies widely (it can increase almost four times between rest and high stress conditions), makes it a convenient parameter for studying the mechanisms of regulation of physiological systems of the body. Therefore, HR is an important indicator of functional activity and strain of the body that most integrally reflects various different components of the strain, such as muscular, nervous, emotional, and thermoregulatory (Rosenblat, Solonin, 1975).

Widespread use of HR in applied physiology is substantiated theoretically and experimentally in numerous publications. Therefore, HR can be rightly considered one of the most well established and proven criteria of physiological strain for muscular activity in a variety of environments. There is a clear correlation between HR and the subjective difficulty of work (Alexeev, 1989).

Simplicity and the accessibility of measuring HR, creating informative advantages of this parameter over many other indicators, is the basis for utilizing it a parameter in classification of work by its workload and intensity (Principles of Integrated Assessment, 1986; Christensen, 1987; Solonin, 1991), and as one of the main indicators for the "Physiological norms of the strain of the body during physical labor" (Solonin, 1980, 1991). In recent years, HR has been used as one of the criteria for assessing a person's physical health (Apanasenko, 1988).

Speaking of the shortcomings of this indicator, it should be noted that HR is only one of the parameters reflecting the minute-to-minute change in blood circulation volume. Although many experiments demonstrated a slight increase in stroke volume during the physical workload, this component of the cardiac outonent of the cardiac work was performed using a respirator, etc. However, we agree with the opinion of most authorities in the field that the main mechanism of amplification of blood circulation at work is the increase in HR (Karpman, Parin, 1980).

The imperfection (emotional impact and short term measurements) of non-standard methods of HR palpation counting widely used in practice imposes a large imprint on its validity, reducing the value of the data. Therefore, the use of continuous and non-burdensome HR registration significantly increases the importance of HR as an informative indicator of physiological strain for many types of human activity.

Without any exaggeration HR can be considered a "core" of any complex physiological, hygiene, or ergonomic research when assessing the gravity of physical labor, the emotional impact of nervous tension, climate impacts; assessing the effectiveness of physiological, hygienic, and ergonomic improvements; and when creating guidelines for acceptable workload.

HR is an important criterion in the evaluation of the effectiveness of ergonomic and hygienic innovations in industry. However, when assessing ergonomic improvements, integrally derived bioeconomic components such as the specific physiological unit cost is a very valuable criterion (where work performance can be taken into account). Data in Table 10.8 demonstrates that bioeconomic indicators show a higher visibility and provides a quantitative estimate of the real increase in work efficiency. Comparisons taken in isolation production and pulse parameters with the integral indicator of specific physiological cosst clearly illustrates the advantage of the latter.

TABLE 10.8

Influence of ergonomic innovations (events) on productivity of labor, HR$_{work}$, and bio-economic index.

Professions	Innovations	Indicators	Average value		Changing, %
			Before innovation	After innovation	
Painter (interior	Efface mechanism	PL	1.00	1.97	+97.0
building		HR	130	114	−12.3
construction)		HR/PL	130	58	−55.4
Aluminum smelter	Automation and	PL	1.00	1.78	+78.0
	personnel	HR	109	117	+7.3
	reduction	HR/PL	109	66	−39.5
Metal sorter	Lifting tables with	PL	1.00	1.62	+62.0
	supplemental	HR	110	106	−3.6
	spring balance	HR/PL	110	66	−40.0
Blacksmith (hot	Complex	PL	1.000	1.078	+7.8
forging)	organizational	HR	105	100	−4.8
	actions	HR/PL	105	93	−11.4
Copper miner	Mandated break	PL	1.000	1.045	+4.5
	time	HR	108	106	−1.8
		HR/PL	108	101	−6.5
Worker replacing	Mandated break	PL	1.00	2.54	+154.0
anode in aluminum	time	HR	121	125	+3.3
production		HR/PL	121	49	−59.5
Worker who repairs	Pneumatic safety	PL	1.00	1.82	+82.0
the tanks	suits	HR	129	121	−6.2
		HR/PL	129	67	−48.2

Notes: Abbreviations: PL—productivity of labor (conditionally units), HR—HR$_{work}$ (bpm), HR/PL—derived bio-economic measure physiological (pulse) cost of unit of production.

In the areas of fitness and sports, HR is the most commonly used indicator for assessing the daily motor activity of different groups of the population; when monitoring sports' training and competitive loads; and in assessing the degree of fitness during endurance exercises (Rosenblat, 1967; Thorevsky V.I., Kramer H., Kalashnikova Z.S. et al. 1986).

REFERENCES

Alekseev, V.I., (1989). The Relationship between Heartrate and the Subjectively Perceived Exertion of the Body during Manual Work: *Human Physiology*, 15(1), pp. 69–74.

Apanasenko, G.I., (2004). Diagnosis of the Individual Health. *Hygiene and Sanitation*, 2, pp. 55–58.

Bedny, G.Z., Seglin, M.H., (1997). The Use of Pulse Rate to Evaluate Physical Workload in Russian Ergonomics, *American Hygiene Association Journal*, 58(5), pp.375–378.

Bedny, G.Z., Karwowski, W., Seglin, M.H., (2001). A Heartrate Evaluation Approach to Determine Cost-Effectiveness of an Ergonomic Intervention. *International Journal of Occupational Safety and Ergonomics*, 7(2), pp. 121–133.

Christensen, E.H. (1987). Physiology of Labor. In: *Encyclopedia of Occupational Safety and Health*, 4(2), Professional Publisher, Moscow, pp. 2712–2715.

Fedorov B.M. (1997). Stress, cardiac aspects. *Human Physiology*. 23(2). pp.89–99.

Gorbunov, V.V., (1997). Terms of the Adequacy of the Use of Indicators to Assess the Heart Rhythm of Operator's Activity in Psychophysiological Tension. *Human Physiology*, 23(5), pp. 40–43.

Karpman, V.I., Parin, V.V., (1980). The Value of Cardiac Output. In: *Physiology of Circulation, Physiology of Heart, Science*, St. Petersburg, pp.271–278.

Koscheev, V.S., (1981). *Physiology and Hygiene of Individual Protection from the Cold. Medicine*, Moscow, p. 288.

Meyerson, F.Z., Pshennikova, M.G., (1988). *Adaptation to the Stressful Situations and Physical Activity*, Medicine, Moscow, p.256.

Pavlov, I.P., (1932). Selected Works (Ed. by Koshtoyants). *Foreign Languages* Publishing House, Moscow, Russia.

Physiology of Man in Space, (1963) (Ed. by Brown, J.H.). National Institute of Health, Bethesda, Maryland, pp. 208.

Physiology of Work. (1993) (Ed. by Medvedev V.I.). Science, St. Petersburg. p. 528.

Principles of Integrated Assessment of Work Intensity and of Work Strain in Mental Activities (guidelines), (1986). Approved by the Ministry of Health of the USSR, Moscow, p. 26.

Rosenblat, V.V., (1967). *The Radio Telemetry Studies in Sports Medicine*. Medicine, Moscow, p. 208.

Rosenblat, V.V., Solonin, Iu. G., (1975). Principles of Physiological Normalization of the Heart Work Based on Pulsometry. In: *Functions of the Organism during Work*. Erismana Institute of Hygiene, Moscow, pp. 31–50.

Scherrer J. (1973). *Physiology of work* (ergonomics). Trans. from French. Moscow: Medicine. p. 496.

Schmidt-Nielsen K. (1987). *The size of the animal: why are they so important?* Trans. from English. Moscow: Mir. p. 259.

Solonin, Iu. G., (1969). On the Estimation of the Heaviness of Labor According to the Heartrate during Recovery Period. *Hygiene of Labor and Occupational Diseases*, 9, pp.23–26.

Solonin, Iu. G., (1980). *Physiological Norms Strain during Physical Labor*, Guidelines Approved by the Ministry of Health of the USSR, Moscow, p. 6.

Solonin, Iu. G., (1991). Physiological Approach to the Standardization in the Work under Strain, *Human Physiology*, 17(2), pp. 141–146.

Solonin, Iu. G., (1991). *Identification and Assessment of the Physical Strain at Work*: Scientific Advice, Komi Science Center, Ural Branch of the Academy of Sciences of the USSR, Vol. 93, Syktyvkar, Russia, p. 24

Solonin Iu.G., Kuznetsov A.A. (1977). *Assessment of severity and improvement of working conditions in press and forging production*. Approved by the Ministry of Health of the RSFSR, Sverdlovsk, p. 20.

Solonin Iu.G., Maslentseva S.B., Kuznetsova Z.M., Kozlovsky V.A., Ustiantsev S.L., (1984). Functional state during work activity, work capacity and human health. *Human Physiology*. 10 (1), pp. 66–71.

S. N. Swisher, J. V. Warren, T. G. Coleman, H. T. Milborn, 1968. In *Physiology in the Space Environment: Volume I*, Washington, DC, The National Academies Press, p. 79.

Thorevsky V.I., Kramer H., Kalashnikova Z.S. et al. (1986). *The dependence of the reactions of the cardiovascular and respiratory systems the size of the active muscles during static operation. Hygiene Labor and Occupational Diseases*. 4. pp. 5–10.

Section III

Management and Education

11 A Point of View on Management from the Systemic-Structural and Applied Activity Theories Perspective

Fred Voskoboynikov
Baltic Academy of Education, St.-Petersburg, Russia

CONTENTS

11.1 INTRODUCTION

Management is the targeted influence of the subject on the object with the ongoing process of developing, adopting, and implementing solutions aimed at achieving a desired result. Ensuring the effectiveness of management requires a new level and character of thinking from the performer, which corresponds to the level of complexity of the activity itself. Characteristic features of new thinking are systematic, flexible, responsive, non-standard approaches to decision making. Management is a multidimensional concept that includes technical, economical, psychological, and social aspects. It is an interdisciplinary field of activity that carries out its important value for the society. Regardless of the type of organization and the field of activity, management functions are similar. In fact, management functions are considered to be universal. Managers plan and organize, coordinate and control, make decisions and handle physical, informational and financial resources, create and communicate, motivate and reward, and all the rest required for the effective carrying out of multifaceted managerial activity.

In the study of management as a science, both qualitative and quantitative methods are used. The qualitative method is based on creativity, intuition, and experience.

The scientific basis for this direction of management is psychology and sociology. The qualitative method is used where the decisions are made under both certainty and risk, in the program evaluation and review technique (PERT), and in the queuing system. Specificity of work considered in this scientific direction is in the psychological aspects of management. The quantitative method is based on objective measurement procedures that are necessary for making rational and logical decisions. These two methodical approaches do not contradict each other and do not supplement each other either. The principle of using the quantitative method was developed within the framework of SSAT (Bedny, 2015). It is used for optimization of performance and the relationship between the information that is directed to the operator, and the ability of the operator to timely react on it. The use of this method is especially important in situations where people operate under the time contraints.

11.2 BASICS OF ACTIVITY THEORY

Analysis of the basics of activity theory and its terminology is important for better understanding how activity theory can be used for the study of human performance. General activity theory (AT) was created and developed in the former Soviet Union in the 1930s–1960s by three prominent scientists, Rubishtein, Leontiev, and Vygotsky. For a long time this theory was used in the study of work activity. However, with the development of mechanization and automation in industry, transportation, and in the military sphere, it became obvious that the direct application of the theory of activity for the effective study of human work was not possible. In response to modern times, a more advanced theory, the applied activity theory (AAT), was created in the 1970s later years by Soviet and Russian psychologists. Thus the theoretical foundation of Rubinstein-Leont'iev-Vygotsky's general activity theory became the philosophical basis of AAT. The effectiveness of AAT was confirmed in ergonomic techniques. The most representative objects of introduction were aviation systems, automated control systems for technological processes, remote control, and software. The further development of AAT led to the creation of the system-structural activity theory (SSAT) as an independent direction in AAT. The emergence of this direction was the reaction of the scientists to the challenge of the development of science, and scientific and technological progress. The main contribution in this field is presented in the works of Gregory Bedny (Bedny and Meister 1997; Bedny and Karwowski, 2007).

SSAT views activity as a structurally-organized goal-directed self-regulating system, rather than as a set of responses to multiple incentives. Such a system is considered to be purposeful and self-regulating if, in changing environmental conditions, it continues to pursue the same goal and can reformulate its purpose. By the systemic-structural activity theory, activity is a coherent system of internal mental processes and external behavioral processes and motivations that are combined and organized by mechanisms of self-regulation to achieve conscious goals. Activity is defined as conscious, intentional, socially formed behavior that is specific to humans. Activity consist of actions playing the role of a mediator connecting personality with the social world. In activity a person develops different strategies that derive from the mechanisms of self-regulation. Self-regulation of activity manifests itself in the way people, through trials, errors, and feedback corrections, create strategies of performance that depend on personality features. It takes place through both the non-conscious (automatic) and conscious levels (Bedny and Karwowski, 2007). At the non-conscious level self-regulation unfolds as an uninterrupted process. Automotive mental operations are not organized into cognitive actions. This can be explained by the fact that the non-conscious level of self-regulation is not subordinated to conscious goals. Activity is triggered automatically and performed through non-conscious automotive reflective processes. The subject is only conscious of the results of this process. The conscious level of self-regulation presents itself not only as a process but also as a system of logically organized actions. Each action is organized according to mechanisms of self-regulation and has a beginning and an end. At the conscious level of self-regulation activity can be considered a hierarchically organized system of self-regulative stages of interrupted reflective processes. Actions can be cognitive/or internal/and behavioral/or external/(Bedny, Karwowski, and Voskoboynikov, 2010).

The fact that activity theory was developed in the former Soviet Union explains why philosophical, cultural, and psychological roots of the theory are significantly different from the interpretations of Western psychologists. Activity theory, and especially its applied fields, utilizes different terminology. The terminology, which is used in activity theory, has totally different meanings compared to the meanings used in the West. This circumstance makes translation of some terms from Russian into English almost impossible when trying to preserve their meaning. Therefore, for presenting activity theory to the West, it was necessary to give the correct interpretation of important terms of activity theory. The thorough analysis and a detailed interpretation of the important terms of activity theory, like activity, action, goal, and task is presented within the framework of SSAT in the publications of Gregory Bedny and his colleagues (Bedny, 2015; Bedny and Karwowski, 2007; Bedny, Karwowski, and I. Bedny, 2015). That's why we will only briefly touch on the subject of terminology here.

For example, the Russian word *deyatel'nost'* as a term was first introduced by Russian philosopher and psychologist Nikolai Grot (1852–1899). It translates into English as *activity*. However, deyatel'nost is a much broader concept than the English word activity, it's what a person does on a regular basis. A builder builds houses, a pianist performs, a scientist conducts experiments and write scientific books and articles, and so on. SSAT considers deyatel'nost to be a *coherent system* of internal mental processes and external behavior actions and motivation that are combined and directed to achieve conscious goals. This term has a very different meaning in the English language as well as the terms action, activity, activities, which are utilized differently in Western psychology. In the US, the term *activities* is utilized in psychology and management. Management is defined as a set of activities (planning, decision making, controlling, etc.) with the aim of achieving organizational goals. This term is frequently used for designating some bits of work. However, work activity has a different and more specific meaning in activity theory. In a similar way the concept of the system goal, the goal of organization, and the goal in human activity each has a different meaning.

To understand activity as a goal-directed system it's essential to understand the concept of the goal of activity. In the West, in cognitive psychology, in social and personal psychology, and in the field of motivation, goal is considered a combination of cognitive and motivational components. For example, Austin and Vancouver (1996) define the goal as "an internal representation of a desired state of the system." That is, they present the goal as it includes both cognitive and motivational features. In activity theory, in contrast, the goal is a cognitive mechanism connected with motives (Bedny, 2015). Thus the main difference is that the concept of goal has to do with human behavior or activity only. The goal always includes conscious components. It is a cognitive representation of a future result of the subject's own activity. Goal is connected with motive and creates the vector *motive(s)* → *goal*. In activity theory the term goal is used only for analysis of human activity. The desired future state can be applied not only to a human goal, it may be the result of events not directly related to human activity. The desired future result becomes the goal only if it directs human activity and is achieved as a consequence of such activity. The goal is a conscious mental representation of the desired result of actions or activity during the performance process. The goal of activity is a combination of imaginative and verbally-logical representations of future desired action result of. At any specific moment a subject can be conscious of some aspects of goal. An image or a verbally logical representation of a future result does not become a goal until it is connected with the motive.

Goal is always connected with motives and creates the vector *motive(s)* → *goal*. This vector gives the goal a directed character of human activity. The more significant the goal is for the subject, the more he or she is motivated to reach it. Further analysis of Vancouver's publications shows that currently there is no clear understanding of goal, task, self-regulation, and other important concepts that are necessary for task analysis. For example, Vancouver reduces the self-regulative process to the elimination of errors that are the result of disturbances, like situations of danger, unanticipated events, emergencies, etc. Self-regulation is a complex process. Our activity does not consists of elimination of errors; it is a self-regulative system. Self-regulation takes place even when there is no disturbance and errors. The self-regulation process is a conscious activity, and first of all, a goal-directed process that allows not only the correction of errors but prediction and prevention of them as well. Because of the disturbances, the self-regulation process becomes more complex and strategies of task performance change. Subjects have to improvise and adapt to the contingency of such disturbances. There are different strategies that are utilized in normal work conditions, in dangerous situations or other disturbances, and transitory strategies, when subjects change their existing strategy to a new one. The foundation for all these strategies is the process of self-regulation that involves goal-formation (Bedny, 2015).

The following is the hierarchically organized elements of human activity: *activity* → *action (cognitive or behavioral)* → *psychological operations (cognitive or behavioral)*. Actions can be broken into smaller elements, namely into mental acts or cognitive operations. Motor actions can be broken into motor operations or motions. Goal of actions integrate cognitive or motor operations into actions. Actions in turn are integrated into some type of activity during the execution of various tasks. From the activity theory perspective, the task is a logically organized system of cognitive and behavioral actions that is integrated by the goal of task.

11.3 THE STUDY OF INDIVIDUAL STYLE OF ACTIVITY IN MANAGEMENT

The study of personality and individual differences is a critically important area of activity theory. The central notion in this area of study is the individual style of activity that connects features of personality with mechanisms of self-regulation and strategies of performance. According to the concept of individual style of activity, people can efficiently adapt to the objective requirements of activity not by utilizing the non-normatively prescribed methods of performance, but by their own individual style of activity. It should be understood however, that the use of the individual style of activity does not mean a violation of the standard job requirements that exist in any type of working activity. Individualization in performance is restricted by a range of tolerance and by managers. Any performer can consciously or unconsciously develop efficient strategies of performance, which are adequate to their personality features, given they do not violate the objective requirements of working activity. Individual style of activity allows different individuals to rely on their personal strength to compensate for their individual weaknesses. People attempt to diminish the impact of their weaker features of personality in a given task situation by utilizing their strong features instead for performing their tasks more efficiently. Quite often people, unaware of the fact that they utilize their own individual style of performance, think that others perform similar tasks the same way they do. Individual style of performance can be developed based on the analysis of individual features of personality and on the obtained training methods that are directed toward the formation of individual strategies of activity. The important role of managers is to understand that in his or her own work and in the work of subordinates, they can rely on an individual style of performance. The formation of the individual style of performance helps reduce the process of selection of people for the tasks of a different significance.

The concept of individual style of activity was first introduced by the Soviet psychologist Merlin (1964) and his follower Klimov (1969). In the subsequent years some other authors studied the effect of individual personality features on performance and associated with it individual style of performance (for example, Bedny and Voskoboynikov, 1975; Bedny and Seglin, 1999; Voskoboynikov, 2014). In SSAT, individual style of activity is considered a strategies of performance deriving from the mechanism of self-regulation, which depends on personality features. Such a strategy occurs at the conscious and unconscious levels, and is based on the principles of self-regulation consciously or unconsciously. Both levels are tightly interconnected and transform from one to another. The process of self-regulation manifests itself in a formation of desired goals, in developing the program of actions that corresponds with these goals, with conditions for achieving the goals, and with a persons' individual abilities. In other words, through trials, errors, and feedback corrections, individuals create strategies of performance suitable to their individuality. For example, people with an inert nervous system develop a predisposition to organize and plan their work in advance, and attempt to utilize a stereotyped method of performance. It should be distinguished however that individual style and methods of performance is not the same. The latter is not dependent upon individual personality features but rather upon organizational factors, imposed supervisory procedures, etc. Sometimes methods of performance that derive from organizational factors may contradict the individual's personality features. And that is not desirable. In cases of an inadequate training that ignores individual features of personality, the subject may acquire methods of performance that contradict his or her individuality. It may negatively affect the performance level and the job satisfaction. Understanding of individual features of personality allows the manager to select the more adequate tasks for individual subordinates, or present individual requirements to the performer of the tasks, and develop individual strategies of social interaction with subordinates.

Empirical facts and theoretical studies show that the adaptive mechanism of personality features can work up to a certain limit. In studying an individual style of activity, it is important to observe how people with different individual characteristics acquire the same knowledge and skills. On the other hand, it is as important to identify how subjects disintegrate into distinct groups with respect to their ability to skill acquisition. Such disintegration takes place as the capacity of some individuals to adjust to the requirements of the activity is reduced due to the increased task complexity according to some specific characteristics. In the tasks of average difficulties all individuals exhibit a similar level of achievement regardless of their abilities. In such situations it is hard to identify individual differences in performance. For example, when subjects take part in the training process, different curves of skill acquisition can be seen. Those trainees, whose individual properties are more suited to the requirements of activity, adapt to the task more rapidly. As a result, individual curves of the skill acquisition have different positions, and their difference are more noticeable at the beginning of the training process. Gradually, differences in performance decreases and skill acquisition falls into a narrow acceptable area. At this stage individual differences of performance cannot be detected. At the final stages of the training almost all participants show approximately the same results. This means that all trainees were able to adapt to the objective requirements of activity.

In our experiment (Bedny and Voskobonikov, 1975), three groups of elementary school students with different levels of mathematical abilities (as evaluated by their teachers) were selected. They were instructed to perform identical calculation tasks multiple times. The time of completion of the tasks was recorded. Surprisingly, at this first stage of the experiment, all students ended up with approximately the same results performing these repeated calculations. Our observation revealed that the similarity of the results were due to each group using different individual strategies. Subjects with weaker abilities used their fingers as well as their external speech. It meant that their mental operations depended on external practical actions. Only after multiple executions of the task with the experimenter's help did they start to rely exclusively on internal mental operations. Subjects with average ability also had an initial tendency to rely on external practical actions to perform the task. Their calculations were accompanied by whispers that were barely audible; that is they facilitated slow mental operations with some external actions. The students with superior abilities from the outset started their calculation without relying on external practical actions. In the second stage of the experiment, when these three groups were given more complex tasks, students with lower mathematical abilities could not develop an individual style of activity in the attempt to adapt to the complex tasks.

Thus, we could observe that in performing relatively simple tasks subjects were able to reduce their differences in performance quite noticeably, whereas increased complexity of the task requirements resulted in distinct groups with distinct levels of performance. This says that though individual style of activity is helpful in the process of adaptation to the objective requirements of activity, it has its limitations. When people demonstrate similar results in their performance, it is important to take into an account how they achieved the required skills and expertise, and how much time and efforts each of them dedicated to that. We can conclude that for managers it is beneficial to rely on subordinates' individual features of personality in the selection process, in utilizing an adequate training for the development of their strategies of performance for specific complex tasks, and for the distribution of the tasks.

11.4   SOME ASPECTS OF DECISION MAKING IN MANAGEMENT

Decision making is one of the most important functions of a manager's work. It is the type of human activity where the subject makes a choice from more than one alternative. In some situations it may require one decision-making action, in others—it's a complex activity. An example of the simplest decision-making action is a situation when the subject needs to either act or not to act on a specific signal or situation. A complex decision-making activity manifests itself in a sequential process that involves performance of a complex multiple-choice task, in which there is more than one alternative course of actions. The decision-making activity includes recognition of the situation, its interpretation, identifying alternatives, then choosing the most preferable course of actions by the subject's opinion. Usually, the decision-making action is performed by the subject because of the motive to achieve specific goals and satisfy life aspirations. In management, the decision-making theory presents the system of knowledge that is required for achieving organizational, group, and individual objectives. This theory integrates diverse disciplines such as economics, psychology, statistics, and others. It includes questions as how to formulate goals of various types of activity and objectives of the organization, and how to select optimal strategies for achieving such goals and objectives. Decisions can be classified by different criteria. They can be an administrative type, which determine the structuring of the company resources, or an operative type, which deals with repetitive day-to-day tasks. Decisions can be routine with a well-defined structure, or non-routine, novel and unstructured, which requires developing certain procedures to handle them. Depending on the complexity and on the number of people involved in the decision making, they can be individual and managerial. Depending on the sphere of interest they can be related to either economics, psychology, politics, and other fields. Decisions can be descriptive and prescriptive.

The study of descriptive decisions is about how people actually make decisions, and how they should make decisions. The purpose of the analysis of such decisions is a discovery and analysis of the factors that lead to successful decisions. The study of prescriptive or normative decisions is to find out how decisions should be made. The development of normative decisions is based on the assumption of rationality. Objectives of the decision should be clearly defined and should be ranked in a defined order according to the preliminary defined criteria. Decisions can be made under the conditions of certainty or uncertainty. In the former situation we talk about decisions with a deterministic outcome. In the latter situation it's about the decision making under the risk with a probabilistic outcome. There is also such criterion as expertise or when the decision maker is familiar with the situation. Naturally, sometimes the subject is unfamiliar with the decision making situation. In such cases it is necessary to determine the differences in possible strategies of decision making by novices and experts. Time also plays an important role in the decision making process because the time pressure is a critical factor in the decision making. In the present, the psychological theory of decision making plays a significant role in management and specifically in economics.

Let's consider some types of decision making in a more detailed manner. In management, intuitive decision making and analytical decision making should be distinguished (Hammond, 1993). There are different types of decisions depending on the conditions in which they should be made. Conditions under which decisions are made in organizations are divided into three categories: certainty, risk and uncertainty. Most decisions in organizations are made under a state of uncertainty (Grrifin, (1999) where the decision maker, with reasonable certainty, knows possible alternatives and in what conditions they may occur. A more common decision-making process in organizations happens under the state of risk. The state of the risk is a condition in which there is a possibility of each alternative and its potential payoffs, and the cost associated with the probability of estimation. Most decisions made in organizations are performed under the state of uncertainty. In such conditions, the decision maker does not know all the alternatives and does not know the level of risk associated with each alternative, nor possible consequences of each alternative.

Original researchers of decision making concentrated their efforts on the study of optimal or rational decision making. The major assumption is that by making the choice managers could either lose or benefit. They can evaluate all aspects of the decision situation logically and rationally by utilizing mathematical models that can help to determine the optimal choice. This approach is used in attempts to prescribe what people should do when they encounter decision choices. Rational decisions are sometimes called normative models of the decision because it describes what people ideally should do in a decision-making situation. In the normative decision approach the central concept is the utility or the value of the choice, or its "worth" for the decision maker. The considered approach manifests itself in the assumption that managers act rationally and logically.

According Griffin (1999), these are the following steps in the rational decision-making process:

1. *Recognizing and defining the decision situation.* For example, when equipment begins to malfunction, the manager must decide whether to repair it or to buy a new one.
2. *Identifying alternatives or courses of possible operating procedures.* There can be standard and creative or innovative alternatives. Sometimes the development or selection of the alternatives takes time and effort. There can be various constrains in the selection or the development of adequate alternatives.

3. *Evaluating each alternative.* Each alternative can be evaluated in terms of its feasibility, satisfactoriness, and consequences. At this stage it's important to evaluate the alternative probability and particularity. When the alternative has passed the test according to feasibility criterion, it should be evaluated from the perspectives of its satisfaction to the conditions of the decision situation. At this stage the probable consequences of a selected alternative must be assessed and the affordability of chosen alternative should be considered. The manager can decide that even if the considered alternative is both feasible and satisfactory, it can still be eliminated because it may turn out to be too expensive for the company.

4. *Selecting the best alternative.* Sometimes even when most alternatives fail to pass evaluation according to such criteria as feasibility, satisfactoriness, and affordability, several alternatives can still remain. Selecting the best one can be a complex task for the manager. Selection can be performed by using some formalized procedures based on the mathematical analysis. However, the manager can often develop his own subjective criteria for choosing the alternative.

5. *Implementing the chosen alternative.* At this stage a manager must decide how to integrate all operating procedures in a new business, including purchasing and distribution into the ongoing organizational framework. It's worth mention that, at this stage of the decision making, a manager should still anticipate some potential resistance in the implementation process.

6. *Following up and evaluating the result* is the concluding step in the decision-making process. The manager should be sure that the chosen alternative is adequate for its purpose. They have to recognize that the selected alternative may be inappropriate for the company.

We considered rational or normative decision making. Decision making under risk and uncertainty is a more specific area of the study from the perspective of psychological analysis. Here, we further consider descriptive decision models. Simon (1957), who was awarded the Nobel Prize in economics, was the first one to recognize that decisions are not always made according to rational and logical principles. Decision makers have the tendency to act according to their individual differences, past experience, and their subjective features of personality in general. People do not usually make absolutely best or optimal decisions. They choose the "good enough" or satisfactory decision. The satisfactory criterion is useful because people have limited cognitive capacities. This criterion is especially useful when a decision is performed in the condition of the time limit. Cognitive resources limitation is a critical factor for the simplification of decision making (Janis, 1982). Simplification heuristics sometimes can lead to misperception and biases. Very interesting data in this area of study was obtained by Tversky and Kahneman (1981). People can make decisions based on such factors as *framing effects* or how the problem is formulated. For example, it was found that engineers and managers made project decisions differently depending on whether the information was framed in terms of a previous team failure or success. The first step in making the decision involves diagnosis of the situation. During diagnosis of the situation, the decision makers can apply various heuristic strategies that are useful but can sometimes lead to an incorrect interpretation of the situation (Kahneman, Slovic, and Tversky, 1982). According to G. Bedny and I. Bedny (2018), this factor is associated with the mechanism called gnostic dynamics, when thinking plays a special role. In a cognitive psychology study of how people understand the situation before making the decision (often called the diagnosis), special attention is paid to perception and memory while thinking is not considered at all. In a similar way Kahneman and Tversky (1984) describe human heuristics at the stage of choosing the course of activity. In SSAT, origins of heuristic strategies is explained by the mechanisms of activity self-regulation (G. Bedny and I. Bedny, 2018). According to this theory there are normatively prescribed methods of task performance and their real individual strategies of performance. This requires analysis of the individual style of activity. The goal of any human is associated with its motivational forces. Hence, there can be a conflict between the various motives and associated goals in decision-making situations. In the example mentioned earlier about the marketing and production departments, there was a conflict of interest. The marketing department staff would like the company to produce a wider variety of products. For the production department staff this means developing a more complex production process. In such a situation, top management has to make a decision in order to minimize conflicts of interest.

When talking about group making decisions, it's important not to fall into the situation of the so-called "Abilene paradox." The Abilene paradox refers to what in social psychology is known as conformity and social influence. Sometimes, in a group making decision situation, people who think contrary to the group opinion feel stressed by not going along with the group. Being afraid that the group may potentially express negative attitude toward them, they go along with the group even though they do not agree with them. If many group members act the same way, it is very likely that the group will make a poor decision. One of the tragic consequences of the Abilene paradox is the disaster of the American space shuttle Challenger in 1986 (Hughes and White, 2010). As the investigation of the causes of disaster showed, just before the launch of the shuttle, the engineers warned their superiors that certain components—particularly the rubber O-rings that sealed the joints of the shuttle's solid rocket busters—were vulnerable to low temperature. If the decision was up to one person, he probably would have decided not to launch the Challenger at subzero temperature because of the risk of a possible accident and the death of seven crew members. The Commission, however, made the decision to go ahead with the launch, and Challenger lifted off. On the 73rd second after lift off, the disaster occurred.

In the opposite to the group making decision described above, where some members of the group go along with the group, there is another well-known non-standard method for decision-making. It's called "brainstorming." In this method, group members are not under group think pressure. Just the opposite—they are encouraged to present their own opinion on the matter. The method was offered by advertising executor Alex Osborn (1888–1966). According to Osborne, the generation of new ideas should occur in a relaxed environment, people should feel free from criticism, everyone should feel equal, and everyone's opinion is considered valuable and is not discussed. Nor should there be a hierarchy of relationships. It is desirable for the participants to be representative of different professions, since ideas from different fields of knowledge may give a boost to unexpected solutions. The most unrealistic idea can generate a chain of associations and lead to an important practical decision. Recommendations in brief for conducting the procedure in sequence are: formulate the problem, announce the time limit, record all the ideas without evaluating or discussing them, and discussion of ideas.

11.5 THE INFLUENCE OF THE GROUP STRUCTURE ON OPERATORS' PERFORMANCE EFFICIENCY

In this section, we will analyze some applied activity theory data of how the group structure influences the efficiency of operators' performance. We utilize the system approach for the study of group performance, which means that particular attention is paid to the relationship between the group members and how this relationship influences functioning of the group as a whole. In our analysis we concentrate on the group's structural organization and its effect on performance. The increasing complexity of social structures and technological progress increases the share of collective knowledge-intensive sectors of industry, which highlights the problem of increasing the effectiveness of individuals' joint activities. In the study of group activities, an object of research becomes a group of people linked by a common purpose. Their activity is connected by the means of carrying out a common task. The concept of a group activity indicates multifaceted phenomenon, which must be differentiated.

Groups are differentiated by a variety of factors. Structural organization is one of the most important factors for group collaboration. Spasennikov (1992) proposed distinguishing five major forms or models of structural organization. He particularly notes the differences in the characteristics of the distribution, alignment, and harmonization of the key elements of the group activity structure, such as objectives, motives, actions and operations, as well as performance. *Multilevel* group activity includes a hierarchical management structure of large systems in which there are several interacting levels of group management and a number of independent objects of management, together constituting a combined object of management. *Interactive* group activity is when the overall objective is performed by each team member simultaneously with other members. This kind of activity takes place when the performance of management functions by one expert is not possible, since the state of the object is determined by many interdependent variables. This interaction is a characteristic of a wide range of different kinds of activity such as assembly work, team sports, group discussion of the problem, etc. *Duplicating* group activity is when several specialists perform the same task simultaneously. It is used for the purpose of improving the reliability and performance in carrying out certain functions. *Co-sequential* group activity takes place in cases where the process of joint activity consists of several sequentially executed tasks, and each member performs a different operation. An example of this type of activity can be an assembly line type of production, or the activity of several specialists in multiphase systems of transmitting and processing information, etc. *Joint-individual* group activity is the most elementary form of group activity. A group receives a general task where each

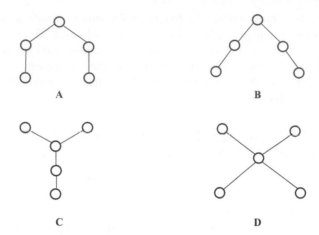

FIGURE 11.1 Types of interconnections within the group. A—complete decentralization; B and C—weak centralization; D—total centralization.

member does its part of the work independently. An example of such an activity is a working crew or a group of designers working on a project, etc.

Graphically, the structure of group organization can be presented in the form of connected lines. These lines are called ribs or interconnections. The total number of ribs (or interconnections) is one of the indicators of the effectiveness of the group structural organization. Examples of types of interconnections are shown on Figure 11.1

One or another kind of group activity is usually determined depending on the specifics of the technological process in an organization. Depending on the nature of the group tasks, different functional organizations of groups are used: *chain, star, circle, network*, or *hierarchical structure*. The group is organized on the principle of *chain* in cases when the process is divided into a number of operations performed sequentially, and each operation is assigned to different individuals. The principle of *star* is used in cases of operations that make up the entire process, where members of the group perform their operations in parallel and independently from each other, and the entire process is planned and coordinated by a certain person. A good example of that is the organization of escorting aircraft objects in automated systems for collecting and processing radar data. A senior operator evaluates the traffic situation and distributes objects among operators (by sector or by regulations of air objects), and each operator watches and accompanies the assigned object. Functional structure of the group on the principle of *circle* is used when the process is organized cyclically, that is, the operations included in the process are performed by different people, and the final operation of one cycle is the beginning of another. The functional structure of the group based on the principle of *network* is used in cases where all group members are interconnected to each other in the process of activity. A network may be complete (each member is connected to all other members of the group in a joint activity) or incomplete (the interconnection between members are realized partially). A hierarchical structure is characterized by the presence of multiple levels of control, and therefore, by a number of independent objects of management, which represent a total management object. In such groups the interaction of operators are organized on multiple levels of hierarchy.

Groups are also differentiated by the size, the duration of existence, and the degree of organization, etc. A *quantitative composition* of a group is small, medium, and large. Human behavior in groups of various sizes can vary significantly. Increasing the size of a group does not necessarily manifest a higher effectiveness and better performance of tasks. Upon achieving a "critical mass" of the group quantity, the efficiency of its operations may cease. It is possible to determine an optimal size for each specific task empirically for achieving the highest efficiency and quality of the group work. *Duration of the existence* of small groups is usually defined by specific goals and objectives for which it is organized. The group can be created for the performance of some specific short term assignments, for example, for a year. Others can be created for a period of one to several years. And, respectively, long term groups can be created, which are practically permanent. By the *degree of organization* groups are divided into several types: *nominal* groups—unified by common goals and the need to perform tasks; *related* groups are also unified by common goals and the need to perform tasks, but they are also united by mutual preferences to each other; *consolidated* groups have the same features as these two, but the task cannot be performed without functional interactions of its members with each other. *Highly organized* groups represent the highest degree of organization. In addition to the characteristics described above, these groups have a higher degree of psychological compatibility between group members and a higher degree of interaction between them, which occurs in the process of communication.

The problem of psychological compatibility has two sides: social-psychological and psycho-physiological. Social-psychological includes the generality of the goals of the group members, the generality of their value orientations, similar attitudes to activities and to other in the group, the motivation for their behavior, as well as the characteristics of their mental state, etc. However, even this may not be enough for successful solution of group tasks by operators. Unlike many other types of activity, the characteristic of the activities of operators often take place within the framework of a single, compressed, and closed sensorimotor field. Sensory and motor skills build on the foundation of our innate abilities. Sensory skills are vision, hearing, touch, smell, taste, vestibular (for the balance and head position in space), and proprioception (coming from muscles and joints). They are responsible for receiving information. The activity of operators includes perception of information and its evaluation, and the decision making for executive actions. It's important to note that these phases of mental and physical actions can be mutually intertwined among different operators. All this can contribute to the formation of a complex of individual and group inhibitory psychological mechanisms and barriers in the interactions of operators. Therefore, for the group of operators, the presence of psychological and physiological compatibility is a very important factor for the successful execution of their professional activity. Psycho-physiological compatibility is understood in terms of the level of physical and psychomotor development, in terms of identical manifestation of the basic mental processes and sensorimotor functions, and in terms of the degree of similarity of various professional skills between the members of the group. Also, consistency in the dynamics of temporal and strength indicators of

higher nervous activity is important as well as other functional systems of the organism, like compatibility in temperament, and the similarity of the dynamic orientation
of emotional-vegetative reactions.

Psychological researchers have shown that the cohesion of the group, psychological compatibility, and the formation of the phenomenon of a group consciousness
depends on the type of functional organization. These trend increases as the transition from a centralized to a chain, and then to a network form of functional group
organization takes place. With all other things being equal, the functional organization of the star type stresses and activates the "individual effect," i.e., contribution
of each group member in the overall result. On the other hand, organizations of the
network type activate the "group effect" to a greater extent. Such groups have an
advantage, as they provide opportunities for the fullest realization of reserves. The
"group effect" is understood not only in a sense of increased productivity, but as an
educational effect as well because it increases social activity of the group as a whole
and of each individual included in it.

The effectiveness of the group's task performance depends not only on the number of ribs (interconnections), but on the degree of influences of operators on each
other as well. The stronger is the degree of mutual influences of operators on each
other, the less effective is the result of the group performance. The degree of operators' influences on each other is estimated by the coefficient of mutual influence. The
coefficient determines how many times the operator's activity has a greater impact
on the result of the partner's work than on his own result. It was experimentally
established that for the same type of group activity $k$add $\times$ N = constant, where
$k$add is the permissible coefficient of mutual influence of operators and N is the
total number of edges of a graph. For identifying the predetermined coefficient, the
device called a "homeostat" is used. The device was originally created by William
Ross Ashby[1], an English psychiatrist and cybernetics pioneer. The device was then
adapted by the Russian psychiatrist and psychologist Feodor Gorbov (1968) in his
work with Soviet astronauts. Gorbov got an idea for the use of the device for his
studies and work by pure chance. Namely, he accidentally observed people taking a
shower in a military medical facility. There were four shower cabins there; the diam-
eter of the water pipe was apparently not sufficient to ensure hot water delivery to all
four cabins. Gorbov noticed that when all four cabins were in use at the same time
the behavior of each person identified a strategy aimed at creating the most favorable
water temperature. When one of them tried to make hot water for himself without
consideration about others, the water flow to other cabins was not hot enough. This
caused immediate reaction by all others; they began rotating their handles trying to
adjust the water temperature, and as a result, that person received either too hot or
too cold water. Only the price of mutual concession eventually allowed adjusting
water to acceptable water temperature for all. In the event when someone quickly
stood out and took the initiative, that is, took the role of the leader, the group could
achieve the desired results faster. If leadership was claimed by two or three people,
it took much longer to have reasonable water temperature for the group. Members of
such group were not able to regulate the water temperature by impeding each other;

---

[1] Ashby, W. R., English psychiatrist and a pioneer of cybernetics created the device *Homeostat* in 1948.

it either took a longer time or they could not perform the task at all. And certainly, quite an impasse took place when one of them did not want to cooperate with others.

Later on, when Gorbov used the "homeostat" for determining the level of compatibility between the group members, he observed similar results. The use of the device is shown to be especially effective for measuring psychological compatibility in environments of extreme conditions. The work of operators of technical systems quite often flows in such conditions. The device consists of the experimenter's control panel and panels for each operator. The task for each operator was to put the arrow indicator on his or her panel into the zero position from a random starting position by using the remote switch. All subjects' panels were connected so each subject's action affecting the position of arrows of others. Each subject could see the indication on his or her device only while trying to adjust the position of the pointer arrow. Sometimes the groups were given unresolved tasks deliberately for the purpose of studying the behavior in conflict situations. Solvable tasks were used to evaluate the ability of the group to rapidly find a successful performance tactic. The groups were able to perform the tasks only if someone took the leadership initiative while the rest of the group acted under his or her influence, often unconsciously.

The deviation of the arrow on the homeostat panel is determined by the *potentiometer,* which is used for measurement of electric potential (in voltage). The degree of influence of the position of the potentiometer on the J-th device reading is set in the appropriate scale link, with the gear ratio Mij. This coefficient characterizes the influence of the I-th operator on the indication of the J-th device. The value of $Mij = 0$ corresponds to the absence of influence between the I-th and J-th operators. The influence of all operators on the J-th device is summed up and affects the position of its arrow. Each operator can see the reading on his device. When $k > kadd$, the group task becomes intractable. These kinds of task requires the distribution of functional roles in the group and identifying the leader, who takes the responsibility as the organizer for resolving the group task. All other operators perform the role of followers under directions of the leader. It was established experimentally that most of the time the leadership ability was observed in those operators whose peripheral indices are of minimal importance. This also confirms the fact that a group with a centralized management system has the best ability to solve complex group tasks in a critical situation in a short period of time by quick redistribution of the organization's resources in the right direction. Leaders should also possess certain personality features that make them stand out from the group. These features include knowledge of a given task, authority, and ability to organize followers for the task performance. Thus, for the successful resolution of the presented task two conditions are necessary: objective requirements (central position in the group) and subjective requirements (certain personality features). Failure to adhere to these conditions leads to the downplaying of the leader's role, and as a consequence, to deterioration of the group's ability to solve joint tasks. In cases when the leader selection is not planned in advance, the role of the leader surfaces spontaneously.

The advantages of a decentralized management system (Figure 11.1A) is that the problems are solved at the level at which they arise. Decentralized management system allows professional growth of employees by their active participation in solving production problems instead of waiting for instructions "from above." This also allows building creatively thinking personnel. At the same time, for groups with a centralized structure (Figure 11.1D), the probability of finding a leader in a node with the smallest periphery is much higher than in a homogeneous graph. The data available on group psychology allowed us to formulate some principles of engineering-psychological design of the group activity of operators. Based on the distribution of functions between human and machine, the tasks solved by the operators are determined. Based on the possible structure of the management process and the considered features of group activities, an acceptable group structure is chosen. When choosing a structure, one should strive to reduce the overall peripheral index and the number of edges of the structural graph. However, in such cases another important thing that should be taken into account is the psychological capabilities of operators in receiving and processing information, because the simplification of the group structure can lead to unacceptable informational load per operator. In addition, it is also necessary to check the magnitude of the coefficients of mutual influences and to provide an increase of inequality $k < k$add.

In conclusion, the position of the leader and the followers in the group is determined, and the main requirements for them are formulated. In some cases, for the purpose of increasing the reliability and speed of the operators' work, duplication of their work should be applied. As a result of the engineering and psychological design of group activities, the task of forming a consolidated group is solved, that is, a group united by a common purpose and having a distribution of functional responsibilities among group members. Such a group has the necessary potential ability for successful resolution of the group tasks. In order to turn these opportunities into reality it is necessary to select the group members accordingly, that is, to create a *gomphoteric* group[2]. These kind of tasks, which designers and engineers of automated production face, are the most important tasks for the effective operators' team performance. The creation of a gomphoteric group involves the solution of two interrelated tasks. First, to assign to each member of the group such a position, which mostly corresponds with the person's personality type in their group activities. Secondly, ensuring the psychological compatibility of the members of the group. The need to solve the first problem is due to the fact that the magnitude and effect of the personal contribution of an individual operator is influenced not only by their business and professional qualities, but also by the communicative features of personality, that is, by the ability to communicate with others.

It is well known that not all people are able to easily understand the context of others. People's communicative properties are different, and as a consequence, their preferences of communication with others are different too. Some people feel comfortable working shoulder to shoulder in groups, others prefer to work on individual assignments, some feel in "their shoes" when they lead others, while others feel comfortable being followers. In the 1970s, psychologists at Leningrad University (now St. Petersburg University, Russia) established differences in people's communicative characteristics of personality. They revealed four fairly distinct types by the abilities of interpersonal communication: *leaders, followers, individualists,* and

---

[2] Gomphoteric group, or a "put together" group, is the term used by Soviet scientists in their work with Russian astronauts.

collaborators. The *leading type* is the type of person with a pronounced emphasis on the leading role in a group environment. These people can work efficiently in the conditions of controlling other group members. *Followers* are the type of people with a tendency toward voluntary subordination. People of this type can successfully participate in the group task under the leading role of the more confident, independent, and competent members of the group. Such people are generally good performers; they are good in performing their work "from—to." Their relations with the group members are based on the group influence; they feel comfortable acting "like others." By agreeing with the group they are free from making decisions, which is not always an easy task. The relationship between the followers in the group is built on the basis of the principle of conformance. Conformity is understood as suggestibility of the operator, susceptibility to his or her group influence, and if necessary, acting like everyone else. This property of the operator can be estimated quantitatively using special techniques. *Individualists* are people of the category that represents the type of behavior with pronounced individualistic emphasis. Operators with such communicative features of personality prefer to work on individual assignments; they do not aspire to leadership and cooperation. They can successfully work on their part of the group task, subject to exclusion from the group in a relative solitude. We can say that for such operators the group environment creates additional exertion. *Collaborating* type are people who aspire to work with other group members; they are open to cooperation and ready to follow others in case of reasonable solutions. Typically, people of the collaborating type are good business partners.

The described communicative differences in people should be taken into account in the process of managing working groups. Imagine a group of operators where two of its members are of a leading type, or a group which consists of followers and individualists. In the former case conflicts between two leading members is very likely; in the latter case—an atmosphere of uncertainty and confusion. In both cases the performance level will be negatively affected. Based on this data, some suggestions for the use of people with different types of behavior in the organization of group activities can be developed. For example, it makes sense to assign persons of the individualist type to distinct tasks with minimal interaction with other group members. An example of that would be a duplication of the activities of operators, that is, parallel independent work of two operators with the comparison of only final advisories. This method is used in order to improve the reliability or speed of the operators. Individuals of the *follower* type will work effectively upon direct assignments "from above." Operators of the follower type are the most prepared to work for retransmission of signals, on processing of commands developed by others for the implementation of executive actions. Persons of the *leading* type should be used, respectively, to carry out the organizational activity, that is, in the nodes of the structural graph that have the least peripheral indices (indicators). *Collaborative* type persons are good for joint activities with other members of the group on an equal footing. This kind of activity is characterized by the work of several operators of similar level in their combined effort to solve a common task.

To conclude, the influence of the group structure on operators' performance efficiency lies in the fact that the individual-psychological approach to its solution is proved untenable. Studies of Soviet psychologists decades ago demonstrated that regardless of how personal qualities of operators were taken into account individually, they do not determine the qualitative and quantitative characteristics of the contribution of the particular operator to the result of group activity. From this it follows that the result of group activity is not a simple sum of the results of the activity of individual operators, but rather a more complex functional dependence. This leads to the necessity of developing special methods for studying group activities, which are different from the methods of investigating the individual characteristics of operators. The object of such studies is not the activities of operators as individuals, but the activities of the group as a whole.

CONCLUSION

In this work we made an attempt to analyze management as a scientific discipline. We described the basics of activity theory in order to understand activity as a structurally-organized goal-directed self-regulating system. We briefly dwelled on the subject matter of activity theory terminology, like activity, goal, task for the purpose of better understanding how activity theory can be used for the study of human performance. We analyzed differences in the meaning of these terms interpreted within the frame of SSAT compared to the way they are interpreted in the West. This approach should be taken into consideration when SSAT is utilized in the study of human activity and in the field of management. The decision making process in general and specifically in management was discussed as well. We described the concept of individual style of activity in management, which is based on personality features. We tried to concentrate in details on the effect of the group structure on the efficiency of operators' performance. We also emphasized the necessity of developing special methods for studying group activities, which are different from the methods of investigating the individual characteristics of operators.

REFERENCES

Bedny, G., (2015). *Application of Systemic-Structural Activity Theory to Design and Training.* Boca Raton, Florida: CRC press, Taylor and Francis Group.

Bedny, G., Bedny, I., (2018). *Work Activity Studies Within the Framework of Ergonomics, Psychology, and Economics.* CRC Press, FL: Taylor and Francis Group.

Bedny, G. Z. and W. Karwowski (2007). *A Systemic-Structural Theory of Activity. Application to Human Performance and Work Design.* Boca Raton, FL: Taylor & Francis.

Bedny, G., Karwowski, W. and Voskoboynikov, F. (2010). The Relationship between External and Internal Aspects in Activity Theory and Its Importance in the Study of Human Work. Bedny, G. and Karwowski, W. (Eds.). *Human-Computer Interaction and Operators' Performance.* Boca Raton, FL: Taylor & Francis.

Bedny, G., Karwowoski, W. and I. Bedny. (2015). *Applying Systemic-Structural Activity Theory to Design of Human-Computer Interaction Systems.* Boca Raton: Taylor & Francis.

Bedny G. and Seglin, M. (1999). Individual style of activity and adaptation to standard performance requirement. *Human Performance.*, 12, 59–78.

Bedny, G., Voskoboynikov, F. (1975). Problems of how a person adapts to the objective requirements of activity. In Aseev, V.G. (Ed.), *Psychological Problems of Personality.* Russia: Irkutsk University Press. 2, 8–30.

Dmitriev, N. E., Gorbov, F.D. (1968). Device for modeling the group interrelated activities (homeostat). Collection of articles *"Problems of Engineering Psychology,"* Issue 3. Moscow, USSR: Publishing house of the Scientific Council on Cybernetics at the Academy of Sciences.

Griffin, R. (1999). *Management*. Mason, Ohio: South Western Cengage Learning.

Hughes, P. and White, E. (September, 2010). The Space Shuttle Challenger disaster: A classic example of Groupthink. *Ethics and Critical Thinking Journal*, Issue 3, 63.

Hammond, K. R. (1993). Naturalistic decision making from a Brunswikain viewpoint: Its past, present, future. In G. A. Klein, Orasanu, R. Calderwood, and C. Zsambok (Eds.). *Decision making in action: Models and Methods* (pp. 205–227). Norwood, NJ: Ablex Publishing.

Janis, I. L. (1982). Decision making under stress. In I. Goldbergand S. Brezniz (Eds.). *Handbook of stress: Theoretical and clinical aspects.* New York: Free Press.

Kahneman, D., Slovic, P., & Tversky, A. (1982). *Judgment under uncertainty: Heuristics and biases*. New York, NY: Cambridge University Press.

Kahneman, D. and Tversky, A. (1984) Choices, values and frames. *American Psychologist*, Vol. 39, No. 4, 341–350.

Klimov, E.A. (1969). *Individual Style of Activity*. Russia: Kazansky State University Press.

Merlin,V. S. (1964). *Outlines of Theory of Temperament*. Perm, Russia: Perm Pedagogical Institute.

Simon, H. A. (1957). *Models of man; social and rational*. Oxford, England: Wiley.

Spasennikov V.V. (1992). *The Analysis and Design of Group Activities in Applied Psychological Research*. Moscow: Institute of Psychology, Russian Academy of Science.

Tversky, A., & Kahneman, D. (1981). The framing of decisions and the psychology of choice. *Science*, 211, 433–458.

Voskoboynikov, F. The Influence of Personality Features on Performance in Work, Study and Athletic Activity. In T. Marek, W. Karwowski, M. Frankewicz, J. Kantola and P. Zgaga (Eds) *Human Factor of a Global Society: A System of Systems Perspective*, 187–192. Boca Raton, FL: Taylor & Francis Group (2014).

Voskoboynikov, F. *The Psychology of Effective Management: Strategy of Relationship Building*. New York, NY: Routledge (2017).

12 A Systemic Approach to the Study of Psychological Factors Applied to Management

Fred Voskoboynikov
Baltic Academy of Education, St.-Petersburg, Russia

CONTENTS

12.1 INTRODUCTION

Systemic-structural activity theory (SSAT) considers human activity as a goal-directed self-regulative system. Activity is defined as conscious, intentional, goal-oriented, and socially formed behavior that is specific to humans. Activity theory distinguishes two types of activity: "object-oriented" and "subject-oriented." The former refers to a subject using tools on material objects with the goal of completing the task and evaluating the results. The latter refers to social interaction between people, which is the most important element in management.

In activity theory, special attention is given to the interrelationship of personality and activity. The subject not only changes the situation or object, but also develops their own personality features, which are formed through activity and social interaction. Activity theory emphasizes fundamental differences between human behavior and animal behavior. Animals are developed according to the laws of biological evolution, while the psychic processes of humans are influenced by the laws of social-historical evolution. Historically, activity theory was developed in the former Soviet Union as a counterweight to the American Behaviorism (Skinner, 1974). In his studies, Skinner portrayed external reality as a variety of stimuli to which a person reacts, thus considering humans as reactive organisms. For example, thinking was presented as a set of verbally motor reactions. The concept of meaning, which plays a leading role in verbal thinking, was totally ignored. This kind of approach is inappropriate in psychology because it eliminates the study of mental processes.

The leading Soviet scientist Rubinshtein (1935) introduced personality principal in psychology, which integrates individual and social aspects in the study of human development. According to this principal human development is the result of the interaction of material and social practice with human subjectivity. Activity plays the role of mediator between social reality and human subjectivity. In activity theory, a person who interacts with a situation is considered the subject. That is, we are talking about cognitive and behavioral actions and not about the stimuli to which the subject reacts. Subjects' actions have a voluntarily goal-directed character. Behaviorism ignores mediated functions of activity that provide the basis for personal development. It denies important elements of activity such as reasoning, judgment, creativity, and concept formation. This is an important point of our discussion. Managers do not react as reactive organisms in emerging situations when making their decisions and formulating their instructions. Subordinates, in turn, do not react as reactive organisms on the manager's directives either, but act as conscious human beings. Even highly automated actions should be distinguished from reactive behavior. For example, a very quick response to an emergency signal looks like a reactive response. However, this is not reaction but a meaningful and purposeful action because it has a specific goal or a desired future result. Usually requirements are presented as instructions. Only after interpreting these requirements can the subject transform them into the goal of a task. During this process the subject compares requirements with the past experience and then makes a decision about how to perform.

The Symposium *Cerebral Mechanisms of Behavior*, held at the California Institute of Technology in 1948, is regarded by many as the end of the reign of behaviorism in psychology and as the beginning of cognitive science as a formal field of study. No stricter stimulus-response explanation of human behavior was considered acceptable. With the rise of cognitive science, human behavior was not looked at as the conditional responses anymore, but rather as the ability of human mind to explore between stimulus and response.

12.2   ABOUT MANAGEMENT

Management is a multidimensional concept that includes technical, economic, psychological, and social factors. The psychological aspect of management is an important area of study in ergonomics. In management, there are always two components—the subject, the one who conducts managerial functions, and the object, to which the managerial actions are directed. The subject and the object of management form a unified management system. Effective management is manifested in transformation of the system from one state to another desired state.

A systemic approach to management manifests itself in a wide range of managerial functions. Regardless of the type of organization and the field of activity, management functions are essentially the same. In fact, management functions are considered universal. Managers plan and organize, coordinate and control, make decisions, and handle physical, informational, and financial resources, create and communicate, motivate and reward, etc. All these functions can be combined into four categories of recourses: physical, informational, financial, and human. The content of the first three categories varies significantly depending on the specifics of the organization or business. For example, a software company differs very much from a restaurant business in regard to the physical supply they need to conduct their businesses. The same kind of differences can be observed in regard to the financial and informational resources needed for conducting their respective activities. Whereas the fourth category of resources, human resources, has a lot of similarity regardless of the specifics of organizations. People are people everywhere with their personality features, needs and abilities, goals and expectations, and so on.

The concept of manager should be understood not only in relation to people who are called managers by their title. Any person who is in a position of managing other people's activities falls into this category. Business owners and plant foremen, supervisors and heads of departments, commanders of military units and coaches of athletic teams, and many others fall into this functional category. Management, first, is interaction with people. A bank manager does not manage computers, safes, and accounts, a construction site manager does not manage machines and equipment, a ship captain does not hold the steering wheel himself but nevertheless gets to the desired destination by managing the ship's crew. Managing people at work is not a part of the management process. Managing people at work is management as a whole.

People live and act in various groups and are influenced by various formal and informal leaders: parents, teachers, managers, coaches, commanders, and so on. Their personal traits and qualities, their behavior and life style as the dominant parties, have a strong impact on people's mentality and play a significant role in their personal development. Many factors influence the psychological environment in the workplace, but the strongest one comes from the manager. The way the manager relates to subordinates affects the whole nature of business communication and largely determines the group moral and the psychological atmosphere in the working environment. If the manager does not project a positive image, it automatically transmits into the relations between the team members. The working environment becomes stressful, people are less inclined to cooperate with each other, they feel uncomfortable and morally vulnerable. To act as a conductor, and not as a drill sergeant, is the key for creating an atmosphere of togetherness and coherency. Submission is quite rarely pleasant, and any overbearing tone is perceived as a suppression of personality, as an encroachment on individual freedom. Demands of blind obedience and underestimating of subordinates' initiative and abilities is perceived as abnormal and determines the corresponding attitude toward the manager. Requirements expressed in the form of proposals are perceived as more acceptable and tend to have a greater effect. By not showing their superiority, discussions are conducted on an equal footing rather than via direct criticism and instructions. By acting in this manner, managers show respect to people, value their competence, and psychologically put them on an equal level. It should not be construed however as an endless idyll. When circumstances dictate, clear and direct instructions are justified in order to achieve the team's objectives.

In the world of management today, many different management styles and their variations have been identified. Some authors identify five styles, others six. Some even describe thirteen management styles. The styles are also named differently by different authors. However, regardless of the number of styles and terminology, the main thing when talking about management styles is to follow how managers make decisions and how they relate to their people. That in turn influences the group dynamic and the working atmosphere. Some managers demand from subordinates obedience to their instructions with no explanation and discussions, whereas others encourage subordinates' initiative and active involvement in problem solving and decision making. That is, we are talking about two opposite approaches (with some variations and the degree of their manifestation) in relation to subordinates. Representatives of the former are characterized by the desire for excessive centralization of power, exaggeration of the role of administrative methods, and the sole solution to most questions without discussing them with subordinates. Representatives of the latter are characterized by allowing group members into coordinated activities and by maximizing their involvement in joint definition of group goals. These managers do not seek concentration of power and contacts, but rather vice versa; they try to delegate responsibility to informal leaders of the group. However, the most important thing about these two management styles is that both styles are appropriate depending on the specifics of the team activity, the situation, and other factors. It can be illustrated by the following example. If a subordinate was competent and well aware of their responsibilities, the use of the authoritative approach would not be justified. Such a subordinate is mostly in a need of support and positive motivation. In the opposite case, if the subordinate does not have the sufficient knowledge and experience, the manager's clear directive instructions and oversight will intensify their activities. If the manager limits his or her actions in relation to this particular subordinate by only friendly support, it will do no good.

There is another management style that differs from the styles described above. It is called liberal or passive by some authors, and chaotic by others. Managers practicing this style are usually people who are not very knowledgeable in the field and often take the position to advance the self-serving tactical goal of "jumping" to a different (often higher) position in the near future. On the other hand, they are people who are conscientious and responsible by nature, but with the features of inertness. The management activity of such managers is reduced to transferring the directives from the top down and to function as observers and collectors of information. They allow subordinates maximum independence in the performance of work. This management style cannot be effective in the production environment, in power structures, in collectives where the activity takes place in extreme conditions, etc. However, this style is quite acceptable in activities where control and guidance is not necessary. For example, activity of the teaching staff in the universities' departments, in scientific institutions, and everywhere else where creativity is the essence of the work.

12.3 FROM EFFICIENT TO EFFECTIVE MANAGEMENT

The early figures who planted the seeds for the employee motivation movement were British reformer Robert Owen (1771–1858) and Scottish doctor Andrew Ure (1778–1857). Robert Owen was one of the first to recognize the importance of human resources and human needs of employees. He is considered a man ahead of his time. Until his era, factory workers were generally viewed much as appendages to machines and equipment. Andrew Ure also was a proponent of humanistic approach to employees' needs and providing workers with good working conditions. Andrew Ure introduced human factors as an additional aspect of manufacturing along with mechanical and commercial ones.

Most attention in any organization is directed toward achieving financial goals, i.e., towards profitability. This is vital for the organization and well understood. However, particularly for this reason, people's interests are not often on the priority list in organizations' affairs. If that is the case, eventually such an approach will backfire and prevent the organization from functioning successfully in the long run. Without taking the the human element into account, achieving the desired objectives and maintaining people satisfaction at work does not seem possible. Psychology has a lot to contribute to management in general, and more specifically to the management of projects. Projects are done in groups; they require team members to communicate, empathize, comprehend, influence, and engage. Delivering successful projects requires an understanding of people and psychology

Until the beginning of the twentieth century, practically no one thought about the system management of an organization from the point of view of psychology, nor from the point of view of technology or management. The concept of management and managers did not exist. Society then was a society of owners and their "helpers." In the twentieth century, our society became a society of organizations. This in turn led to the demand for people who were neither owners nor helpers. The new society needed people who practiced professional management: planning, organizing, integrating, developing people, and so on. Thus, management as the specific activity and the area of study has been developed. The term *management* was first popularized by Frederick Winslow Taylor (1856–1915). He is considered the earliest advocate of scientific management and a principal innovator in industrial engineering, especially in relation to improving efficiency and utilizing time and motion studies. Taylor introduced the scientific management theory in his book *The Principals of Scientific Management* (1911). The major postulate of his theory was an assumption that individual workers would be willing to work hard for monetary rewards. Taylor believed, that in order to improve efficiency, managers were required to do all the thinking, leaving workers with the task of implementation. "In our scheme, we do not ask the initiative of our men. We do not want any initiative. All we want of them is to obey the orders we give them, do what we say, and do it quick." (Taylor, 1911, p. 11).

Taylor has been instrumental in the development of modern management. He was "the first man in history who did not take work for granted, but looked at it and studied it" (Drucker, 1974). Taylor introduced wage incentive schemes so that workers could be paid more for the increased production. He insisted that workers would be satisfied by this economically motivational method, and as a result, cost per production unit will be reduced. To offer more money was the only method at that time to motivate workers for better performance. Scientific management can actually be condensed into five simple principles (Morgan, 1997): 1. Shift all responsibility for the organization of the work from the worker to the manager, 2. Use scientific methods to determine the most efficient methods for completing the work (while specifying the precise way in which the work is to be done), 3. Select the best person to perform the 'designed' job, 4. Train the worker to do the work efficiently, and 5. Monitor worker performance to ensure the appropriate method is followed and that the appropriate results are achieved. Despite the fact that the wage is still considered the main motivating factor for increasing productivity, motivation by only economic incentives works up to a certain point. People see more than just earnings in their work, they also work for reasons other than money. People are filled with thoughts and ideas, and want to see them implemented along with receiving monetary rewards.

Through a focus on efficiency, Taylor abandoned the strengths of human nature and capability, ignoring workers' individuality, their health, their motivation, and decision-making skills. By ignoring human factors, Taylor, the man most dedicated to performance efficiency, actually created the most significant barriers to business efficiency and decision making. One of the early critics of Taylor's work, who initiated the human factors, was Lillian Gilbreth (1878–1972). In her book *The Psychology of Management: The Function of the Mind in Determining, Teaching and Installing Methods of Least Waste* (Lillian Gilbreth, 1914), she propagated the exploring of the psychological element within management, and incorporated concepts of human relations and worker individuality into management principles. She expressed the view that scientific management should be built on the principle of recognition of the individual not only as an economic unit but also as a personality, stressing the importance of including the human element in management, which will complement and enlarge the scientific management perspective. Lillian Gilbreth is deservedly considered as one of the first propagandists of the human relation approach in management. She made important advances by progressing the search for efficiency to bring human subjects into consideration. Other advances around human relations, human activity systems, ergonomics, human centered design, and human interaction followed over time.

In our recently published book *The Psychology of Effective Management: Strategies for Relationship Building* (Voskoboynikov, 2017) we attempted to advance the agenda of the psychological aspects of management by encouraging a deeper consideration for the human element in management. Dalcher (2018), in his comments on the content of the book, writes that the author "acknowledges the dramatic changes in human work, which in contradiction to Taylor's view, increasingly require greater reliance on human intelligence, knowledge and insights. In order to address such a wider agenda new fields such as work physiology and occupational psychology are needed to integrate human capabilities and improve performance. Excelling and enhancing performance emerges from an understanding of the participants, and their strengths and capabilities. The implication is that in order to improve performance and deliver, organizations need to employ conductors who are able to bring different skills and expertise together, rather than efficiency experts and drill sergeants who endeavor to optimize individual tasks and minutiae. Voskoboynikov is therefore able to progress the discussion from one concerned with mechanistic *efficiency* of operations and reduction of waste, towards one that addresses the challenges of modern life by embracing and acknowledging the role of *effectiveness*, and fitness for purpose management."

Motivation is perhaps the most important factor in human activity, as it involves the action phase in human behavior, which has a direct impact on the outcome of the activity. Two things motivate people more than anything else—achievements and acknowledgment of achievements by their superiors. The late American philosopher John Dewey wrote, "The deepest urge of human nature is the desire to be important" (Schul, 1975). People desire recognition; they want to experience their own importance. They want to have their ideas considered and want to feel a real sense of accomplishment. Namely, everyone has a natural need for working activity and not just for generating income to meet consumer preferences, but for the work that brings satisfaction from the work process and from achieved results. Not only monetary incentives important to people, but moral and psychological appreciations and support are important as well. People thrive in the atmosphere of acceptance and recognition, and sometimes open some gifts and talents in themselves, the presence of which was not even suspected. To make all possible efforts to note people's merits when they expect it is called positive reinforcement, which tends to increase the probability that the act will occur again.

Motivation is considered a source of energy that drives activity. An interaction between such components of activity as needs, motives, goals, and objects constitute the inducing aspects of activity. Needs are treated as states of individuals, the desire for some objects that are required for survival and growth, that become the ground for activity. With tools, humankind changes objects and modifies them in accordance with its needs and goals. During the satisfaction of human needs they change and develop. Needs become motives for activity as they motivate an individual toward a goal of activity. In general, the goal of activity is a conscious future result of an individual's own actions or activity. The relationship between motive and goal determines the directedness of activity. While motives are *energetic* component, goals are a *cognitive* element of activity. The interrelationships between goals and motives are dynamic and complex, and may vary over the course of activity (Bedny and Karwowski, 2007).

12.4 PERSONALITY FEATURES AND PERFORMANCE

12.4.1 Temperament Types and Some Other Personality Features

Personality manifests itself in the relatively stable set of psychological attributes that differentiate one person from another. Personality traits are understood as the most fundamental sets of individual differences. They affect people's behavior and performance, as well as their attitudes toward the production unit and the organization in total. Whether a person's individual differences are good or bad should not be considered in their totality, but rather in relation to the specific performance. A person may be dissatisfied and negative in one job setting but very satisfied and positive in another. In addition, the "good or bad" depends on the circumstances where performance is unfolding. In other words, whenever there is a need to assess individual differences of subordinates, the situation in which behavior occurs must be considered.

In the most general terms, every human being is characterized by two components: *biological* and *social*. Genetics is a biological or innate component of personality, whereas social is an acquired component of personality. The biological component includes physical characteristics and physical peculiarities, which practically do not change during the life of the individual. These include the size and body shape, height, hair and eye color, etc. To the same extent, it applies to the makings, temperament, the nervous system, cognitive processes, and so on. The latter are of the greatest importance for our consideration as they manifest themselves in activity. Some people are quick and agile, while others are slow and calm. Some have a well-developed logical memory; others are good at remembering things mechanically. Some have the imagination of an artist, others, developed abstract thinking. Generally, individual characteristics of a person are a product of his or her heredity, physical being, and the acquired experience of mankind.

Temperament is an important characteristic of personality, which manifests itself in activity. It is well known that people react differently to the same impact. When we say a is short tempered or "temperamental," we mean the person is quick and emotional with distinct facial expressions and gestures. If another person does not clearly manifest these kind of reactions, we say that this person is a "non-temperamental." If one person can quickly put aside some unpleasant comments directed towards him or herself, another one won't be able to sleep well the following night and will keep dwelling on it. Where one person calmly reacts to criticism, another one will "jump through the roof." His eyes, color of his face, his body language etc., will display his emotional stage. We say that people react to the same impact with different "psychological colorings." That is, temperament characterizes people's behavior from the position of force with which they respond to the same stimuli. Soviet psychologist Merlin (1964), most known for his research of temperament, emphasized that temperament characterizes people only by the dynamic of their reaction on the impact, and does not predetermine their mental ability or social significance. It should be noted that some people, by acquiring knowledge and skills, and by their life experiences in general, are able to conceal their real features of temperament. However, these kind of behavioral reactions are actually "masked" reactions, and are possible only in non-extreme situations. In extreme situations, in contrast, a reduction of the acquired life experience manifestation of behavior take place, and the natural qualities of the individual's temperament are revealed.

There are four known conditional temperament types. We will restrict ourselves to a very brief description of the types. People of *sanguine* temperament have a strong nervous system. They are steady in their feelings and actions, sociable, talkative, and easily converge with new people. The *choleric* type is an individual with a strong nervous system, quick-tempered, straightforward, and aggressive. They are characterized by stable aspirations and persistence in achieving their goals and are capable of overcoming great difficulties. People of choleric temperaments are characterized by prevailing of excitation over inhibition. They are the ones who, we can say, have "bad brakes." To put such an individual on the front line of communication with customers, where "the customer is always right," is hardly a good idea. The *phlegmatic* type is an individual with a strong nervous system. They are balanced, diligent, patient, and peaceful; tend to be self content and kind, relaxed and rational.

People of the *melancholic* type are individuals with a weak, easily vulnerable nervous system, capable of sustaining a short term stress. Melancholic type persons are usually perfectionists, can sustain monotonous work, possess an ability for paying close attention to details, which is quite important in a number of professions.

This brief description of four temperament types should be understood as general types of temperament, which are quite rare in a "pure" form. Indeed, many of us exhibit some mixture of temperament characteristics; all human beings have some degree of each of these four types within them. This is how it can be explained. In any classification, the type is characterized by the severity and the ratio of its constituent properties or other characteristics. Theoretically, the degree of severity of the properties may vary indefinitely, thereby creating an endless number of possible types. However, in reality, there is no need for such theorization and the type approach can be used for practical purposes. Singling out the most prominent feature of temperament attributes a person to one or another type. In summary, each temperament type has its strength and weaknesses, advantages and disadvantages. While one temperament type better relates to some kinds of activity, another type is good for some other kinds of activity. All four types have both good and bad qualities in relation to different kinds of activity. To take subordinates' temperament type and other personality features into account is helpful in the process of implementing managerial functions.

Extraversion—introversion and neuroticism—stability is another feature of personality, which manifests itself in activity. Hans Eysenck (1970), best known for the study of personality, viewed individual differences as a result of two independent variables: neuroticism (weak emotions, unstable) versus stability (strong emotions), and extraversion and introversion. Extraversion characterizes people as outgoing. Extraverts are in constant need of "psychological food" from the social environment. They are characterized by high motor and speech activity, easily respond to a variety of proposals and are actively involved in their implementation, but on the other hand, they may easily lose interest and switch to a new activity. Introversion is understandably the opposite to extraversion. Introverts are thoughtful, rational, inclined to plan their activity, inclined to self analysis, and looking "inside themselves." Generally speaking, extraversion-introversion has a weaker effect on behavior, while neuroticism-stability has a stronger effect. For example, manager-introverts can learn to behave in an extraverted manner if circumstances require. Neuroticism can be described as an enduring tendency to experience negative emotional states, such as anxiety, guilt, and depression. They draw some negative scenarios in their mind; their body language reflects their emotional state, their physiological reactions—blood pressure, pulse count, etc., evidence the same. Those that score high on neuroticism tend to respond poorly to stress. McCrae and Costa (1986) pointed out a very important difference in coping with stress by people with high and low scores of neuroticism. In their research they found that people with high neuroticism scores are "emotion-focused" on stress, and they interpret situations as threatening or hopelessly difficult, while people with low score in neuroticism (stable) are "problem-focused" and tend to ignore the source of stress.

Bagretzov and L'vov (2001) studied the reliability of operators, who often make decisions under the impact of stress factors. They were able to establish four types of operators by the ability to resist stressing factors based on their personality characteristics: *stress-resistant* personality type, *labile* personality type, *inert* personality type, and *stress-volatile* personality type. Operators of *stress-resistant personality* type have a strong balanced type of higher nervous activity and high emotional stability. They are characterized by rapid entry into the labor process, and long-term efficiency in solving intellectual problems at a fast pace. Operators of a *labile personality type* are characterized by low emotional and volitional resistance and quick excitability. They usually do not maintain long-term stress resistance; they make mistakes when solving problems in a time limit. However, in favorable conditions they are able to work with high intensity. Operators of the *inert type* of personality have a strong balanced type of higher nervous activity. They are characterized by a pronounced inertness of nervous processes. Emotional and volitional resistance is seen at the medium level, tend to work at a uniform medium pace, and it is difficult for them to switch from one activity to another under time limit conditions. Representatives of the *stress-volatile type* are people with a weak type of higher nervous activity. They are characterized by low emotional-volitional resistance, high anxiety, inadequate self-esteem of their state, and high exhaustion of mental functions in the solution of intelligent, heuristic tasks at a fast pace. The authors' analysis of the described personality types suggests that for each of them, the defining nature of the behavior and changes in performance, is a certain group of characteristics that reflect the dependence of the intensity of their work on the correctness of the tasks performed.

12.4.2 INDIVIDUAL STYLE OF PERFORMANCE

Any kind of human activity presents more than one objective requirement to people in order to perform. It allows different individuals to rely on their personal strength to compensate for individual weaknesses. The study of personality and individual differences is a critically important area of activity theory. The central notion in this area of study is the individual style of activity that connects personality features with mechanisms of self-regulation and strategies of performance. All kinds of work, learning, and athletic activities are characterized by the interaction of subjective personality features and objective requirements of activity. This interaction goes in two directions. The first one is the adaptation of objective requirements of activity to the subjective properties of the individual, the second is the adaptation of subjective properties of the individual to the requirements of activity. Hence, there are two ways of ensuring the effectiveness of human performance. One of them is by professional selection, the so-called "screening out" of individuals with specific attributes. The other one is individual training methods directed toward the formation of individual strategies of activity based on the personality features of the individual in the process of adaptation to the objective requirements of activity (Voskoboynikov, 2014).

While in the West the selection method was used more intensively, in the former Soviet Union the attention was mostly directed toward the development of methods for individual training. The concept of individual style of activity was first introduced by the Soviet psychologist Merlin (1986) and by his follower Klimov (1969). In subsequent years, some other authors also studied the effect of individual personality features on performance (Bedny and Voskoboynikov, 1975). The outcome of these studies was establishment of the fact that different individuals can perform with *equal efficiency* using their own strategies of performance that are more suitable to their personality features. That is, people attempt to compensate for individual weaknesses with their personal strength in a given task situation. By implementing the individual style of activity on performance, they diminish the impact of their weaker features of personality.

Individual styles of activity should be considered strategies of performance deriving from the mechanism of self-regulation, which depends on personality features (Bedny and Seglin, 1999; Voskoboynikov, 2014). Such a strategy occurs at the conscious and unconscious levels, and is based on principles of self-regulation. Both levels are tightly interconnected and transform from one to another. The process of self-regulation manifests itself in the formation of desired goals, in developing of a program of actions that corresponds with these goals, with conditions for achieving the goals, and with a person's individual abilities. In other words, people through trials, errors, and feedback corrections create strategies of performance suitable to their individuality. For example, people with an inert nervous system develop a predisposition to organize and plan their work, and attempt to utilize a stereotyped method of performance. Based on the individual style the subject can adapt to the situation more efficiently. It should be distinguished however that individual style and methods of performance is not the same. The latter is not dependent upon individual personality features but rather upon organizational factors, imposed supervisory procedures, etc. Sometimes methods of performance that derive from organizational factors may contradict with individual personality features, which is not desirable. In cases of inadequate training that ignores individual features of personality, the subject may acquire methods of performance that contradict with their individuality. It may negatively affect performance level and job satisfaction.

To identify the individual style of activity of a particular individual who interacts with the task situation is important. Through individual style, the subject can adapt to a situation more efficiently. Empirical facts and theoretical studies show that the adaptive mechanism of personality traits can work up to a certain limit. In studying the individual style of activity, it is important to observe how people with different individual characteristics acquire the same knowledge and skills. On the other hand, it is as important to identify how subjects disintegrate into distinct groups with respect to their ability to acquire skills. Such disintegration takes place as the capacity of some individuals to adjust to the requirements of the activity is reduced due to increased task complexity. In simple situations, individuals exhibit similar levels of achievement regardless of their individual style of performance. In such situations, it is hard to identify individual differences. In the early stages of task performance almost all participants show approximately the same results. As the task gets more and more complicated, the differences in results will be demonstrated. People who are able to perform the more complicated tasks are assigned to a higher professional category. Those whose individual style of activity does not provide adaptation to the objective requirements of activity will remain in the category of lower requirements. Individuals begin to vary more and more in their performance level (Bedny and Voskoboynikov, 1975; Bedny and Karwowski, 2007). We consider the individual style with some more details in our other article in this book.

12.4.3 PROFESSIONAL SUITABILITY AND RELIABILITY

Reliability in general terms refers to consistency in performance. In technical systems a qualitative characteristic of reliability is "the ability of the system to perform its required functions at a given time interval and under the conditions of overloads for which the system is designed" (G. Bedny, W. Karwowski and I. Bedny, 2015). Human error in the failure of the technical system can occur for various reasons. It can take place because of inadequate equipment design in relation to human factors, which leads to erroneous human actions. It can also be due to a mismatch between the personality properties of the human and the activity requirements, which also leads to erroneous human actions. For example, for effective performance in some dangerous professions the operator must possess a high level of emotional stability. It manifests itself in the ability to demonstrate learned skills under adverse conditions, such as stress, fatigue, danger, time limits, etc. In order to perform any professional activity (or any kind of activity for that matter) a person must meet certain objective requirements that vary with professions. It is implied that any normal human being can be trained to learn any professional trade. Human's innate qualities possess a great plasticity that allows a vast majority of people to master different professions and perform quite successfully. There are many specific skills that can be learned if a person wants or has to. However, the question is not that it can be learned, but how much time and effort it will take for those whose innate features are not up to the requirements of activity. For example, when two people perform the same work equally successfully we don't know how each of them came up to the required level of knowledge and skills, or how much time and effort each of them attributed for achieving the needed level of performance.

Gurevich (1970) studied this phenomenon. He writes, "Even if to assume that some special training methods could be found, a person with limited innate properties would have to spend unreasonable amount of time and energy for mastering a chosen profession. The period of active human life is limited and unproductive bleak activity would not only bring personal unhappiness, but ultimately will negatively affect the society as a whole." He proposed the division of professions into two types. Professions of the first type are the ones that present excessive requirements to some individuals. He called these requirements *non-compensable*. That is, due to the certain incongruity of innate features of some individuals to the objective requirements of activity, they either cannot master the profession, or even if they can, mastering it may have a negative impact on their health. Below is an example of how a mismatch between the psychological and physiological features of personality on the one hand and objective requirements of activity on the other hand negatively impacted human health (Marischuk et al., 1969). This real life experiment took place in the early sixties at the Russian Aviation Academy in Leningrad (now St. Petersburg). Upon applying to the Academy all applicants went through the required physiological and psychological tests. Some of them did not do very well, but demonstrated a strong desire to become pilots. Exceptions were made for those applicants and they were accepted. Within the first and second years of study and training some of that particular group dropped out because they could not keep up with the requirements of the training program. The remaining part of the group continued their study and training demonstrating a strong will to master the profession despite their inadequate physiological and psychological properties to the demanding activity. They completed the full program and graduated as pilots. However, in a few years all of them were discharged from service due to different chronic health disorders as a result of what was for them, a highly stressful activity. Prolonged resistance to the impact of the excessive objective requirements of activity created a state of stress that made them vulnerable to diseases, and eventually led to the breakdown of the weakest link of the organism. Some of them developed disorders of the gastrointestinal tract, others, high blood pressure, and some others disorders of a similar nature. Indeed, "a chain is as strong as its weakest link."

All other professions which present *compensable* requirements are the professions of the second type, the mass professions. Most people can master mass professions and perform successfully without harming their health. The major role in adaptation to the objective requirements of activity in mass professions is the individual style of activity. It manifests itself in the specific strategy that people consciously or unconsciously use in order to adapt to the objective requirements of activity. It's worth noting that professional suitability manifests itself in the specific professional activity, whereas the individual style of activity manifests itself in various kinds of activity.

12.5 THE EFFECT OF A GROUP ENVIRONMENT ON INDIVIDUAL PERFORMANCE

12.5.1 INDIVIDUAL AS A COMPONENT OF THE SYSTEM

A systemic approach in management includes concentrating on leading the group as a whole, rather than concentrating on the leading of individual subordinates. This in turn means paying attention to the group dynamics and roles distribution in the group. In particular it has to do with the presence of informal leaders in the group, i.e., the most respected group members.

It has long been observed that people behave differently in a group setting compared to behavior in private. This is because an individual in a group appears in a new capacity—as a component of the system "individual–other individuals." Groups have properties of their own; they are different from the properties of the individuals who form the group. In most cases, the group environment has a positive effect on the individual's behavior and performance. People's relationship in groups can be on the level of functional business contacts and on the level of psychological human contacts. If a group is formed for a certain purpose—production, education, military, recreational activity and so on—they are called *formal* groups. Examples of such groups are a factory shop, a production unit, a school class, an aircraft or a ship crew, an athletic team, and so on. Other groups are formed based on personal relations, subjective feelings, sympathy, trust, common interests, and so on. These groups are called *informal*. For example, a group of friends is a group of persons who are pulled together by common amateur interests. The structure of any informal group in general is as follows: leader, followers, and isolated (rejected by the group or rejecting the group). From the point of view of management, it is important to understand that psychological connections between people take place not only within informal groups. People built informal connections by functioning in formal groups that are formed for certain organizational goals. That is, in every working unit there are two structures—formal (official) and informal (unofficial), and each structure has its respective formal and informal leaders. The formal structure is based on formal rules and written instructions, where the circle of obligations for each employee is clearly defined. The informal structure within the formal groups is a system of psychological connections between the group members. This structure is particularly important for our discussion.

In every production unit are people who are in the process of joint activity project psychological feelings towards each other—sympathy or antipathy, convergence or divergence in tastes, personal interests, ethnic preferences, amateur affiliations, and so forth. The manifestation of these feelings results in either mutual attraction or repulsion. Precisely these factors affect emotional well being of the team members, which has a direct influence on their satisfaction in the workplace, and ultimately, on their productivity. People come to work expecting a colleague-friendly work atmosphere. The sense of belonging to the informal group gives a worker certain status and recognition, creates the feeling of the relation to others; the feeling that she is somebody, even though in the formal structure she is just one of many. The effectiveness of the group and individual performances is largely determined by the conformity of formal and informal structures. Simply put, if people at work evoke positive perceptions of one another, the process of activity will run more effectively.

12.5.2 INDIVIDUAL APPROACH

To work on creating a *cohesive team* is one of the most important things a manager must to do in order to achieve the teams' objectives. From the social psychology perspective, a team is a small social group of persons who unite and cooperate to achieve a common goal. For the successful functioning of the team, two factors are of utmost importance: team members must possess the needed technical skills and experience in the field, and they must complement each other. The first factor is usually taken into account by managers, but the second one is not always paid much attention to. To create the cohesiveness of the team is easier when team members are at his or her respective place according to their best qualities. Hence, to take an individual-focused approach is perhaps the most crucial role of the manager. Each person perceives the world through the prism of her or his unique individual personality and life experience. Each person is different in his or her own unique way. Some people are quick and can easily adapt to the changing environment, while others are slow and not as dynamic. Some individuals can sustain tough impacts while others can't, but the latter are able to navigate in slightly noticeable changes of the surroundings, which enable them to react more keenly. Some feel comfortable in performing monotonous work, while others "fall asleep" doing the same thing. Some people are happy to work in a group environment; others prefer to work on individual assignments. That is, some people are good in some things while others are good at something else.

This brings us to the necessity of the individual-focused approach. To see employees as individuals, and recognize their abilities and desires helps to bring out their best. As has been mentioned, any activity requires more than one quality of people in order to perform. That allows people with different qualities to adapt to the requirements of activity by relying on their strong qualities as compensation for the weaker oncs, and as a result, to perform equally effective. It suggests that to rely on people's strong qualities is more effective than to insist on fixing the weaker ones. As a result, managers will best benefit from what people are capable of and they will experience satisfaction by their performance. Respecting people's individuality and using it the best possible way will eventually benefit the team and the entire organization. Hence, the golden rule in dealing with people is not to try to change them, but rather to build on what they are and compensate for what they are not. To give a person the wrong role is like asking them to be what they are not. When one is pressed to be what they are not, they do not feel good and will not perform as effectively. However, when placed in the comfort zone, where an individual feels "in their shoes," everything changes—they feel good, productivity increases, and all the rest that comes with it. People around that person are amazed about the changes. However, the individual has not changed, he or she has simply become themself. Morris Viteles, who is considered one of the fathers of industrial psychology and an enthusiast of taking the human element into the practice of management, wrote, "It is important that a man be kept out of a job for which he is not fitted. It is even more important that he be placed in a job where he can be efficient and happy" (Wallace, 1996). Such an approach should be a sort of guiding star for managers in their work with people. Assigning tasks to people according to their personality features elevates people's satisfaction in the work place and leads to better performance.

12.5.3 The Phenomenon of Compatibility

In any group activity the question of compatibility arises. For the effective execution of the group tasks, the compatibility factor approach is important. The presence of needed specialists and their technical skills only constitutes the necessary condition for the desired performance. The sufficient condition for the effective execution of group tasks is the degree of compatibility between them. Depending on the degree of compatibility, the result of the group performance may be either equal to the sum of the results of individual performances, or greater or lower than the sum. This suggests that the group is not the arithmetic sum of separate individuals, but rather a single organism—the whole—and the result of the group performance is not always a positive sum of the results of individual performances by its members. People in the group act and behave in a new capacity, as a component of the system "individual—other individuals." Representatives of various professions and other kinds of activity such as polar explorers, mountain climbers, commanders of aircrafts and ships' crews, coaches of athletic teams, and many others experienced in real life that not all people are equally fit for complex teamwork. That is, speaking in the language of psychology, "two plus two is not always four."

An example of the incompatibility can be a working crew, where there is a significant difference in skills of workers requiring coordinated application of muscular effort or relatively accurate movements. This kind of incompatibility is called *physiological*. In this example, such physical parameters as physical strength and motor skills are described. To note such differences in people is not that difficult, and it's unlikely that anyone will instruct people with such differences to perform a task where these differences present hindrances. People always experience a certain flow of feelings toward others within a group. These feelings are based on the commonality or differences of the psychological nature, such as temperament, character, social orientation, amateur interests, religious and ethnic peculiarities, and others. They may be positive, negative, or neutral; they can be weak or strong in intensity; they can be mutual or non-mutual and therefore conflicting. Differences of this kind are not always obvious and apparent. However, particular differences of this kind quite often have a decisive impact on compatibility, and in turn, on the implementation of the group task. The incompatibility by the described differences is called *psychological*. The presence of psychological incompatibility is a major obstacle for the effective group performance.

Psychological incompatibility has its negative influence not only on the group's performance, but also on human health. Unfriendly, uptight relationships between group members in the working environment cause negative emotions. In mass professions, where there are no expressed extreme conditions, people can perform productively under the influence of negative emotions for a fairly long time. However, it's important to understand that it flows at the expense of unnecessary stress, "until then, until the time" so to speak. Many can recall depressing mental states due to incompatibilities with colleagues or bosses at a current or previous job. Working activity on the background of negative emotions for a long period of time may cause pathological developments in the central nervous system, which could lead to various diseases of a neurotic order. People become irritable, experience headaches, insomnia, blood pressure disorders, dysfunction of the gastrointestinal tract, and other deviations in health conditions. Typical medical approaches for the treatment of such conditions does not always give positive results. There are statistical data in different countries on the loss of a huge number of person-hours as a result of nervous breakdown due to psychological incompatibility. Psychological compatibility is the most crucial factor when activity takes place in the extreme conditions. Examples of such conditions among others are situations of danger, time limits, extreme cold or hot temperatures, limited activity space, and so on. In such conditions fuzzy coordination between the group members due to the insufficient compatibility may endanger people's lives. The activity of aircrafts and submarines' crews, polar explorers, and mountains climbers flows in such conditions.

12.6 THE IMPORTANCE OF CLEAR AND SYSTEMATIC COMMUNICATION

By working and socializing, humans developed a special ability to communicate with each other. This ability is language in its various forms: audio, written, and sign. Language that is understood as consisting of words is the most common way to communicate. Computer language and other symbols can also be described as languages of communication. Other forms of communication include gestures, facial expressions, signals, art forms, and music. Famous Russian physiologist Ivan Pavlov (1951) called the ability to communicate by the means of language the second signal system (or the signals of signals) comparing it to the first signal system. The first signal system is signals in the form of external and internal irritations perceived from the real word by visual, auditory, and other receptors of the body. The real world for animals is perceived only by the first signal system. The availability of the second signal system allows humans to transfer knowledge and to profit from the knowledge of others.

Communication is a two way process where both parties want to communicate. The communication process takes place when each party makes an effort to understand what the other party is trying to communicate. In some activities the value of clear communication cannot be overestimated, for example in communication between pilots and air traffic controllers. If a command or confirmation of the command is not understood correctly by either of them, it may lead to serious and sometimes tragic consequences. Here is a real world example of how miscommunication between the air traffic controller and the pilot led to a tragic end. On March 5, 1973, an aircraft Boeing 747-249F of the Flying Tiger Line was on the way to the Kuala Lumpur Subang airport (KUL). The air traffic controller gave the command to the pilot of the approaching aircraft, "Get into echelon (go down) "two-four-zero-zero" ("2400"). Because of the similarity in the pronunciation of the numeral two and the preposition "to" in English, the pilot confirmed the command incorrectly. He responded, "OK, to four-zero-zero" (that is "to 400," which sounds similar to "two-four-zero-zero"). The airtraffic controller did not catch the mistake and the aircraft crashed into a hill. Among recently introduced FAA rules and regulations some commands were modified in order to prevent possible miscommunication. For example, they changed the air traffic controller's command "affirmative" to "affirm," because possible background noise in the microphone may muffle the first part of the word, and the pilot may hear the "tive," which could be understood as "negative." Some other commands were also modified (Makarov and Voskoboynikov, 2011).

This example is of an extreme situation, but in ordinary everyday business affairs clear and proper communication is as important. In 2009, after the financial crisis of 2005–2007, the US government launched a program to help homeowners with their mortgages. Under the program, banks reduced the interest rate to qualified borrowers and thus their monthly mortgage payments were reduced accordingly. Homeowners sighed with relief. Unfortunately, that was not the end of the story. Some banks departments that offered mortgage relief to homeowners failed to communicate with the department in charge of implementing foreclosure. As a result, there were cases where the foreclosure departments, seeing that homeowners were making lower than their original monthly mortgage payments, assumed that they defaulted and seized their properties.

Understanding is subjective; it can occur in the receiver's mind. The fact that the communicator transmitted the message and the receivers heard it does not mean that communication took place. Even when the receiver understands the instruction or information the way it's intended to be understood by the communicator, it does not yet constitute a completed communication process. Communication is not only the receipt and understanding of information, it is also an acceptance and action. One of the most common blunders in communication management is when the manager assumes that everything is going well because he or she has not heard anything. Such an assumption is clear evidence of the manager's rare and inefficient communication with subordinates. If subordinates are not recieving regular check ins regarding their work, they won't know if everything is going in the right direction. If the manager, on the other hand, is not getting regular feedback from subordinates, it will be difficult for her to track the progress of the work. Maintaining regular and clear communication with all people involved in the ongoing projects is essential for ensuring a smooth work flow progress. That in turn allows anticipating possible missteps and making the necessary corrections in advance. In order to maintain such a business environment managers must provide appropriate and timely information to subordinates so they will know what they should do, when they should do it, and what is expected of them in general in the framework of the work requirements.

A systemic approach in management includes focusing not only on communications with subordinates in their own production unit, but with their bosses, managers of neighboring production units, as well as with people outside the organization—suppliers, customers, and others with whom managers have to deal on a regular basis. Management is, first, an interaction with people. In managers' relation to subordinates, it all comes down to managing people. A ship captain does not hold the steering wheel him or herself but gets to the desired destination by managing the ship's crew. Managing people at work is not a part of the management process. Managing people at work is management as a whole.

CONCLUSION

In the material here, we introduced the basis of activity theory as a counterweight to behaviorism, which portraits humans as reactive organisms. In activity theory, a person who interacts with a situation is considered the subject; that is, we are talking about external behavioral and internal cognitive actions as components of cognition. Self-regulation of activity manifests itself in the way people, through trials, errors, and feedback corrections create performance strategies. How performance strategy depends on personality features was also considered. In particular, the individual style of activity allows the subject to adapt to the situation more efficiently and to achieve conscious goals. We analyzed and described some important factors of the psychological nature that can be applied to the practice of management. The influence of a group environment on individuals' behavior and performance, and the phenomenon of compatibility was analyzed as well. The importance of systematic communication between the manager and subordinates, and with all those with whom managers have to deal on a regular basis, was considered. Activity theory distinguishes two types of activity: "object-oriented" and "subject-oriented." The former refers to a subject using tools on material objects with the goal of completing the task and evaluating the results. The latter refers to social interaction between people, which is the most important element in management.

REFERENCES

Bagretzov, S. and L'vov, B. (2001). *Diagnostics and forecasting of functional states of the operators in the activities.* Moscow Radio, p. 158.

Bedny, G. Z. and W. Karwowski. (2007). *A Systemic-Structural Theory of Activity. Application to Human Performance and Work Design.* Boca Raton, FL: Taylor & Francis.

Bedny, G., Karwowoski, W. and I. Bedny (2015). *Applying Systemic-Structural Activity Theory to Design of Human-Computer Interaction Systems.* Boca Raton, FL: Taylor & Francis.

Bedny G. and Seglin, M. (1999). Individual style of activity and adaptation to standard performance requirement. *Human Performance.*, 12, 59–78.

Bedny, G.Z., and Voskoboynikov, F. (1975). Problems of how a person adapts to the objective requirements of activity. In Aseev, V.G. (Ed.), *Psychological Problems of Personality.* Irkutsk, Russia: Irkutsk University Press. 2, 8–30.

Dalcher, D. (2018). What has Taylor ever done to us? Scientific and humane management reconsidered. In Dalcher, D. (Ed.) *Managing Projects in a World of People, Strategy and Change.* Milton Park, Abingdon, Oxon and New York, NY: Routledge, 29–38.

Drucker, P. (1974). *Management.* New York: Harper & Row.

Eysenck, H. J. (1970). *The Structure of Human Personality.* London: Methuen.

Gillbreth, L. M. (1914). *The Psychology of Management: The Function of the Mind in Determining, Teaching, and Installing Methods of Least Waste.* New York: Sturgis and Walton.

Gurevich, K.M. (1970). *Professional Suitability and Basic Features of the Neural System.* Moscow: Pedagogical Academy of Science.

Klimov, E.A. (1969). *Individual Style of Activity.* Kazan: Kazansky State University Press.

McCrae, R. R. and Costa, P.T. (1986). Personality, coping, and coping effectiveness in an adult sample. *Journal of Personality,* 54, 385–405.

Merlin, V. S. (1964). *Outlines of Theory of Temperament.* Perm, Russia: Perm Pedagogical Institute.

Merlin, V.S. (1986). *Outlines of integral study of individuality.* Moscow: Pedagogy.

Makarov, R., Voskoboynikov, F. (2011). Methodology for teaching flight-specific English to nonnative English-speaking air-traffic controllers. In G. Z. Bedny and W. Karwowski (Eds) *Human-Computer Interaction and Operators' Performance,* 277–304. Boca Raton, FL: Taylor & Francis Group

Marischuk, V. L., et al. (1969). *Psychic tensions in flight.* Moscow: Voenizdat.

Morgan, G. (1997). *Images of Organization* (2nd ed.). CC: Thousand Oaks, Sage.

Pavlov, I.P., (1951). *Complete collected works,* Volume 3, book 2. Moscow: Academy of Science.

Rubinshtoin, S. L. (1935). *Foundation of Psychology,* Moscow: Pedagogy Publishers.

Skinner, B. F. (1974). *About Behaviorism.* New York: Knopf.

Schul, B. D. (1975). *How to be an Effective Group Leader.* Chicago: Nelson-Hall.

Taylor, F. W. (1911). *The principals of Scientific Management,* New York: Harper and Brothers.

Thorndike, E., (1920). Intelligence and Its Use. *Harper's Magazine,* 140, 227–235.

Voskoboynikov, F. (2014). The Influence of Personality Features on Performance in Work, Study and Athletic Activity. In T. Marek, W. Karwowski, M. Frankewicz, J. Kantola and P. Zgaga (Eds.) *Human Factor of a Global Society: A System of Systems Perspective,* 187–192. Boca Raton, FL: Taylor & Francis Group.

Voskoboynikov, F. (2017). *The Psychology of Effective Management: Strategy of Relationship Building.* New York, NY: Routledge.

Wallace, A. (1960). Morris Viteles, Industrial Psychologist. *Philadelphia Enquirer.* (December Issue).

13 The SSAT Paradigm as an Effective Activity Design Guide for Efficient Human Performance in Organizations

Mohammed-Aminu Sanda[1,2]
[1] University of Ghana Business School,
P. O. Box LG 78, Legon, Accra, Ghana
masanda@ug.edu.gh
[2] Luleå University of Technology, SE, 97187, Luleå, Sweden
mohami@ltu.se

CONTENTS

13.1 INTRODUCTION

Human work has dominated cognitive psychology study for a long time. In recent times, there is a growing interest among researchers and practitioners in the application of activity theory to the study of human work. In cognitive psychology, the basic concept is information. External activity is considered to be completely dependent on internal mental activity. The dependence of inner mental activity from the external practical has not been studied. Rubinshtein (1922; 1935) was the first to draw attention to the dependence of mental activity from external, practical. A critical feature of activity is its relation to consciousness. Later Leontiev (1978) also began to develop this idea. Human psychological processes are unique in terms of their social aspects. Psychological processes of humans are products of social-historical development. Important data in the study of this aspect of human activity was obtained by Vygotsky (1960/1978). His major idea was that signs as mental tools are a major factor in human mental development. In contrast, Rubinshtein and Leont'ev empha-size the importance of material activity and its interaction with material objects in mental development rather than social interaction.

In this regards the work of Rubinshtein has much in common with the work of Leont'ev. However, there are important differences in understanding the relationship between inter-nal cognitive and external behavioral activity. Leont'ev considers the formation of internal activity as a process of internalization. Rubinshtein did not accept the concept of internal-ization. According to him, a subject does not internalize ready made standards derived from practices. Subjects utilize exploration and interaction with the objective world as a source of reflection and developments of his or her consciousness. Rubinshtein intro-duced basic principles of activity, which he called the principle of unity of conscious-ness and activity. From this principle it followed that not only cognitive activity regulates external practical activity, but external practical activity shapes internal mental activity. Human practical activity plays a leading role in mental development. At the same time both scientists in their studies underestimate the role of interaction of object-practical activity and social interaction between people during mental development, and the role of such interaction in mental development. This aspect of human development was studied more successfully by Vygotsky (1960/1978). According to SSAT (Bedny, Karwowski, 2007), the term of internalization can be preserved, but its understanding is different. This process should be understood as a self-regulative process. Based on feedforward and feedback interconnectiona between external and internal components of activity, a subject can develop both types of activity. Internal activity at the first stage is performed with the support of external activity, and at a later stage the subject can perform only mentally. Self-regulation provides flexibility of internal mental activity, and psychological processes start to perform reflective functions. Psychological reflection is a process of capturing external reality. Information reflection is studied from semantic, pragmatic, and quantitative perspectives. Semantic refers to the qualitative meaning, pragmatic to the utility, and quantitative of information available. The psychological reflective process at the one level includes automated psychological operations that are triggered by external stimuli. The other level of reflection includes voluntarily goal-directed actions that can be cognitive and behavioral. We consider an understanding of all these basic concepts from work psychology perspectives in the following section.

The degree of specification of any theory should be adequate for its application. The general activity theory developed by Rubinshtein, Leont'eve, and Vygotsky is not well specified and adapted for study human work. Terminology is not clearly defined for such purposes. For example, in general activity theory there is no clear method of description and classification of cognitive and behavioral actions, and how they can be extracted from the process of activity. There is no clear description concept of task. Davydov, Zinchenko, Talyzina (1983), leading specialists in general activity theory, wrote that in activity theory "differences between action and activity is not clear, and movements between them are possible in both directions." This statement, similarly to other examples, clearly demonstrates that degree of specifica-tion of terminology in general activity theory is not adequate for applied studies. For example, Strekov (2007) from Moscow University, erroneously described a pilot's task as an action. Leont'ev (1977, p. 107, see also Leont'ev, 1978) made mistakes when he described a metalworker production operation (task) and conflated it with psychological motor operations. As an example, we present his explanation of cutting a material object:

For example, material object may be physically taken apart by means of various tools each of which determines the method of carrying out the given action. Under certain conditions, let us say, an operation of cutting will be more adequate, in others, an operation sawing; it is assumed here that man knows how to handle the corresponding tools, the knife, the saw, etc.

The cutting of material objects by a knife or saw are not psychological operations, which can be considered as elements of separate actions. For example, if we desire to cut a brad we have to perform a number of motor actions. We move out hand and grasp the knife (the first motor action), than move the knife above the brad to a specific position (the second motor action), then we start to move the knife back and ahead, and at the same time push it down (the third motor actin). The last action we have to perform several times until pieces of the brad separate. In the first motor action we have two motions: move hand to knife (the first motion or operation, and grasp knife, the second motion or operation). In the second motor action we extract the following operation: move knife to the brad will be the first operation, and putting it in the exact starting position will be the second operation. The third action includes moving the knife back and down (combined motion)—the first operation, move knife ahead and down the second operation. In a similar way we can describe cutting the metal object by using a saw. In his example, Leont'ev mixed the psychological term "operation" with the term "operation" in a technological meaning. Cutting of material objects with various instrument are examples of different technological operations.

Terminological inaccuracy is characteristic not only of Leont'ev but also for his colleagues who attempt to study human work. As an example, consider the work of Strelkov (1990) who worked at Moscow State University and presented his study in aviation. Strelkov's studies in the field of aviation are of interest because they demonstrate the inapplicability of Leont'ev's general activity theory to the analysis of the operator's performance. There is no real understanding of human actions in the study of human work. Strelkov could not distinguish between the flight navigator's task and the psychological operations during performance of various tasks. For example, he wrote that the content of the flight navigator's actions during flight included: 1) perceptual operations of controlling various displays, 2) mental operations such as correction of navigation systems, 3) standard computing operations performed with the aid of special tools, 4) manual operations that involve entry of various data, 5) negotiations with terrestrial services, and so on. In addition, the author noted the pilot's single actions such as reducing an aircraft's altitude during flights lasting up to 20 minutes. It is important to indicate that "reducing an aircraft's altitude" is not a single action but rather a complex *task* that includes various cognitive and behavioral actions. Moreover, a single action cannot take such a long time. A duration of up to 20 minutes can be related to a *task* performance time. All these examples clearly demonstrate that some scientists in applied activity theory do not have a clear understanding of what such terms (task, actions, operations) mean when they apply their theoretical knowledge to a practical situation. The accurate description of the flight navigator's task should be as follows: 1) perceptual *actions* of controlling various displays, 2) mental *actions* such as correction of navigation systems, 3) standard computing *actions* performed with the aid of special tools, 4) manual *actions* that involve entry of various data, 5) verbal *actions* that include negotiations with terrestrial services, and so on. All of these actions should be extracted and described as a human algorithm of task performance.

A few more attempts to study human work were made at Moscow State University and most of them were unfitting. This was because the authors could not understand the technical content of human work. For example, Reshetova (1985) in her book, tried to develop instructions that describes a method of treating a cylindrical shaft with a manual feed. However, instead of the term "cylindrical shaft" she erroneously utilized the term "cylinder." According to the technical terminology, a cylinder is a hollow part and has a completely different processing technology. The manual presented by Reshetova describes the sequence of the steps for part machining on a lathe. However, the entire sequence of steps for part machining was incorrect. For example, in the instructions the author describes the following sequence of steps for selecting the cutting procedure for the considered part: determinination of the cutting speed, spindle speed determination, determination of cutting depth, and determination of the number of passes. However, the described sequence of steps is totally incorrect. According to technological requirements the sequence of steps should be the following: determination of cutting depth, determining feed of lathe, determining the cutting speed, determining frequency of rotation, etc. (Denezhnyj, et al. 1976, p. 27). There are multiple other incorrect and meaningless steps of part machining described in Reshetova's instructions and we will not discuss them here. Before any psychologist begins an analysis of a particular type of work he or she have a minimal knowledge of the technical aspects of the considered work. For example, such American and Russian scientists as Norman (1986), Wickens, Hollands, (2000); Kotik, (1978); Ponomarenko, tc. (2006) start to a study pilot's work, and they spend a lot of time and effort obtaining the required basic knowledge in this field. Any knowledge in activity theory can be useless without such background.

We have to note that we conducted an analysis of some other publications from Moscow State University where the authors tried to study human work from the perspective of Leont'ev's concept of activity. In general, these studies were not successful. This is not an accident. For example, in this theory, there is a lot of discussion about cognitive and motor actions, but we can never find their specific description and classification. A general AT does not have a clear developed terminology, which would allow the effective use of this theory in the study of human work.

General activity theory contains a lot of interesting data. However, it does not provide rigid principles and methods of study of human work. General activity theory provides only a philosophical and general psychological foundation for the study of human performance. General psychological theories developed by Vygotsky, Rubinshtein, and Leont'ev were important for the development of applied and systemic-structural activity theories, which were adapted for the study of human work.

13.2 ANALYSIS OF LANGUAGE AND BASIC CONCEPTS OF ACTIVITY THEORY IN THE WEST

A fundamental difficulty associated with activity theory in the West is associated with the specificity of Russian language, and historical and cultural roots. For example, there are no clear equivalents between Russian and English words. Such words as *deyatel'nost'* (translated as activity), *dejstvie* (translated as action) are examples of this. *Deyatel'nost'* in broad meaning and in common sense is the performance of some work, learning, play, etc., that involves human consciousness and language. This makes such an active state different from an animal's behavior. *Deyatel'nost'* unfolds in time and has a longer duration. Practical experience demonstrates that there are some smaller units of active process that consciously transform reality. Such units in the Russian language are designated as actions. Moreover, such actions before designate only motor actions. *Dejstvie* and *deyatel'nost'* became basic psychological concepts when psychology and philosophy later emerged as science. Later these concepts become more clear and specific in Soviet psychology. English speaking scientists have a lot of difficulty translating these two basic concepts of activity theory. As just stated, *deyatel'nost'* is translated as activity and *dejstvie* as an action. However, *activity* and *action* have different meaning in English language. They can designate active or passive states, and have similar or different meanings in psychology. For example, action theory translated from German psychology has exactly the same meaning as activity theory in present US. Diaper and Lindgaard (2008) wrote that translation problems between different cultures involves complex social processes that can last decads before agreement about concepts can be reached. There is another example that demonstrates translation issues from English into Russian (Bedny, Harris, 2008).

No exact equivalent of the English word "purpose" exists in the Russian language. The Russian concepts of *goal* and the English *purpose* carry different meanings within activity theory and in considered languages. The word *tcel'* translates as *goal* and *namerenie* as *intention* or *purpose*. It is interesting that a recognized leading Western activity specialist such as Engestrom (2008, p. 257) cannot understand the exact differences between a goal and a purpose. He wrote "what exactly the difference is between a goal and purpose is not made clear—except that purpose is somehow larger or at a higher level than goals." However, anyone who knows both English well and the Russian language can find differences in these two words without being a specialist in activity theory. Thus, there are difficulties in understanding Russian words. On the other hand, there is a problem derived from necessity to correct interpretations of basic theoretical concepts of activity theory. We consider below some examples of incorrect interpretations of basic concepts of activity theory in the West.

It is interesting to consider examples that describe Diaper (2008) in his work where he uses data from other work. On page 263 he considers the example "requesting X-rays." According to Diaper, "requesting X-rays" can be considered an action and following it from there are two goals of this one actions: 1) obtain information from a diagnostic test, and b) avoid future litigation. However, "requesting X-rays" is not an action. This is an activity during performance of a task. The author did not describe the considered situation in a detailed manner. Hence, we can hypothesize that task performance includes some hypothetical steps that consists of some actions: 1) discussion with the patient about health issues, and 2) examining a patient's health state. Based on obtained data the doctor decides what area to X-ray (decision-making action), takes a pen (motor action—move the hand and grasp), takes a prescription form (motor action), and writes a diagnosis (can be considered a sequence of actions). Usually for task analysis it is sufficient to know the duration of writing. Here we note that regular writing requires the third level of attention concentration. The last action is "give the prescription to the patient." The last motor action includes two motions: 1) move prescription to the patient, and 2) release prescription. This can in an approximate manner describe a considered hypothetical task. The goal of a task is *make a diagnosis and if required, to give the patient a prescription*. The motives can be: 1) providing health care for the patient; 2) earn a salary; and 3) protect my job. Hence "requesting of X-rays" is a *decision-making action* that is performed in the context of a diagnostic task and 1) providing health care for the patient; 2) earn a salary; and 3) protect my job are not some goals but simply possible verbalized motives. In our work (Bedny, 2015, p. 32) we argue with Diaper in a similar way when he tries, based on such an analysis, to eliminate the concept of goal altogether. In this example he described a chemical plant operator's task. Diaper makes general conclusions that these are not motives, and calls them unrelated goals for a single action and therefore goal concept is redundant in psychology.

The subject of course can give various reasons for activities or performed actions; this may not be the actual motive. This can be explained by the fact that some motives are unconscious.

Moreover, there are defense mechanism such as rationalization, and a person can try "to act rationally." As a result, a subject can incorrectly to describe their motives. This is a well known fact in studies of motivation within the field of psychology. In activity theory unrelated goals do not exist. As we already wrote, it is not Diaper's fault that he confused the goal with motives, which can with some approximation be verbalized by a subject. In cognitive psychology and ergonomics, goal and motives are not distinguished. We have one goal of a task or one goal for each action. At the same time, our behavior is poly-motivated and can includes a number of motives. In cognitive psychology a goal is considered with a motive as a unitary concept. If the goal includes various motives, each of them can be considered as unrelated goals and therefore Diaper is right. In SSAT, it was clearly demonstrated that goal and motive(s) are different concepts. Motive(s) and goal should be considered as a vector motive(s) → goal. This concept is described in a totally different manner in comparison to general and applied activity theories. Goal and motives and their relationship are critically important concepts in activity theory in general. However, there are some differences in understanding this concept in general, applied, and systemic-structural activity theory.

In SSAT goal considered a combination of imaginative and verbally-logical components. Hence, the goal is broader than its verbal equivalent. An imaginative or verbally logical representation of a future desired result emerges as a goal only when it is joint with motives. For example, a student has some desire to obtain a good grade. Such components as "desire" and "obtaining a good grade" are two components that can be transformed into motive and goal when they create vector "motive—goal."

Desire is transformed into motive, and the mental representation "obtaining the good grade" becomes the goal.

Goal can be independently formulated or prescribed by instruction during task performance. At the first stage, the prescribed goal of a task emerged for the subject as requirements of task. Presented requirements should be interpreted and accepted by the subject. This means that the objectively presented goal (prescribed goal) and subjectively accepted goal are not always the same. We also need to distinguished the overall goal of task from partial goals of actions. Goal of actions after their execution can be quickly forgotten during task performance.

It is also possible to have different relationships between the same goal and different motives. In this situation goal-directed activity acquires an ambivalent character. If according to some, motives in motivational process appear to a person to be undesirable, motivation will emerge as avoidance of the goal. In contradictions between motives, volitional process is important. The goal of task plays a critical role in activity self-regulation. This is not a homeostatic process of self-regulation. Self-regulation acquires goal-directed character. Based on self-regulation, subjects develop different strategies of task performance. Informational or cognitive (including goal) and energetic components of activity (emotionally-motivational processes) are tightly interconnected. There are several types of such interconnections (Vekker, Paley, 1971). An example of the first type of interconnection was described in psychophysics and psychophysiology when intensity of external stimuli results in increasing the sense experience. Another type of interconnection is related to the reticular activating system of the brain. This interconnection plays an important role in controlling the state of arousal and awareness. A third group of interconnections is linked to the emotionally-motivational components of activity. This fact is often overlooked in cognitive psychology. For example, human informational models, models of separate cognitive processes, ignore energetic components of activity.

Emotionally evaluative mechanisms are also critical factors in activity regulation and the specificity of the functioning information processing system. For example, significance of the goal or some components of the situation demonstrates how the goal or such components has value for the person. Factors of significance play an important role in the selection of information, performance of diagnostic tasks, and situation awareness (SA) in general.

Understanding the relationship between emotionally-motivational (energetic) and goal (informational) components of activity can interpret the theory of motivation in different ways. Western theorists (Locke and Lathman, 1984; Pervin, 1989) postulate that the goal has both cognitive and affective features. Pervin stated that the goal can be weak or intense, and Locke and Lathman (1990) consider the goal as a motivational component of behavior. Presumably, the more intense the goal, the more a subject strives to reach it. Hence, the goal "pulls" the activity. In contrast, according to Kleinback and Schmidt (1990), goals push activity and therefore also should be considered a motivational factor.

From this discussion one can see that there can be several motives but only one goal of activity during task performance, and therefore in SSAT there are no unrelated goals. Awareness of the goal is critically important. The other aspects of the activity such as motives or some aspects of performance can be conscious or unconscious. The goal of task or sub-goals of task are higher-order hierarchical goals in comparison to goals of individual actions. They are more often developed consciously. The goals of individual actions are often formed unconsciously or automatically. Such goals can be quickly forgotten after their achievement.

Activity differs from the purposive behaviorism of Tolman (1932), where goal directness was not distinguished from purposive behavior. Goal directed human activity always includes conscious understanding of the goal's intent. Purposiveness of animals' behavior does not suggest conscious understanding of future desired results.

13.3   BASIC CONCEPTS OF SYSTEMIC-STRUCTURAL ACTIVITY THEORY

Currently, we need to distinguish general (AT), applied (AAT), and systemic-structural activity theory (SSAT). Applied activity theory is not a unitary theory. There is no standardized terminology, units of analysis, and procedures. Publications in this field were practically unknown in the West until we conducted some translation and interpretation of data in this area of study (see for example some material in Bedny, Meister, 1997). At present, the most advanced and developed activity theory that studies human work is SSAT (Bedny, Karwowski, 2007; Bedny, Karwowski, I. Bedny, 2015; Bedny, 2015). This is general theory or framework for studying ergonomics and work psychology with well-developed terminology, and required basic concepts adapted for practical application. Withing framework of this theory developed not only qualitative and experimental, but also analytical and quantitative methods of study. According to this theory activity integrates external behavior, internal mental processes, and emotionally motivational processes into a system that has goal directed character. (Bedny, Karwowski, 2011). This is a structurally organized goal directed self-regulated system where cognition, behavior, and motivation influence each other.

According to SSAT the task is a basic component of an activity, and lives of an individual can be presented as a continuing attempt to solve various tasks (Bedny, 2015). The activity during task performance consists of conscious cognitive and behavioral actions that in turn depend on psychological operations. The goal of actions is conscious until information about it is kept in working memory. Actions are integrated by conscious goal of task. Thus, activity is a hierarchically organized system. This is a historically developed phenomenon that evolved over time within a culture. The social and physical environment determines what types of tasks are performed by people and how they interact with others. Any activity includes the objects, tools, procedures, and norms. This means that people acquire work activity that is prescribed by society and is divided into various work tasks. Any task includes work activity and technological components. From an activity perspectives, task is situation-bounded activity directed to achieve a conscious goal of a task. Analysis of interrelationships between activity and technology components is important for developing efficient work methods, design of equipment, and computer interfaces. Nardi (1997, p. 73) and Kaptelinin and Nardi (2006, p. 32) wrote that the basic unit of analysis is activity. However, activity is an object of study and basic units of analysis are cognitive and behavioral actions that are elements of activity. A production process is a collection of tasks that are organized in a particular order. The general hierarchical scheme of activity includes four levels: *work activity, task, action,* and *operation*. Hence, Kaptelinin and Nardi's (2006, p. 119) statement, "There is an evident difference between high-level actions and low-level actions (or tasks)" contradict general psychology (Leont'ev, 1978) and SSAT (Bedny, 2015). Task consist of logically organized system of actions that are integrated by the goal of the task. Activity during task performance is the object of study. Such activity is a logically organized system of actions, and actions integrated into activity by goal of the task. Cognitive and behavioral actions integrated into activity are units of analysis. This understanding of activity coincides with the historical development of the word *deyatel'nost'* in the Russian language.

Activity (*deyatel'nost'*) is a sociohistorical category that is studied by various sciences and not only by psychology.

One such direction is the analysis of efficiency of human labor and time study in manufacturing, which was developed in the early twentieth century. This direction of study emerged in Russia under the influence of work by Taylor (1911) and Gilbreth (1911). Specialists in this field were had engineering backgrounds. They developed their classification of elements of human activity during performance without any relationship to psychology nor knowledge of work by Rubinshtein, Vygotsky, or Leont'ev. According to an engineering approach, there are the following hierarchical relationships between production operations and their elements in study the human labor: *production operation–motor action–movement*. Production operation according to contemporary terminology is a task that is performed in the field of manufacturing. Of course, there are certain shortcomings in this area. They studied mass production operations where elements of task are performed in the same sequence. Specialists in this area cannot clearly classify and describe motor elements of the activity. These professionals totally ignore the concepts of cognitive actions and mental components of activity in general. Consider that this approach cannot be adapted for contemporary variable human activity, etc. At the same time, specialists in activity theory who study human work are not sufficiently familiar with engineering approaches that for a long time have studied human labor. This explains the fact of insufficiently effective application of the general activity theory in the study of human work. This also explains why many specialists in the West face great difficulties in applying the general activity theory to studying human work.

For example, Kuutti (1997, p. 33) presents examples of actions such as "writing a report," "arranging a meeting," "transporting merchandise," etc. In this figure he also wrote that building a house is an activity and fixing a roofing is an action. Let us as an example describe the first stage that should be performed when a manager decide to organize a meeting. The first stage usually involves the following steps.

1. scheduling the meeting
2. determining the meeting time
3. determining the list of participants
4. reserving a meeting room
5. making sure all participants are available at the set time
6. notifying all parties of the meeting

We describe only the first stage of the possible task cConducting the meeting." As can be seen, the manager tried to formulate the purpose of the meeting, secured the room for the meeting, called subordinates on the phone, wrote down some information, etc. This is a complex manager activity. It includes a number of cognitive and motor actions. For example, the manager can move his or her hand and grasp a paper list in front o themselves, and release the paper. This is the first motor action. Then he or she can move their hand and grasp a pen, and put their hand into a starting position for writing. This is the second motor action. If in front of him or her is a list of subordinates' phone numbers, reading some of them will involve perceptual actions. During connections with others the manager can perform various decisions. These hypothetical examples give no description of each action. More precise descriptions of various cognitive and behavioral actions and their logical organization can be found as examples in Bedny (2015).

Kuutti (1997), citing Leont'ev work, wrote that *building a house* is an example of activity and is an action. However, both examples label different parts of the production process that includes a technological process and a work process (Bedny, Harris, 2005). They could be further divided into production operations. Fixing the roofing includes worker activity frequently associated with hammering nails. Hammering a nail is completed when the nail is flush with the roof board. Suppose that in order to hammer a nail until it is flush, the worker performed on average five strokes with a hammer. The worker raises the hammer and strikes the nail. He on average does this five times. The worker performs two motions in each motor action. The hammer is up in an approximate position and the hammer is down in the exact position under visual control. These two motions, which are integrated by one goal, hit nail, is one motor action. Thus, at present, a specialist in the West cannot explain simple questions about what activity, action, and operations (in psychological meaning) are. This is not accidental. Most general activity specialists in Russia do not know either. All required information in this field can be obtained in SSAT (G. Bedny, I. Bedny, 2018).

There are some other examples that demonstrate the difficulty of understanding basic concepts in the West. Kaptelinin and Nardi (2006, pp. 67) present some examples of human actions. They wrote that "making a hunting weapon is an action." However, this is a complex work activity that includes a number of cognitive and behavioral actions that have not only hierarchical but also logical organization. Sometimes several simple actions can be integrated into a more complex one. However, such integration is limited by capacity of the short-term memory. The complex action subjectively can be perceived as a unitary one.

According to SSAT (G. Bedny, Karwowski, I. Bedny, 2015), production and operationally monitoring processes consists of tasks. The first one in turn includes in its content production operations or tasks. The second one includes in its content operationally-monitoring tasks. From this it follows that the task is a basic component of any type of work. As an example we present an object of study and units of analysis of production process.

Object of study   Units of analysis

Production process → tasks → cognitive and behavioral actions → psychological operations

Similarly, we can present an operationally-monitoring process.

Activity is a complex goal-directed system. Thanks to self-regulation, a subject can develop various strategies of task performance. In SSAT, models of orienting activity self-regulation and general model of activity self-regulation were developed. Orienting activity precedes to the stage of activity execution. The model of orienting activity shown in Figure 13.1 below (Bedny and Karwowski, I. Bedny, 2015) describes how a person creates a goal, a subjective mental model of the situation, which type of the exploratory actions and operations are utilized, what types of possible mental models are developed, and how a subject selects preferable mental models, among others. This means that the product of orienting self-regulation activity is not the transformation of the real situation, but formation of a mental representation of a situation that precedes decision making and follows performance of executive actions, the purpose of which is to transform a situation or execution. This model in more detail manner describes what is called situation awareness or SA that is described by (Bedny, Karwowski, I. Bedny, 2015). SA is only one mechanism in the model of orienting activity, which is presented in function block 9. In each function block at any particular time, cognitive processes are integrated to achieve a specific purpose of activity self-regulation, and this integration serves as the basis for the formation of functional blocks or mechanisms of self-regulation.

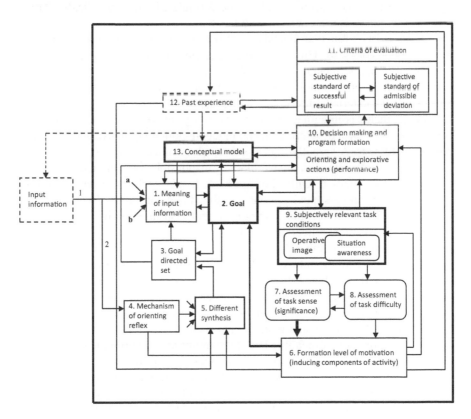

FIGURE 13.1   Model of the self-regulation of orienting activity. From Sanda, Johansson, and Johansson, 2011.

When specialists study activity from a self-regulation perspective, a functional mechanism or function blocks is the main unit of analysis. Any function block includes various combination of cognitive processes. It can also be described in term of cognitive and behavioral actions. Each function block has the same purpose in the structure of activity self-regulation, but its content varies depending on specificity of a performed task. During task analysis, each function block determines the range of issues that are connected with this block, and should be considered when examining the role of this function block in activity regulation. The most important function blocks for each particular case should be considered. It is also important to pay attention to the relationship between and the mutual influence of the function blocks. As an example we consider in brief manner some function blocks.

- The first functional function block that we describe is the goal of the task (block 2). Overall, the goal of the task performs integral functions. It integrates all the mechanisms of self-regulation into a holistic, self-regulative system of activity. There is no activity without a goal. When the goal is not clearly formulated externally, the subject can formulate it him or herself. Even though a goal lacks clear definition, it still exists for a subject. During task performance, such a goal can be formulated more specifically. This stage of analysis is associated with block 2.
- There exists a function mechanism that is called subjectively relevant task condition (block 9). This mechanism is responsible for development of stable or dynamic mental models of task. At this stage, a subject creates a subjective representation of the task. This mental model of the task can vary even if the external representation of situation does not change. The mental model includes imaginative and verbally logical sub-mechanisms or sub-blocks. A subject develops a mental model in both imaginative and verbally logical forms. A verbally-logical sub-block is called a situation awareness sub-block. An operative image block is responsible for imaginative reflection of the situation. Only those parts of a sub-block that overlap with a situation awareness block can be verbalized and conscious. Therefore, not only logical or conceptual components but also imaginative components provide a dynamic reflection of reality.
- Functional block number 7 is responsible for emotionally evaluative aspects of activity where the subject evaluates the personal significance of the goal or task itself. The block "sense" or "significance" is an important mechanism that influences interpretation of meaning and facilitates the creation of the mental model of situation. Features relevant to the task goal, the significance of situation, and motivational state of the subject are selected for the interpretation of meaning out of the multitude of its semantic features.
- Motivational aspects of activity, functional block 6, depend on the mechanism of sense. First, the motivational block is an inducing mechanism of activity. Existence of blocks 6 and 7 in the self-regulation process makes it clear that not only cognitive but also emotionally motivational components affect activity regulation. It is important to note, according to the concept of self-regulation, that in SSAT, the goal is cognitive, and motivation is an

energetic component of activity. Activity during task performance involves multiple motives. However, there is only one overall goal of task and a number of different sub-goals of separate underlying actions required to complete that task.

- The purpose of functional block number 8 listed above is assessment of task difficulty. This block is involved in the evaluation of the objective complexity of the task. The task difficulty is a subjective characteristic of the task. The individual can under- or overestimate the objective complexity of the task, and as a result, select inappropriate strategies for goal attainment. Evaluation of task difficulty is a complex process that involves comparison of a subject's own abilities with demands of the task. The process may also involve taking inventory of other people's ability to compete this task. If an easier task is subjectively perceived as a more difficult one, the subject can reject the task or change requirements of the task. Relationships between blocks 7 (significance) and block 8 (difficulty) can influence motivational block 6. For example, the less significant the task for subject, and the more difficult for him or her, the less the subject is motivated to perform the task.

For purposes of illustration, we briefly considered only some functional blocks or mechanisms in the presented model. However, even this description illustrates some possibilities in the description of activity strategies during task performance. Analysis of activity self-regulation is called functional analysis of activity.

13.4 QUALITATIVE AND FUNCTIONAL ANALYSIS OF MINER WORK ACTIVITY

Functional analysis can be combined with traditional qualitative analysis of activity. Below we present these two methods of study applied to the drilling activity work of two experienced miners (Sanda 2015).

Qualitative data was collected through interviews as well as direct observation and video recording of the miners' activities on two consecutive occasions for periods of four hours. Using the systemic analytical approach as the basic paradigm for the critical analysis of positioning actions (Bedny and Karwowski 2011) for drilling activity in deep mines, both morphological and functional analyses were conducted (Sanda 2015). In the morphological analysis, the constructive features of the rock drilling activity, entailing the logical and spatio-temporal organization of the cognitive behavioral actions and operations involved, were described (Sanda 2015). In describing the structure of the rock drilling activity, the work process was subdivided into tasks. These tasks were analyzed individually in terms of mental and motor actions and operations (Sanda 2015). In the functional analysis, potential strategies of activity performance associated with the miners' actions and their corresponding operations, identified as constituting functional blocks (see Figure 13.1) were analyzed qualitatively using systemic principles (Bedny and Karwowski 2007). This allowed for the evaluation of varieties of performance indicators, such as time and errors, and the selection of the most efficient strategy (Sanda 2015). Functional analysis was conducted based on studying the miners' mechanism of activity self-regulation, which is critical for understanding the drilling activity as a system (Bedny and Karwowski 2011, Sanda 2015). The following functional blocks (see Figure 13.1 above) were considered in analyzing the actions and operations undertaken by miners in the rock drilling activity:

- Block 2: goal
- Block 12: past experience
- Block 11: criteria of evaluation block include two sub-blocks: "subjective standard of successful result" and "subjective standard of admissible deviation"
- Block 10: making a decision and program formation: orienting and explorative actions (performance)
- Block 9: subjectively relevant task conditions responsible for development of stable or dynamic mental models of task or situation
- Block 8: assessment of the task difficulty
- Block 7: assessment of the sense of task or task significance
- Block 6: formation of the level of motivation

In the functional analysis, each functional block is viewed as a self-regulative system (Bedny and Meister 1997, Bedny and Karwowski 2007) in which the miners use their procedural knowledge in various ways. The units of analysis, viewed within the Vygotskian perspective (Bedny and Karwowski 2007), are the miners' cognitive and behavioral motor actions and operations. The cognitive and behavioral actions and operations in the object-oriented activity are analyzed from the perspective of the individual miner using technological tools in production work (drilling) to achieve results. Observations and video recordings of the miners' engagement with the drilling activity show that a miner's individual object-oriented activity consists of both physical and mental actions and operations whose characteristics are influenced by past experiences (Sanda 2015). The functional appraisal of this activity shows that the object of a miner's self-regulation of orienting activity to be twofold. The object, in terms of the miner's physical activity, is the conduction of production (rock) drilling operations, based on informed decisions and programs to orient explorative actions toward performance enhancement, sense-making in task performance, and determination of motivational level (Sanda 2015). In terms of the miner's mental activity, the object is the simultaneous observation of the production drilling work, assessment of task difficulty, and listening to communication models in order to enhance the development of stable or dynamic mental models of task or situation for enhancing the relevant task conditions (Sanda 2015). The object also includes the setting of a subjective standard of "successful results" and "admissible deviation" (Sanda 2015).

Analysis of the miners' mental assessment of task difficulty showed that they view the use of tractor technology with more than one robotic arm (i.e., boomer) for production drilling tasks as excessive for one operator to handle (Sanda 2015). The miners related the perceived task overload to the difficulty an operator encounters in his or her ability to focus on the computerized programming command for the automated rock drilling actions and operations using the multi-boomers (Sanda, 2015). The miners viewed such situation as distracting them from developing the requisite dynamic mental models of task or situation that are required for enhancing the quality of the relevant task conditions (Sanda, 2015). The miners view that even in situations where high technology robotic loaders that are remotely controlled from safe distances are used for important task components in production drilling activity, the guide cameras used to manipulate the robots' movements do not guide their operations as efficiently as the direct use of the operator's eyes. As such, the operators, based on acquired experiences, mostly find ways to guide the technology for optimum performance (Sanda, 2015). A sense of this, provided qualitatively by the miners, is as follows:

> *The technology does not always get it right. Most of the time, I use the experience I have acquired over the years to guide the technology for optimum performance. I have also developed enormous knowledge on manipulating the technology to make my work activity easier (Sanda 2015).*

Researcher: So at times the work can be frustrating.
Miner: Oh yes.
Researcher: How do you feel now? The machine is not behaving well.
Miner: Yes, because it is fighting a hard rock, which is difficult to penetrate.
Researcher: How do you know that the rock is difficult to penetrate?
Miner: From the noise produced. The sound is different when the rock is not diffi-
cult to penetrate
Researcher: How did you know these differences?
Miner: I learned it from experience. Look at the drilling of this here, bad! It took
me almost six minutes.
Researcher: So, using the robots here does not solve all your problems?
Miner: Yeah, at times. We need to use techniques we have developed from our expe-
riences to help the machine work better.
Researcher: So at the end of the day, will you let someone know that you did this
during the work?
Miner: No.
Researcher: Why?
Miner: Because I don't think about it. Sometimes you want to go faster, they will ask
you to wait. You will wait and wait and …
Researcher: I know you want to go faster. Going faster means you are helping man-
agement to make more money.
Miner: Yeah! The more the machine works, the more money…
Researcher: So, don't you think that there has to be a way for managers to appreci-
ate what you have just done, that is, to know that when you are working
and something happens, you repair it yourself?
Miner: Yes I understand. You have to do it.
Researcher: Maybe management also doesn't know that you do this kinds of repair
work here.
Miner: No! They know, but they don't care.
Researcher: So, how do you feel, when you get the sense that you are sacrificing, but
they (i.e., management) don't care
Miner: I feel bad! You know the song "bad to the bone."

A short qualitative and functional description of a miner's work shows that the miners possess procedural knowledge developed over time in various objective activities, but which remained shared (Sanda, 2015). This strategy of task performance is used by the operators to negotiate technology-based standardized task patterns in bids to overcome task repetitiveness and to increase their productive capacities in terms of waste removal during production time (Sanda 2015). Based on the notion that technologies do not always get it right, the miners also use their individual strategies of task performance to overcome subjective perceptions of technological shortcomings in their task undertakings (Sanda, 2015). In some instances, during the bolting operations, operators negotiate tasks to make them move faster. For example, it was found that operators of the high technology machine in the roof drilling and bolting operations in the deep mine used techniques enhanced by old mining culture to negotiate and accelerate the tasks (Sanda 2015). They use their understanding of the sound produced in the machine-rock interaction to instinctively detect and negotiate around an impenetrable rock section (Sanda, 2015). A sense of this is provided by the following extract of the direct conversation between researcher and miner (Sanda, 2015);

Researcher: So like this, is it something you developed from your skills over the
                years?
Miner: Yeah. It is instinct.
Researcher: So it is instincts?
Miner: Yeah!
Researcher: But this instinct. Do you think it can be used in designing a better technology? That is your instinct. If something doesn't work, I have to do this.
Miner: Yeah, you can wake me in the middle of the night, and I can do it like this.
Researcher: If management comes to you today and says, "We now know that you
                have good ideas so write them down on paper and give them to us," what
                will you tell them?
Miner: I will write it down and give it to them
Researcher: For free or you will ask them to pay you for your knowledge?
Miner: They should pay me for my knowledge.

Based on the analysis above, Sanda (2015) argued that since organizations possess technologies (i.e., techniques for processing raw materials and/or people) for accomplishing work, and activity during team performance then emphasizes a work system design in which technology affects social relations in organizations by structuring transactions between roles that are building blocks of an organization (Sanda, Johansson, Johansson and Abrahamsson, 2014). In this respect, Sanda, Johansson, Johansson, and Abrahamsson (2014) argue that application of systemic-structural activity theory stands to provide an understanding of the various processes that are entailed in digitized human work that can be used to design a harmonious work environment integrating the human, technical, and the social system toward increased productivity in the deep mine industry. The findings from the morphological analysis showed that relationship between the miners' external behavior and internal psychological functions are mutually regulated (Sanda, Johansson, Johansson and

Abrahamsson, 2014). This shows a degree of commonality in the cognitive characteristics of the miners' internal mental activity and the regulative nature of their external behaviors toward the drilling activity undertakings (Sanda, 2015). This finding
is underscored by the argument supporting the mutual interdependence of mental
development, semiotic mediation, and external practical activity, which do not exist
separately (Bedny and Karwowski, 2007; Bedny and Harris, 2005). The sense here
is that an inter-subjective aspect of activity, manifested as innovative skills for task
negotiation, is observable in a miner's individual activity (Sanda, 2015).

Findings from the functional analysis support the miners' notion of time reduction in some task undertakings using individual strategies of performance to determine penetrable and non-penetrable sounds differences to negotiate an action in an
activity (Sanda, 2015). It also portrayed the structure of the miners' activity during
task performance as a logically organized system of cognitive and motor actions,
and operations (Bedny, Karwowski, and Sengupta, 2008) that enhances innovation
(Sanda, Johansson, Johansson, and Abrahamsson, 2014). This shows that the specificity of the drilling activity is underlined by the interdependence of the miner's
practical activities and symbolic activities, each of which is in constant transformation of the other (Sanda, Johansson, Johansson, and Abrahamsson, 2014). Since the
practical actions in the drilling activity have a clearly defined object (Sanda 2015), it
then entails semiotic mediation (Bedny and Karwowski, 2007). The occurrence of
such mediation is highlighted by the conscious goal (including planning and understanding of the possible outcomes) with which the object-practical actions in the
rock drilling activity are undertaken (Sanda, 2015; Sanda, Johansson, Johansson,
and Abrahamsson, 2014). Thus, for the miners engaged in the drilling activity, their
active exposure to diversity derived from years of practice and experience appeared
to have increased the number of conceptual categories (Sanda, 2015) they have
developed for storing information (Sanda, Johansson, Johansson, and Abrahamsson,
2014). They also appeared to have procedurally developed new ways for integrating conceptual data in the actions and operations associated with drilling, and as a
result provided more insightful knowledge of the complex problems and solutions
in the drilling activity (Sanda, 2015). The implication here is that the miners, being
cognitively complex persons, tend to be open in their beliefs and relativistic in their
thinking, as well as having a dynamic conception of their work environment (Sanda,
2015; Sanda, Johansson, Johansson, and Abrahamsson, 2014).

The study above has shown that the object-practical activity of the rock drilling
activity is determined by the genesis and content of the miners' mind. In this respect,
the functional efficiency and effectiveness of the drilling activity could be increased by
identifying performance enhancing strategies that are based on mechanisms of activity
self-regulation. Such strategies are used by workers to facilitate the social collaboration
between them and the technology they use in order to enhance their productivities.
Identification of such performance must require an understanding of the interrelationship of internal and external activities, which stands to determine the miners' practi-
cal-external activity and the corresponding external tools that they need to enhance
their mental activities toward developing successful performance enhancing strategies
in the drilling activity. Such understanding of performance enhancing strategies could
be integrated in the design of efficient and effective work systems and/or technology.

13.5   CONCEPT OF MEDIATION AND
TRIANGLE MODEL OF ACTIVITY

Based on Vygotsky's concept of mediation, Engestrom (2000) developed his version of activity model (Engestrom triangle model). His model can be viewed as some expansion of the basic idea of mediation through the addition of the elements of community and rules division of labor while emphasizing the importance of analyzing their interaction with each other. Engestrom (1999) sees joint activity or practice as the unit of analysis for activity theory. However, according to SSAT this is not a unit of analysis but an object of study. He also takes into account the conflictual nature of social practice. There is a lot of contradiction in the understanding of basic terminology and concepts of activity theory in SSAT, and model of activity suggested by Engestrom. However, we do not discuss this in our work. Only some required improvements in the triangle model will be considered. A simplified triadic scheme of activity system according to Engeström is presented below.

This schema was developed by Engestrom based on Vygotsky's idea about semiotic mediation according to which human activity is mediated by a material and mental tool (external tool–material object and internal tool mental objects such as signs, symbols, words, etc.). For our analysis we consider only the upper part of this Figure 13.2. The bottom line of this part of the figure presents the relationship:

$$\text{Subject} \rightarrow \text{Object} \rightarrow \text{Outcome}$$

A number of researchers have interpreted the term "object" in this schema as being synonymous with "objectives." This interpretation will always engender difficulties when attempting to apply activity theory in practice. For example, Kaptelinin and Nardi (2006, p. 66) wrote, "A way to understand objects of activities is to think of them as objectives that give meaning to what people do." Western psychologists interpret the object in this scheme as objectives. However, it is the object of activity. According to general and systemic-structural activity theories, it may be a physical or mental one (sign, symbol, and image). Subjects in accordance with required goals should transform or modify objects of activity according to the goal of activity.

Below we outline a somewhat more elaborate scheme (Bedny, Karwowski, 2007). In this scheme we exhibit not only subject → object interaction, but also subject ↔ subject interaction (see Figure 13.3).

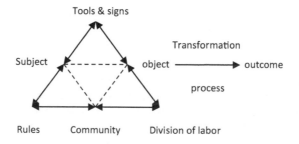

FIGURE 13.2   Triadic scheme of an activity system according to Engeström. From Sanda, 2016.

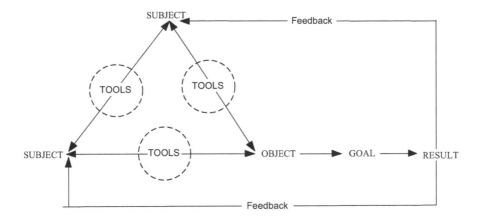

FIGURE 13.3 Triadic schema of activity. From Bedny and Karwowski, 2011.

This schema is intended to emphasize our point that the notion of "objectives" relates to the goal, rather than the object of activity. The broken circles in the figure indicate that subject-object interaction may be either direct or use external mediating instruments. In such a situation, subjects utilize only internal mental tools. In this scheme, the object and goal are treated as distinct components. From this scheme, it may be inferred that the notion of "objective" is relevant to the goals, not to the object. The schema also demonstrates that subjects can interact with each other either directly by utilizing only mental tools or indirectly through instruments. This brings us to the concept of collective activity. Under collective activity we understand the system of task and their actions coordinated in space and time by different subjects toward achieving a common goal.

The schema demonstrates that subjects interact with object. It is possible that subjects have their individually related objects or shared object. In the latter case, subjects during the performance of individual tasks can transform the same single object according to the goals of their task. This requires coordination of actions in space and time during the performance of individual tasks. For example, during the fixing of a roof, two workers take a heavy board on both sides and place it in the required place. Each of them takes a hammer by the right hand and brings it into the required position before hitting. Then they take a nail by the left hand and put it into position before hitting it by the hammer. They hit the nail on the head, until the nail is completely in the board. Of course, this is a simplified description. In this example, board is the shared object, hammer is the individual material tool, and nail is the individual material object used by each worker. In our example, each worker performs a similar task and if they did their task correctly the high order general goal will be achieved. The general goal was to fix the board to the roof. In this considered example, joint activity is the object of study. The example demonstrates the situation when both subjects have the same object and common goal of activity. From the figure we also can see that object and goal can be different. Through feedback, they coordinate their activity. It should be taken into consideration that both tasks can be different. In such situations the goal of each task, and the object of activity during task performance are also different. It is also important to understand that tasks performed in a collective activity should be coordinated in space and time.

Figure 13.3 demonstrates that social interaction can be transferred into object-oriented activity and vice versa. We also use the term result instead of the term output as it is usually utilized in activity theory. Results of separate actions and tasks are evaluated based on feedback.

Finally, this scheme possesses feedback influences, implying that an activity is organized according to the principles of self-regulation. Subjects can utilize some individual strategies of task performance if these strategies do not violate technological and safety requirements.

Circles in this figure exhibit that subjects can interact with an object either through instruments or directly. Similarly, subjects can interact with each other either directly or indirectly through instruments.

However, there is another possibility for performing a collective activity. For example, the object of activity for each subject can be different. Each subject can pursue his or her particular goal during transformation of his object. At the same time, each subject should coordinate this transformative process in time with other subjects. This is a common goal of joint or collective activity. Therefore, the major criteria for common activities are requirements to coordinate joint activities performed by different subjects and the existence of a common goal. Each subject in a collective activity evaluates their result according to the following criteria: 1) how an object is progressively transformed according to the goal of activity, and 2) how the process of the transformation is coordinated with the other subject. From a self-regulation perspective, collective activity requires coordination of activity strategies of different subjects. In Engestrom's model of interacting activity, the major concepts are interaction and contradictions. They are important for the study of any system. However, without the concept of self-regulation and feedback, individual and collective activity cannot be understood. Therefore, one cannot agree with Engestrom (1999), who states that in a collective activity subjects must always share the same object of activity. Further joint activity is not a unit of analysis but an object of study.

CONCLUSION

Systemic-structural activity theory suggests a different way of looking at work psychology and ergonomics. Activity is considered not simply as a process as in cognitive psychology, but as a structure that unfolds in time as a process. Activity is considered as a system that integrates cognitive, behavioral, and emotionally motivational components. This system is goal directed. Cognitive psychology and fields that study motivation postulate that goals have both cognitive and motivational features, whereas SSAT states conscious human goals can be only cognitive in nature. This enables activity to be described as a goal-directed, self-regulative system. In systems of activity, regulation plays not only an important motivational role, but also emotionally-evaluative mechanisms, which are involved in evaluating the significance of various elements of the situation during performance.

From a morphological perspective it is a system of logically organized cognitive and behavioral actions, and from a functional perspectives it is a self-regulative system. When subjects operate with material, or real objects, actions are called motor. Transformation of images, concepts, or propositions in our mind are called mental actions. Thus, for the purpose of study, specialists can utilize various methods of activity analysis. Various methods of quantitative analysis were developed within the framework of systemic-structural activity theory.

REFERENCES

Bedny, G. Z., Harris, S. R. (2005). The systemic-structural theory of activity: Applications to the study of human work. *Mind, Culture & Activity*, 12, 128–147.

Bedny, G. Z., Karwowski, W. (2007). *A Systemic-Structural Theory of Activity: Applications to Human Performance and Work Design*. Boca Raton: CRC Press.

Bedny, G. Z., Karwowski, W. (2011). Analysis of strategies employed during upper extremity positioning actions. *Theoret. Iss, in Erg, Sc.*, iFirst (2011), 1–20.

Bedny, G. Z., Karwowski, W., Sengupta, T. (2008). Application of systemic-structural theory of activity in the development of predictive models of user performance, *Int. J. Human-Computer Interaction*, 24(2008), 239–274.

Bedny, G. Z., Meister, D. (1997). *The Russian theory of Activity: Current Applications to Design and Learning*. New Jersey: Lawrence Erlbaum Associates.

Bedny, G. Z., Karwowski, Bedny, I. (2015). *Applying Systemic-Structural Activity Theory to Design of Human-Computer Interaction Systems*. CRC Press, Taylor & Francis Group, Boca Raton, London, New York

Bedny G. Z. (2015). *Application of Systemic-Structural Activity Theory to Design and Training*. CRC Press, Taylor & Francis Group, Boca Raton, London, New York.

Bedny, G. Z., Bedny, I. S. (2018). *Work Activity Studies Within the Framework of Ergonomics, Psychology, and Economics*. CRC Press, FL: Taylor and Francis Group.

Davydov, V. V., Zinchenko, V. P., Talyzina, N. F. (1983). The problem of activity in the work of A. N. Leont'ev's. *Soviet Psychology* 21(4): 31–42.

Denezhnyj, P. M., Stiskin, G. M., Txor, I. E. (1976). *Turning profession*. Higher Education.

Diaper, D. (2008). Reactionary reactions to altering activity theory. *The Interdisciplinary Journal of Human-Computer Interaction*, 20(2), 267–271.

Diaper, D., Lindgaard, G. (2008). West meet East: Adapting activity theory for HCI and CSCW applications? Interaction with Computer, *The Interdisciplinary Journal of Human-Computer Interaction*, 20(2), 240–286.

Engeström, Y. (1999). Expansive Visibilization of Work: An Activity-Theoretical Perspective, *Computer-Supported Cooperative Work*, 8, 63–93.

Engeström, Y. (2000). Activity Theory as a Framework for Analyzing and Redesigning Work. *Ergonomics*, 43(7), 960–974.

Gilbreth, F. V. (1911). *Motion Study*. Princeton, NJ: D van Nostrand Company.

Kaptelinin, V., Nardi, B. A. (2006). *Acting with Technology. Activity Theory and Interaction Design*. The MIT Press, Cambridge, Massachusetts, London, England.

Kleinback, U., Schmidt, K. H. (1990). The translation of work motivation into performance. In V. Kleinback, H.-H. Quast, H. Thierry, and H. Hacker (Eds.).*Work Motivation*, pp. 27–40, Hillsdale, NJ: Lawrence Erlbaum Associates, Publishers.

Kuutti, K. (1997). Activity theory as potential framework for human-computer interaction research. In B. Nardi (Ed.). *Context and Consciousness. Activity Theory and Human-Computer Interaction*, pp. 17–44.

Kotik, M. A. (1978). *Textbook of Engineering Psychology*. Tallin, Estonia: Valgus.

Leontiev, A. N. (1978). *Activity, Consciousness, and Personality*. Englewood Cliffs: Prentice-Hall.

Locke and Lathman, (1990). Work motivation: The high performance cycle. In V. Kleinbeck et al. (Eds.) *Work Motivation*, pp. 3–26. Hillsdale, NJ: Lawrence Erlbaum Associates, Publishers.

Nardi, B. A. (1997). Some reflection on application of activity theory. In A. B. Nardi, A. B. *Context and Consciousness: Activity Theory and Human-Computer Interaction.* Cambridge: MIT Press, 235–246.

Norman, D. A. (1986). Cognitive engineering. In Norman, D. and Draper, S. (Eds.) *User Centering System Design* (pp. 31–61). Erlbaum: Hillsdale, NJ.

Pervin, (1989). *Goal Concept in Personality and Social Psychology.* Mahwah, NJ: Lawrence Erlbaum Associates, Publishers.

Ponomarenko, V., Bedny, G. Z., Makarov, R. N. (2006). *Activity Pilot in Flight and its Imaginative Components.* Maastricht, the Netherlands: Triennial International Congress.

Rubinshtein, S. L. (1922/1986). Principles of creative independent activity. *Questions of Psychology*, 4, 101–107 (reprinted from the first publication).

Rubinshtein, S. L. (1935). *Principles of General Psychology.* Moscow, Russia: Pedagogical Publishers.

Sanda, M. A. (2015). Application of systemic structural theory of activity in unearthing employee innovation in mine work. *Procedia Manufacturing*, Vol. 3C, pp. 5147–5154. Doi: 10.1016/j.promfg.2015.07.546.

Sanda, M. A., Johansson, J., Johansson, B. Abrahamsson, L. (2014). Using systemic structural activity approach in identifying strategies enhancing human performance in mining production drilling activity. *Theor. Issues Ergonom. Sci.*, 15(3), 262–282.

Strelkov, U. K. (1990). Operationally-semantic structure of professional experience. *Scientific news of Moscow University. Psychological series.* Moscow University, 3, pp. 50–55.

Strekov, U. K. (2007). Action as unit of analyzes of operators work activity. *Psychological foundations of professional activity.* PER CE; Logos, pp. 808–814.

Taylor, F. M. (2011). *The Principles of Scientific Management.* New York: Harper and Brothers.

Vekker, L. M., Paley, I. M. (1971). Information and energy in psychological reflection. In B. G. Anan'ev (Ed.) *Experimental Psychology*, Vol. 3. Leningrad, Russia: Leningrad University, pp. 66–66.

Vygotsky, L. S. (1960). *Develop High Order Psychic Functions.* Moscow, Russia: Academy of Pedagogical Science of the RSFSR.

Vygotsky, L. S. (1960/1978). *Mind in Society: The Psychology of Higher Mental Functions.* Cambridge: Harvard University Press.

Wickens, C. D., Hollands, J. C. (2000). *Engineering Psychology and Human Performance*, Third edn. New York: Harper-Collins.

Index